# Recent Advances in Natural Products Science

This book provides a summarized information related to the global herbal drug market and its regulations, ethnopharmacology of traditional crude drugs, isolation of phytopharmaceuticals, phytochemistry, standardization, and quality assessment of crude drugs. Natural products science has constantly been developing with comprehensive data contemplating different parts of natural drugs, such as global trade, quality control and regulatory concerns, traditional medicine systems, production and utilization of drugs, and utilization of medicinal and aromatic plants. This broad information about crude drugs gives rise to a subject that is now recognized as advance natural products science. By contemplating all of this thorough knowledge of the areas, this book is intended to provide considerably to the natural products science. The area of natural products science involves a broad range of topics, such as the pharmacognostical, phytochemical, and ethno-pharmacological aspects of crude drugs. Each chapter gives a sufficient understanding to academicians and researchers in the respective topic. This book includes 40 illustrations and descriptions of roughly 80 medicinal plants used for herbal medicine. The book is an imperative source for all researchers, academicians, students, and those interested in natural products science.

## Features

- Includes advance knowledge and detailed developments in natural products science
- Discusses the most important phytopharmaceuticals used in the pharmaceutical industry
- Explores the analysis and classification of novel plant-based medicinal compounds
- Includes standardization, quality control, and global trade of natural products
- Gives a deep understanding related to recent advances in herbal medicines to treat various ailments
- Discusses national and WHO regulations and policies related to herbal medicines
- Covers the complete profile of some important traditional medicinal plants, especially their historical background, biology, and chemistry

# Recent Advances in Natural Products Science

Ahmed Al-Harrasi
Saurabh Bhatia
Tapan Behl
Mohammed F. Aldawsari
Deepak Kaushik
Sridevi Chigurupati

**CRC Press**
Taylor & Francis Group
Boca Raton London New York

CRC Press is an imprint of the
Taylor & Francis Group, an **Informa** business

First edition published 2023
by CRC Press
6000 Broken Sound Parkway NW, Suite 300, Boca Raton, FL 33487-2742

and by CRC Press
4 Park Square, Milton Park, Abingdon, Oxon, OX14 4RN

**Library of Congress Cataloging-in-Publication Data**

Names: Al-Harrasi, Ahmed, author. | Bhatia, Saurabh, author. | Behl, Tapan, author. | Aldawsari,
Mohammed F., author. | Kaushik, Deepak, author. | Chigurupati, Sridevi, author.
Title: Recent advances in natural products science / Ahmed Al-Harrasi, Saurabh Bhatia, Tapan Behl,
Mohammed F. Aldawsari, Deepak Kaushik, Sridevi Chigurupati.
Description: First edition. | Boca Raton, FL : CRC Press, 2022. | Includes bibliographical references.
Identifiers: LCCN 2022004323 (print) | LCCN 2022004324 (ebook) |
ISBN 9781032227764 (hardback) | ISBN 9781032227771 (paperback) |
ISBN 9781003274124 (ebook)
Subjects: LCSH: Materia medica, Vegetable. | Medicinal plants. | Herbs--Therapeutic use.
Classification: LCC RS164 .A3635 2022  (print) | LCC RS164  (ebook) |
DDC 615.3/21--dc23/eng/20220217
LC record available at https://lccn.loc.gov/2022004323
LC ebook record available at https://lccn.loc.gov/2022004324

ISBN: 978-1-032-22776-4 (hbk)
ISBN: 978-1-032-22777-1 (pbk)
ISBN: 978-1-003-27412-4 (ebk)

DOI: 10.1201/9781003274124

Typeset in Times
by KnowledgeWorks Global Ltd.

# Contents

# Preface

For many decades, pharmacognosy and natural products have played a very important role in health-care and the prevention of diseases. It is estimated that about 70% of the supply of herbal raw materials for Ayurveda and other homeopathic medicines in India comes from the wild. The global market for botanical and plant-derived drugs was valued at $23.2 billion in 2013 and $24.4 billion in 2014. This total market is expected to reach $25.6 billion in 2015 and nearly $35.4 billion in 2020, with a compound annual growth rate (CAGR) of 6.6% from 2015 to 2020. Developments in crude drugs lead to the exploration of more quality and standardized products in the form of phytopharmaceuticals. Standardization of isolation procedures for phytopharmaceuticals can offer better quality products. Industrial pharmacognosy is one of the emerging branches in natural products science, which keeps being updated with the latest developments in natural-product-derived nutraceuticals, phytopharmaceuticals, and cosmeceuticals. Industrial pharmacognosy is a broad coverage of these three defined areas along with its control, standardization procedures, and regulatory guidelines that keep changing every year. Recent innovations in industrial pharmacognosy have incurred certain biopharmaceuticals that require new regulation set up for trading and other post- and pre-marketing practices. Crude drugs have become an important component of complementary and alternative drugs in the globe to treat various ailments. Majority of these crude drugs are derived from plants and formed essential part of human health-care system. These crude drugs have been practised since ages in different traditional systems of medicine of China, Japan, Korea, India, and other Asian countries. Nevertheless, due to variances in genetic characters, geographical, cultivation, processing, harvesting conditions, exogenous impurities, etc., the quality difference of crude drugs is considerable. Due to this, quality difference of chemical profile and therapeutic properties of plant is badly impacted, and this has become one of the major differences among crude drugs and chemical medicines. Also due to the complex nature of multicomponent systems, there is always challenge in developing the suitable quality control and standardization protocol for respective product under examination. Achieving quality uniformity has become one of the main characteristics of herbal medicine. Variation in the quality can result into natural medicine with substandard and compromised quality which can affect its safety and efficacy. Apart from the quality control measures, it's important to identify those medicinal plants with high therapeutic efficacy and have more understanding related with its recent development in an order to develop safe and effective products. This book includes six chapters: the first chapter of the book includes details about the herbal drug market and its regulation at global level, whereas the second chapter is primarily focused on the isolation and production of important phytopharmaceuticals at industrial level. Chapters 3–5 summarize the most important crude drugs used in herbal drug industry with a recent knowledge of their profile, phytochemistry, and ethnopharmacological uses. Chapter 6 includes an updated information about quality control as well as the standardization of herbal products. This new volume plans to cover all of these potential areas in separate chapters, which will be arranged systematically to provide current information about latest developments in this field. In contrast with allopathic medicine, herbal medicine also requires stringent guidelines to regulate its unauthorized use in different nations or to make broader communities aware of the significance of using standardized products rather than in crude form. The book will discuss recent amendment in WHO guidelines regarding these guidelines. These chapters could be more interesting for pharmacognosists, pharmacologists, traditional practitioners, natural products chemists, academicians, agronomist, and ultimately all academicians, researchers as well as students in the area of natural products science. The authors are enormously thankful to the publisher for their support, guidance, and cooperation, especially Dr Renu Upadhyay of CRC Press, in publishing this book. Recommendations and comments will continuously be solicited to further enhance the quality of the book. Valuable suggestions from the readers will always be appreciated by the authors to further improve the quality of the book.

# Acknowledgements

We would like to dedicate this book to all students, researchers, academicians, and all the scholars who have sincerely contributed to the area of natural products, depicted in the book, at national as well as international level. We would also like to acknowledge the University of Nizwa, Sultanate of Oman to extend its support for accomplishing this book project successfully. We are grateful to Natural and Medical Sciences Research Center (NMSRC) housed at the University of Nizwa for providing central resources of advanced analytical instrument for our research work and for promoting interdisciplinary research studies. NMSRC is a research centre where several scholarly and research activities, including natural product research studies, are highly encouraged. This centre has given a valuable base to this book project. The authors are also thankful to NMSRC for offering excellent facilities required for the completion of this book. Last but not least, we show our sincere gratitude to the whole team of the CRC Press for furnishing their active cooperation and support.

جامعة نزوى

University of Nizwa

مركز أبحاث العلوم الطبيعية والطبية
Natural & Medical Sciences Research Center

**Prof. Ahmed Al-Harrasi**

Professor
Natural and Medical Sciences
Research Center
University of Nizwa, Oman

**Dr. Saurabh Bhatia**

Associate Professor
Natural and Medical Sciences
Research Center
University of Nizwa, Oman

# Authors

**Ahmed Al-Harrasi** received his BSc in chemistry from Sultan Qaboos University (Oman) in 1997. Later he moved to the Free University of Berlin from which he obtained his MSc in chemistry in 2002 and then his PhD in organic chemistry in 2005 as a DAAD fellow under the supervision of Prof. Hans-Ulrich Reissig. His PhD work was on the New Transformations of Enantiopure 1,2-oxazines. Then he received the Fulbright award in 2008 for his postdoctoral research in chemistry for which he joined Prof. Tadhg Begley's group at the Cornell University where he worked on the synthesis of isotopically labelled thiamin pyrophosphate. After completing the postdoctoral research stay at the Cornell University in 2009, he started his independent research at the University of Nizwa, Oman where he founded the Chair of Oman's Medicinal Plants and Marine Natural Products merging chemistry and biology research that became a centre of excellence in natural and medical sciences. He is currently a professor of organic chemistry and the vice-chancellor for graduate studies, research, and external relations at the University of Nizwa. The budget of his interdisciplinary-funded projects exceeds $15 million. He was a chair and invited speaker in many international conferences. He is a referee for more than 15 international chemistry and biotechnology journals. He has authored and co-authored over 500 scientific papers, 5 books, and 15 book chapters. He taught chemistry courses both at undergraduate and postgraduate levels. He has been included in the Stanford University's global list of top 2% Scientists in the years 2019–2020 and 2020–2021.

**Saurabh Bhatia, PhD,** is an Associate Professor at the Natural and Medical Sciences Research Center, University of Nizwa, Oman, and an Adjunct Associate Professor in the School of Health Services School of Health Science, University of Petroleum and Energy Studies, Dehradun, Uttarakhand. He graduated from Kurukshetra University, India, in 2007, followed by postgraduation in 2009 from Bharati Vidyapeeth University, India. He successfully completed a PhD programme in pharmaceutical technology at Jadavpur University, Kolkata, India (2015). He has 12 years of academic experience, and his areas of interest include nanotechnology (drug delivery), biomaterials, natural products science, biotechnology (plant and animal), microbiology, modified drug delivery systems, analytical chemistry, parasitology (leishmaniasis), and marine science. He has promoted several marine algae and their derived polymers throughout India and has published more than 10 books and 67 articles in many areas of pharmaceutical science. He has participated in more than 30 national and international conferences.

# 1 Global Herbal Drug Market and Its Regulations

*Ahmed Al-Harrasi*
Natural and Medical Sciences Research Center, University of Nizwa

*Saurabh Bhatia*
Natural and Medical Sciences Research Center, University of Nizwa
School of Health Sciences, University of Petroleum and Energy Studies

*Sridevi Chigurupati*
Qassim University

*Tapan Behl*
Chitkara University

*Deepak Kaushik*
M. D. University

## 1.1 INTRODUCTION

Herbal drug is defined as plant-based products which have been transformed into processed form of crude drug-based raw materials after subsequent procedures of collection, drying, storing as well as packing. However, this definition also includes certain unorganized form of crude drugs such as essential oils, latex, mucilage, gums, waxes, and extracts, which generally require less processing. Herbs are natural products, and their chemical nature differs from species to species depending on numerous factors including environmental conditions (biotic as well as abiotic factors). Also, plant type, harvesting method, processing and storage conditions, extraction methods, and type of phytochemicals significantly impact shelf life of the crude drugs. These all factors significantly contribute to the phytochemical profile of plant extract which determines therapeutic efficacy of crude drugs. Thus, official standards related to phytochemical profile, extraction, storage as well as processing conditions must be referred to maintain the consistency of phytochemical profile as well as therapeutic efficacy. Due the complex nature and multicomponent system of plant-based products, it's important to maintain the consistency in context to the phytochemical profile to maintain the quality as well as the same therapeutic efficacy among different batches of crude drugs [1]. Crude drug-based knowledge as well as practices have been adopted since ages; thus, due to their established ethnotherapeutic effects, consumers have strong belief on therapeutic as well as safety profile. In each nation, traditional crude drugs have been utilized owing to their fruitful therapeutic outcomes and, thus, form an important part of indigenous system. World Health Organization (WHO) has been consistently working in setting up guidelines for crude drugs in terms of their dosage, safety as well therapeutic efficacy part to assess the effectiveness of traditional medicine [1]. It is a branch of pharmacognosy that deals with/evaluate the commercial viability/importance of natural products that are already established in pharmacognosy. In the 19th century, the term *"materia medica"* was used for the subject known as pharmacognosy. While studying the drug called Sarsaparilla, it was Seydler,

DOI: 10.1201/9781003274124-1

a German scientist, who coined the term Pharmacognosy in 1815 during the title of his work *"Analecta pharmacognostica"*. Pharmacognosy is derived from two Greek words

$$Pharmakon\,(a\,drug) + Gignosco\,(to\,acquire\,knowledge) \tag{1.1}$$

Pharmacognosy is a branch of bioscience that deals with comprehensive study of crude drugs obtained from plant/animal/mineral or any other natural origin. WHO has registered 21,000 plants for medicinal purposes across the globe, out of which 2500 species are accessible in India. Out of these 2500, 150 species have been used at commercial scale in India. Still 75–80% of the people rely on herbal medicine, mostly in the developed nations, for various ailments primarily owing to the more compatibility with the human system, better cultural acceptability, and relatively better safety profile. Nevertheless, major changes have been observed in last few years regarding their utilization in developed countries. In countries like Germany and France, various plant-based products have been considered as prescription medications. Their approximate trades among European countries were US$6 billion (1991), and currently it would be greater than US$20 billion. In contrast, crude drugs in the United States are presently traded as dietary supplements in grocery stores with a business of around US$4 billion (1996), which will further increase in future [2]. Indian crude drug-based market is around US$1 billion with an export of about US$80 million [3]. Crude drugs also include nutraceuticals which represent the trade of approximately US$80–250 billion in the United States [4]. Asian countries like India, China, and Japan have their own indigenous system of medicine which have been practised from several years. Nevertheless, except China and India, other nations have not explored their traditional knowledge in terms of using herbal medicines more frequently for different ailments. This can be done by utilizing herbal preparations for different disease, especially for those for whom medicines are not available. Such herbal products with high medicinal value must be available in such nations where other treatments are not available. Advantageously, several Asian countries are practising their respective indigenous system of medicine; however, their rate of utilization of crude drugs may vary among different nations. Currently, Asian countries like India, China, and Japan became a major exporter of herbal drugs across the world. However, there are some important parameters for the nations to enter international trade of herbal drugs:

- Crude drugs must be free from foreign materials such as pesticides, heavy metals, etc.
- Evidenced traditional or ethnopharmacological knowledge or database
- Stability as well as safety profile
- Quantitative as well as qualitative analysis based on phytochemistry

In addition, mechanism, safety profile, pharmacokinetic-based information, dosage as well as pharmacological properties including the molecular targets of phytochemicals in clinical studies is required to establish its therapeutic efficacy.

## 1.2   HISTORY OF HERBAL MEDICINE

Plants had been used for therapeutic reasons long before recorded history. Ancient Chinese and Egyptian papyrus writings explain medicinal utilization for plants as early as 3000 BC. Native cultures (e.g., Native American and African) used herbs in their healing rituals, while others developed traditional systems (e.g., Ayurveda, Siddha, Unani, and traditional Chinese medicine [TCM]) in which herbal therapies were used [5]. TCM is one of the excellent examples to present how crude drugs have been utilized from primordial time (more than 3000 years) till today in healthcare systems. The presence of Chinese manuscript entitled *The Devine Farmer's Classic of Herbalism* (2000 years ago) is considered as the ancient script in the sphere; however, accumulated information has now divided into different herbal pharmacopoeias and numerous profiles on single herb. Recently, utilization of plant-based medicines and other botanicals in the West has augmented

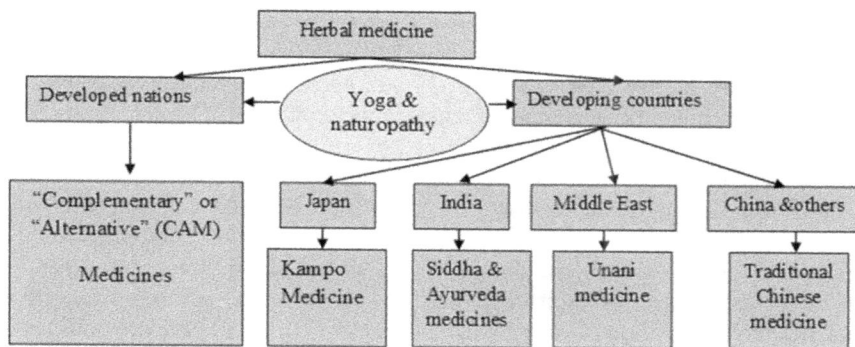

**FIGURE 1.1** Worldwide utilization of herbal medicines.

manifold. In the 19th century, our medicinal practices were largely conquered by plant-based medicines. Nevertheless, the therapeutic use of herbs went into a rapid decline in the West when more conventional synthetic drugs were made frequently available. On contrary, various developing countries continued to promote from the rich data of medical herbalism, e.g., Kampo Medicine in Japan, Siddha and Ayurveda medicines in India, Unani medicine in the Middle East and South Asia, and TCM, are still utilized by a large majority of population (Figure 1.1) [6].

To meet the healthcare requirements, especially in the remote locations where there is no access to main course allopathic medicines, both India (up to 70%) and Africa (up to 90%) are dependent on traditional medicine. China has overall 40% of accountability for traditional medicine in healthcare system, and majority of the hospitals have division for traditional medicine [7]. With more advancement and more awareness in herbal drug technology, herbal drugs utilization among the developing nations has been drastically increased these days. Based on data retrieved in 2007, from developed countries like the United States, utilization of herbal drugs was found to be 38% among grownups and 12% among kids. It was also found that practising natural products except vitamins as well as minerals considered as alternative medicine, and almost 18.9% of population use alternative medicine for different ailments. A study performed over Chinese population (2003) suggested that 40% of the individuals presented noticeable belief in TCM then Allopathic medicine. In a report, it was found that out of adults (n = 21,923) in the United States, some (12.8%) use herbal supplement and others use dietary supplements.

### 1.2.1 Traditional Medicine

Traditional medicine is defined as the health ways, knowledge, approaches, and way of life by utilizing natural remedies, manual techniques, mystical treatments, naturopathies, and trainings, in single or in combination to extravagance, identify, and prevent diseases. In developed nations, adaptation of herbal preparation is known as "Complementary" or "Alternative" (CAM). Currently, there is a switch from synthetic to herbal medicine, which signifies "Return to Nature". Natural products owing to the presence of potent therapeutic phytochemicals have been utilized to treat various ailments. Nature has given India a huge diversity of medicinal plants which is referred to as the *"Medicinal Garden of the world"*. Nations with ancient cultures have their own indigenous system or ways to utilize herbal medicine in different forms for different ailments. Out of all nations, India has its own traditional systems of medicine including Ayurveda and Siddha which have been often used for the healthcare of people. Because of the safety profile of crude drugs, it's widely consumed among both countryside and city-based population of India. Additionally, due to cost-effectiveness, patient compatibility, less toxicity, and easy availability, natural products have been utilized throughout the world for the prevention as well as treatment of various ailments [8]. Utilization of natural product is high in a region where the population have less access to main course medications, especially in rural or tribal regions for primary healthcare [9]. It's due to the common conviction that crude drugs cause less toxicity, are inexpensive, and easily available, they are widely utilized [10]. As per WHO, the utilization of natural medicines across has been drastically increased than synthetic medications which are available [11]. The utilization of plants for therapeutic uses predates human history and forms the origin of much modern medicine. Various conventional drugs are developed from plant sources: a century ago, most of the few efficient drugs were plant-based, e.g., dioxin (from foxglove), quinine (from cinchona bark), aspirin (willow bark), and morphine (from the opium poppy) [12]. Morphine is considered as the first therapeutically active and pharmacologically relevant phytochemical which was obtained from poppy plant 200 years ago. Herbal-based treatments show fruitful results and now it has been used alternatively with main course medication to treat various ailments such as cancer [13]. As per the previous report, out of 177 drugs accepted globally to treat cancer, over 70% are obtained from natural resources and using these lead phytochemicals with combinatorial chemistry many analogues have been developed. Various anticancer drugs of natural origins have been obtained from natural resources such as:

- paclitaxel from *Taxus brevifolia*
- camptothecin from the *Camptotheca acuminata*
- camptothecin further used to synthesize irinotecan and topotecan

It was also found that approximately 25% of the main course drugs approved globally are obtained from natural resources, and around 121 phytochemicals have been used often. Between 2005 and 2007, 13 drugs derived from plant sources were approved in the United States. More than 100 natural product-based drugs are in clinical investigations, and of the total 252 drugs in the WHO's essential medicine list, 11% are exclusively of plant origin. Owing to the less profit generated from herbal drug market, use of herbal medicines was overcome by newer synthetic drugs in the mid-20th century [14]. Since technologies and science became more advance in the 20th century, herbal drug-based treatment was criticized for its potency against newer synthetic drugs. Owing to the iatrogenic and other undesirable effects caused by the herbal medicines, their utilization was highly impacted. Also, regulation of the herbal medicine to ensure the quality, safety, and efficacy is still a major concern, due to which the undesirable components such as heavy metals and proportion of the active component remain uncertain. Owing to emerging interest in natural products due to the conception of using medicinal products those are close to nature, herbal market again retained its existence and value. In 1992, alternative medicine division has been formed by the National Institute of Health (the United States) with the prime objective to monitor and regulate the development of herbs and herbal products. National Center for Complementary and Integrative Health has also been established to create the database for herbs and herbal products. Globally, crude drugs utilization has been increasing since WHO has started boosting developing nations to utilize the natural products, especially herbal medicine to achieve effects or therapeutic responses not produced by allopathic medicines [15]. Major areas of herbal industry are mentioned in Figure 1.2.

- Phytopharmaceuticals: Plant-based pharmaceuticals which are present in isolate and combination forms. Various industries at domestic and international levels producing phytopharmaceuticals at larger level, e.g., reserpine, quinine, diosgenin, and sennosides.
- Cosmeceuticals: Cosmeceuticals refers to the combination of cosmetics and pharmaceuticals. A cosmeceutical is essentially a skincare product that contains biologically active compound that is thought to have pharmaceutical effects on the skin. Rhealba®, cosmeceutical, oat plant extract soothes and restores fragile skin in acne by reducing inflammation and bacterial adhesion. Thus, herbal cosmetics are those beauty products (skin care) which possess desirable physiological activities such as skin healing, smoothening and appearance enhancing, and conditioning properties, e.g., aloe vera, amla, almond oil, brahmi, bhringraj, castor oil, coconut oil, eucalyptus oil, henna, neem, marigold, and sandalwood oil.

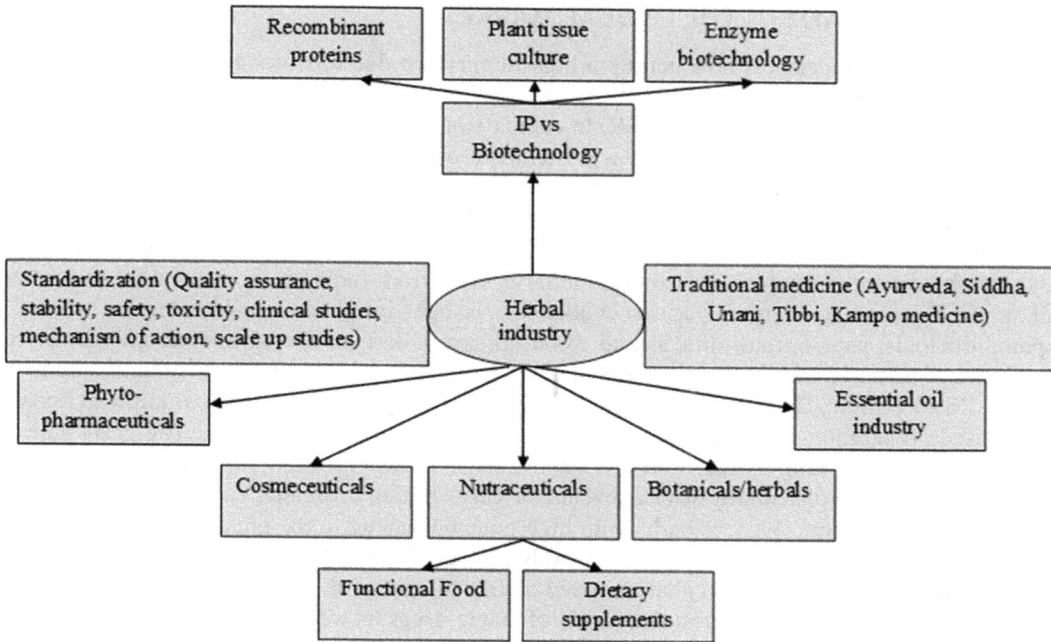

**FIGURE 1.2** Major areas of herbal industry.

- Nutraceuticals: Nutraceuticals are the amalgamation of the words "nutrition" and pharmaceutical. The term nutraceutical is defined as parts of food that have a medical or health benefit, including the prevention and treatment of diseases. There are two major types of nutraceuticals: dietary supplements and functional food.
- Botanicals/herbs evaluation: In botany, herb is a non woody, seed-producing plant that dies to the ground at the end of the growing reason. In culinary practices, herbs are vegetables that are used to add flavour or aroma to food. In medicine, herbs are crude drugs of vegetable origin that are used in the treatment of disease states, often of a chronic nature, or to attain or maintain a condition of improved health. Exploration or estimation of herbs/botanicals determines the regional estimate of number herbs present in area. That's how balance can be maintained between endangered and new species. This estimation also determines number of botanicals and its scope of industrialization of its chemical components as herbal/synthetic medicines.
- Traditional medicine: Each nation has its own way of consuming drugs. Consumption of crud drugs as medicine is widely observed in developing nations. By following their ancient civilization, each nation somehow is able to preserve their culture in the form of traditional medicine, thus each nation has its own traditional system which is now transforming theoretically into pharmacopoeia (monographs) and practically into national offices, policies, and expert committees to regulate the manufacturing of herbal products and classify them in suitable category by registration of products with its regulation under national legislation/act.
- IP vs biotechnology: This branch explores/updates the latest ongoing innovations in pharmacognosy with aid of biotechnology, e.g., recombinant proteins and enzyme production, transgenic plants, and edible vaccines.
- Quality control/standardization: This area ensures/determines the stability, quality, and safety profile of herbal medicines.
- Essential oil industry: Estimate of total number of plants containing essential oils and requirement of their subsequent cultivation with suitable isolation/extraction procedures.

## 1.3  HERBAL TRADE IN THE GLOBAL MARKET

Utilization of herbal drugs is now getting a high momentum due to which herbal drug market is growing and expanding drastically. As it was estimated, the overall annual revenue of the Indian herbal market is around 2300 crores (INR) in comparison to profit generated from pharmaceutical industry's 14,500 crores (INR) [16]. India is the major exporter of the herbal drugs in the global market and thus holds good position in the international herbal drug trade. India holds a good position in producing as well as exporting several crude drugs such as castor seeds the annual production of which was estimated around 1,25,000 tonnes. Some of the most important crude drugs and the phytochemicals which are exported at large scale from India are agar wood oil, cinchona alkaloids, diosgenin, guar gum, ipecac root alkaloids, isabgol, jasmine oil, mehndi leaves, menthol, opium alkaloids, papain, rauwolfia, sandal wood oil, senna derivatives, solasodine, gudmar herb, and vinca extract [17, 18]. Overall profit generated from herbal preparations in India is approximately US$ 1 billion. These preparations include over-the-counter preparations, traditional herbal drug-based preparations along with herbal-drugs-based traditional crude drugs. The profit generated from the crude drug extract was estimated around US$80 million [19]. Indian herbal drug market is growing exponentially with a revenue of US$1 billion estimated earlier, which has now possibly increased more. Various herbs with high potential owing to the presence of high level of therapeutic compounds have been used extensively (Table 1.1) [19, 20]. In India, it's anticipated that there are almost 25,000 approved pharma based on Indian system of traditional medicine. Because of this vast set up, there are almost thousands of single drugs as well as around 3000 compound preparations listed. Some of the most important preparation which have been extensively utilized are highlighted in Figure 1.3.

India-based herbal medicine industry so far utilized 8000 plants with medicinal value. Additionally, there almost 8000 producers of herbal medicines out of which 20 produce herbal drugs as well as its preparation at large scale. Overall, annual profit generated from these herbal preparations in India was estimated around US$300 million while revenue generated from Ayurvedic and Unani preparations was around US$27 million (Figure 1.4). Revenue generated from Ayurvedic and other herbal preparations had reached to US$31.7 million during 1998–1999; however, this value had been considerably increased during the period of 1999–2000 and reached up to US$48.9 million. Overall export of the herbal preparation was estimated around US$80 million [21]. Utilization of crude drugs as well their respective formulation in India is more common, and it was found that almost 960 different plants have been used by the Indian herbal industry. Out of these plants, 178 plants are produced more with more than 100 metric tonnes annually.

### TABLE 1.1
### Medicinal Plant Parts Exported from India, Imported Medicinal Plants and Their Parts

| Exporting of Herbals | | Importing of Herbals | |
|---|---|---|---|
| *Argemone mexicana* | Fruit | *Adhatoda vasica* | Bark and leaf |
| *Curcuma longa* | Rhizome | *Curcuma aromatica* | Rhizome |
| *Acorus calamus* | Rhizome | *Aloe vera* | Dried leaf |
| *Curcuma aromatica* | Wild turmeric | *Garcinia indica* | Fruit |
| *Cassia lanceolata* | Leaves | *Gloriosa superba* | Tuber and seed |
| *Glycyrrhiza glabra* | Root | *Juniperus communis* | Fruit |
| *Withania somnifera* | Vegetable rennet | *Myrica nagi* | Bark |
| *Piper longum* | Fruit | *Phyllanthus amarus* | Fruit |
| *Swertia chirata* | Whole plant | *Ocimum sanctum* | Leaf and essential oil |
| *Zingiber officinale* | Rhizome | *Vinca rosea* | Leaf, seed, and stem |

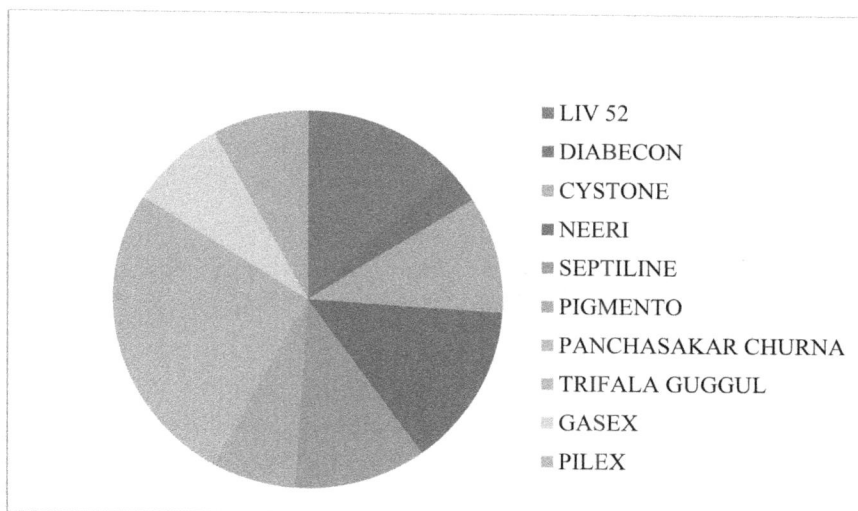

**FIGURE 1.3** Best formulations which are utilized in developing countries.

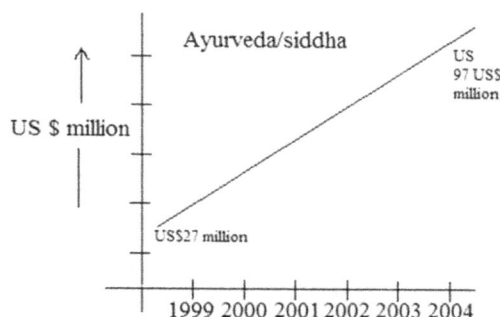

**FIGURE 1.4** Contribution of Ayurveda/Siddha in herbal industry having revenue of approximately US$300.

## 1.4 HERBAL MARKET AT INTERNATIONAL LEVEL

It is quite challenging to evaluate the annual trade value at a global level as due to many factors such estimation is always miscalculated. This is because of the different ways of utilization of herbs in different forms such as food products, multivitamins, raw form, and energy drinks. As per the estimation done by WHO, US$83 billion is the overall trade that has been estimated at the global level [22]. At the national level, the overall revenue obtained from crude drugs has been estimated based on marketing and sales data. Nevertheless, in some countries, it's estimated based on how much quantity of raw material has been produced domestically in a cost-effective manner.

### 1.4.1 INTERNATIONAL TRADE OF HERBAL PRODUCTS

The annual herbal products trade was estimated around US$60 billion at the global level with an annual growth rate of 6.4% [23, 24]. Because of the several favourable factors (listed below) especially inclination of majority of the population towards herbal products, international market as well as trade of herbal products is expanding exponentially:

- Owing to the side effects as well as toxicity caused by the current medications synthetic drugs are losing faith.

- On the other side, high expenditure required to set up the formulation and development, research and development, quality control, manufacturing unit, packaging unit as well as production unit make the whole process more expensive and time-consuming.
- Inclination of end consumers towards natural products.
- Preference of populations for preventive medicine due to increasing population age.
- As these products have been used from long time, thus majority of the population is compatible with this type of treatment.
- Inclination towards self-medication.
- Since herbal products have been utilized from several decades, population has faith on herbal medicines in regard to its efficacy and safety.

As per WHO, herbal medicines are profitable globally and they symbolize a market value of about US$43 billion per year. It was estimated in 1991 that the herbal medicine market in the European countries was about US$6 billion, e.g., France had a revenue of US$1.6 billion, Germany accounting for US$3 billion, and Italy US$0.6 billion, whereas in other nations it was found to be US$0.8 billion. In the late-19th century, US$10 billion revenue from herbal medicine was estimated in Europe while US$4 billion was estimated in America, US$1 billion in India, and in other nations it was found to be US$5 billion. In the late-19th century, revenue obtained from European market raised up to US$7 billion. Out of all the European nations, Germany covers almost 50% of herbal product-based market in Europe with a revenue of US$3.5 billion. This is followed by the other nations such as France (US$1.8 billion), Italy (US$0.7 billion), the United Kingdom (US$0.4 billion), Spain (US$0.3 billion), and the Netherlands (US$0.1 billion). In the United Kingdom generally for the product registration, application of the medicinal dossier must be submitted to the regulatory body, in order to meet the safety as well quality standards. In America, CAM (complementary and alternative medicine system) has been often used with traditional medicine to improve the efficacy of whole therapy such as acupuncture has been used to decrease the pain. In late-19th century, revenue generated from Chinese herbal market was estimated around US$2.5 billion [25]. In other countries as well, during the period of 2003–2004, revenue obtained from herbal drug market has been increased considerably such as Europe (Western part) raised up to US$5 billion. During the year of 2005, Chinese annual sale trade of herbal products reached up to US$14 billion, whereas profit generated from Brazil herbal market was found to be US$160 million in 2007. At the global level, overall revenue generated from herbal medicine was almost found to be US$60 billion. China and India are the major players which contributes 30 and 1%, respectively, out of whole US$60 billion. Most of the Chinese medicines are standardized which increases its authenticity in the market. Total global revenue of herbal industry across several nations per annum is mentioned in Table 1.2.

## TABLE 1.2
## Total Global Revenue of Herbal Industry across Several Nations per Annum

| Countries | Revenue per Annum | | | |
|---|---|---|---|---|
| | 1991 | 1996 | 1997 | 2006 |
| European countries | US$6 billion | US$10 billion | US$7.0 billion | US$10.0 billion |
| France | US$1.6 billion | US$1.7 billion | US$1.8 billion | US$1.6 billion |
| Germany | US$3 billion | US$3 billion | US$3.0 billion | US$3.0 billion |
| Italy | US$0.6 billion | US$0.5 billion | US$0.7 billion | US$0.6 billion |
| United Kingdom | US$200 million | US$250 million | US$400 million | US$400 million |
| Spain | US$250 million | US$350 million | US$300 million | US$300 million |
| Netherland | US$60 million | US$60 million | US$100 million | US$100 million |
| United States | US$4 million | US$4 million | US$400 million | US$4 billion |
| India | US$1 billion | US$1 billion | US$1 billion | US$1 billion |

### 1.4.2 Herbal Medicine and Its Market

Utilization of plant-based medicine has been drastically increased in Europe (US$6 billion, 1991 followed by the US$4 billion, 1996). Almost 50% of the revenue received in Europe was because of the high sale in Germany (US$3 billion). Other countries such as France (US$1.6 billion), Japan (US$1.5 billion), and Italy (US$0.6 billion) also contributed to this sale. Recently, US$250 billion sale of herbal drugs has been reported [26]. Indian herbal market has evidenced an estimated sale of US$1 billion with an overall export of extracts found to be US$80 million. Out of this US$80 million, 50% has been contributed by Ayurvedic traditional preparations. Utilization of herbs-based products has been also evident in Russia as well as Germany. Majority of the Indian as well as Chinese herbal preparation have been imported by other nations and regulated as per their herbal drug national regulation or policy or standards.

### 1.4.3 Development in Herbal Medicine Industry with Reference to Trade

Due to the inexpensive nature, easy availability, strong ethnobotanical background, low toxicity, and broad therapeutic effectiveness, herbal drugs are always preferred over synthetic medicines [27]. Medicinal as well as aromatic plants have been considered as an important source for these herbal drugs and these herbal preparations have been known for their medicinal value due to the presence of active phytochemical content. These crude drugs have been also utilized in raw form as well as various herbal preparations have been made from these crude drugs by using classical or advanced procedures. Some of these herbal preparations have also been considered as food supplements as nutraceuticals and some have been also utilized as cosmetics. Consumption of nutraceuticals is more common among American as well as Japanese communities. In America nutraceutical market is growing rapidly with an annual revenue between US$80 and 250 billion with an annual growth rate of US$1.5 billion [28]. This herbal market based on nutraceuticals has expanded because of Dietary Supplement Health Education Act, United States, in 1994. As per this act, chemical components or extracts or any other fraction isolated from plants as well as animal's sources can fall under nutraceuticals. Indian herbal market found a new space to market the herbal products under the nutraceuticals [29]. However, herbal products including organized as well as unorganized retain great importance in terms of their therapeutic efficacy, safety profile, and unique phytochemical nature [30]. Recently, most of the herbal products-based companies are emphasizing on standardization as well as quality control to determine quality, purity, safety as well efficacy of herbal products. Most of the companies identify biomarkers or the active compound present in the crude drugs and determine their actual content in the crude drug to check its quality. Some of the major pharmaceutical companies which are more into development of herbal formulations as well as its standardization and quality control are as follows:

- Zandu Realty Limited
- Dabur Ltd. (have 29 patents in the United States)
- The Himalaya Drug Company
- Hamdard Laboratories
- Maharishi Ayurveda Products Private Limited

## 1.5 HERBAL MEDICINE IN THE TREATMENT OF CHRONIC DISEASE

Since decades crude drugs have been utilized as an ultimate source to treat various ailments. The primary utilization of natural products in the form of medicine has been evidenced 5000 years ago [30] in China. Similarly, by using the traditional knowledge of Ayurveda, several herbal medicines have been utilized in India to treat various ailments for more than 5000 years. These herbal medicines are the most important element of Ayurveda-based treatment [31]. Still Ayurveda has been practised not even in the nation but across the world as its benefits have now become more visible to different populations from different countries. In the United States, utilization of Ayurveda in the form of CAM to treat different ailments has become more evident these days [32]. It has been also observed that aged group relatively use more herbal medicine as the aged individuals are more likely to have chronic disease, which doesn't every time require treatment by synthetic drugs. It was also found among this age group that the effectiveness in terms of herb vs illness as well as herb vs drug interactions was found to be more fruitful. Due to lack of proper standardization, quality control measures, regulation as well as clinical studies, herbal medicines still face criticism in context with its therapeutic effectiveness as well as safety profile. Many nations use herbal medicines along with main course medication to effectively treat respective ailments; however, some nations use herbal medicines solely as a source to treat different ailments, and some are using it alternatively to treat different diseases. However, still more research is required to assure its efficacy, potency, and safety profile.

## 1.6   PRESENT STATUS OF HERBAL MEDICINE

Utilization of herbal medicine is not limited to developing countries, since it has been projected that 70% of all medical practitioners in France and German regularly recommend herbal medicine [33]. Number of patients those are using herbal medicines to treat different ailments have been increased considerably [34]. By following US FDA regulations for selling herbal products [35], majority of the drug makers produce, sell as well as distribute crude drugs accordingly. Current US market is flourished with several products with high revenue in contrast to the herbal medicine market in 1991; in the countries of the European Union, the revenue was about US$6 billion (may be over US$20 billion now) and in Germany it was accounted for around US$3 billion, France US$1.6 billion, and Italy US$0.6 billion. Whereas in 1996, the US herbal medicine market was accounted for around US$4 billion, which have doubled by now. Out of the US$1 billion sale of herbal extracts, it was found that almost US$8 million revenue was generated from export of herbal extracts in India [36]. As current generation is now more familiar with the utilization of herbal drugs, most of the natural products are now widely utilized at large scale. The current focus on herbal medicine is based on the investigation to evaluate the authenticity of its claims for herbal cures and herbal medicine and it has been found to have some remarkable credentials. Developed empirically by trial and error, many herbal treatments were nevertheless remarkably effective [37]. Recent studies [38] estimated that 39% of all 520 new approved drugs in 1983–1994 were natural products or derived from natural products. Out of this 60–80% of antibacterial and anticancer drugs were derived from natural products [39]. One of the most popular drugs, penicillin, that replaced mercury in the treatment of syphilis and put an end to various deadly epidemics, comes from plant mould. Bellodona still act as a potential source for different chemicals utilized in ophthalmological preparations and in antiseptics to treat gastrointestinal disorders. *Rauvolfia serpentina,* which is known as Sarpagandha in India, contains vital components such as reserpine. This is considered as the main component of different tranquilizer which were used primarily in the mid of 19th century to treat different types of emotional and mental issues. Nevertheless, reserpine is rarely used today for this purpose; its discovery was a step forward in the treatment of mental illness. It is also the main element in a variety of modern pharmaceutical preparations for treating hypertension.

## 1.7   PRESENT STATUS OF HERBAL MEDICINE IN INDIA

India has a wealthy custom of herbal medicine as apparent from Ayurveda, which could not have boomed for 2000 years without any scientific reports. Ayurveda which accurately means knowledge (Veda) of life (Ayur) had its commencement in Atharvaveda. Charak Samhita and Sushruta Samhita are the two main renowned treatises of Ayurveda and various other were accumulated over the centuries such as Kashyap Samhita, Agnivesh Tantra, Bela Samhita, Vagbhata's Ashtanghridaya (600), and Madhava Nidan (700 AD) [40]. There are various preparations based on vegetable listed in *Materia Medica,* India, due to which widespread utilization of plant-based products came into picture. Sushruta and Charak as well as Vagbhata listed 700 crude drugs along with their identification features as well as therapeutic efficacy. Because of the clinical efficacy, 50 clusters of drugs have been made such as laxatives, anti-asthmatic, antiepileptic, digestive, stimulant, anti-haemorrhoid, anti-emetic, anti-diarrhoea, appetizers, anti-pyretic, anti-helminthes, haemopoietic, haemostatic, analgesics, sedative, anti-inflammatory, anti-pruritic, rasayana, rejuvenation, complexion, voice, and strength promoter to improve the secretion of milk, to improve the production of sperm as well as semen, to heal wounds as well as fractures, to cure renal stones, etc. [41]. Utilization of crude drugs as well as their respective preparations in the complications associated with liver such as liver inflammation caused by virus has been treated in India by herbal preparations since Vedic times. As per the survey, almost 170 phyto-chemicals were isolated from 110 medicinal plants which contain hepatoprotective compounds. It was also found that there are almost 6000 herbal preparations claimed for liver protective effects that have been sold out commercially. Out of all these hepatoprotective preparations, India holds patents of 40 herbal preparations containing mixture of 93 different plants [40]. Nevertheless, some of the most potent formulations reported so far are listed below:

- Silymarin-based hepatoprotective formulations
- *Picrorhiza kurroa*-based hepatoprotective formulations
- *Phyllanthus niruri* and *Phyllanthus amarus*-based hepatoprotective formulations
- Glycyrrhizin-based herbal preparations for peptic ulcer as well as hepatic complications

Glycyrrhizin-based herbal preparations known as stronger minophagen C has been used in Japan for liver inflammation caused by virus and known for their high efficacy. Similarly, hepatoprotective herbal preparations like Liv 52 has been extensively used in India which contains extract of various plants listed in Ayurveda for improving biochemical parameters among rodents and patients with liver complications [42]. A mixture of various crude extracts and Ayurvedic medicine known as Arogyavardhini has been clinically investigated for its efficacy against liver complications caused by virus [42]. Other herbal extracts derived from *Mucuna pruriens, P. amarus*, and *Tinospora cordifolia* have been also tested clinically against Parkinson's disease, hepatitis, and jaundice [42]. Some of the most common herbal products are cited in Figure 1.1.

## 1.8 WHO GUIDELINES FOR HERBAL PRODUCTS

In the late-19th century, WHO frame regulations for the assessment of herbs-based preparation and approved those regulation in 6th Int. Conf. of Drug Regulatory Authorities, Ottawa. Some of the most important criteria led down by WHO are:

- Evaluation of herbal drug quality including raw plant-based crude drugs or their extracts or products derived from either raw material or extracts.
- Guidelines associated with stability of herbal preparations to estimate their shelf life.
- Safety profile of raw material and preparations obtained from plants and assessment criteria based on ethnopharmacological background as well as toxicity studies.
- Evaluation of therapeutic efficacy based on in vivo or in vitro studies conducted on animals and human beings.

### 1.8.1 Last Amendment in the Guidelines

Following new points are included:

- Limit of the pesticides: Mainly pesticides like dichloro diphenyl trichloroethane (DDT), benzene hexachloride (BHC), toxaphene, and aldrin can cause serious side effects in human beings if the crude drugs are mixed with these agents.

- Limits of microbial contamination: Since all drugs are present in dried form so there might be chance of the retention of water, and just because of water and other nutrients microbe growth can occur.
- Radioactive contamination.

## 1.9 MAJOR OBSTACLES IN HERBAL PRODUCTS

Owing to multicomponent as well as complexity associated with phytochemicals, it's difficult to corelate active phytochemical profile with toxicological, pharmacological, safety, and clinical profile unless active chemical compounds have not been isolated, identified, and characterized properly. The following key issues remain:

- Uncertain clinical outcome
- Constraints with clinical trials and people available
- Tedious isolation procedures for phytochemicals
- Challenges associated with the characterization as well as identification
- Uncertain toxicological, pharmacological, safety, and clinical profile
- Unavailable clinical and pharmacovigilance data
- Standardization as well as quality control
- Herb-herb interaction in case of compound formulation

## 1.10 THE EVALUATION OF NEW HERBAL PRODUCTS CONSISTS

### 1.10.1 Assessment of Recent Herbal Preparations Comprises Following Steps

- Ethnopharmacological background or data analysis
- Phytochemical data analysis and stability profile
- Therapeutic profile (via involving both animals and human subjects)
- Clinical studies

## 1.11 WHO STRATEGY FOR TRADITIONAL MEDICINE

In 2002, WHO released its complete traditional medicine approach to help various nations. This approach was mainly based on the safety, efficacy, and quality of the herbal products, especially those are used in CAM and as traditional medicine. Following strategies have been introduced by WHO for the effective use of herbal medicine:

- Requirement of regulations or policies at the national level to assess and regulate traditional medicines as well as complementary and alternative medicine methods
- Encourage the effective therapeutic utilization of traditional as well as complementary and alternative medicines
- Record or data maintenance of herbal medicine
- Warrant accessibility and affordability of traditional medicines as well as complementary and alternative medicine

WHO has been also reported for aiding in performing clinical investigations on herbal drugs against malaria in three different nations from Africa. It was found that herbal drugs showed good clinical outcomes. Recently, the potential of herbal drugs against HIV, diabetes mellitus, anaemia (sickle cell), and malaria has been clinically tested in Tanzania by forming an association between WHO and China. Also, they are giving support to produce *Artemisia annua* (an antimalarial drug). This approach has reduced the cost of medicine from US$6–US$7 to US$2 per dose.

## 1.12 IMPORTANCE AND THE STATUS OF HERBAL MEDICINES AND COSMETICS

Herbal medicine has been around for centuries and dates to the Neanderthal period [11]. Herbal medicines have been utilized more among those patients those who usually wouldn't like to visit to clinic and have strong believe on herbal drug's efficacy as well as safety profile. Medicinal as well as aromatic plants have been used from decades and it has been anticipated that almost 35,000–70,000 plants have been used traditionally for treating various ailments. The majority of the population in developing countries still believe that conventional medicine has failed them and most of the patients use herbal medicines for seasonal mild infections, mild illness, mild injuries, and when either immediate access to medicine or clinic is not available and high expenses of medicine. Also, in tribal and rural regions, people have strong belief on herbal medicine as they are using them since ages and because of this reliability, traditional knowledge of practising herbal medicine travels from one generation to the other. Different clinics as well as healthcare centres available in village lack sufficient facilities to provide the full treatment to patients. Thus, tribal as well as rural population is dependent on traditional practices to cure various diseases. However, challenges associated with herbal medicine such as dose error while self-treatment, toxicity, and unknown mechanisms of action brings criticism over its utilization. Recently, majority of the clinics used herbal medicine along with allopathic medicine to sometime boost the effect of the primary medicine owing to the synergistic effect or to reduce or prevent the ill effects (e.g., toxicity) of primary medicine and to rejuvenate or to induce beneficial effects such as immunomodulatory, antioxidant, anti-inflammatory effects, etc.

## 1.13 INCREASING USE AND POPULARITY

Owing to the various benefits, plant-based products have been used from the centuries to treat various diseases. Plant-based products broadly cover various domains, out of which herbal medicine is considered as one of the most important components which was neglected due to the soon arrival of allopathic medicine. Herbal medicine-based treatment is considered as one of the most easily available and affordable treatment which is available in remote areas also. Due to the lack of the availability of modern medicine in such remote regions, herbal medicines have been extensively utilized. Plants are not only source of providing raw material as crude drugs, but also provide an array of therapeutic phytochemicals which have been utilized as lead compounds to synthesize derivatives in the laboratory. Thus, due to these certain herbal manufactures encourage (or invest in) cultivation of medicinal or aromatic plants to get the sufficient raw material to isolate phytochemicals. Hence medicinal as well as aromatic plants and traditional medicines are considered as an important component of pharmaceutical market. In the late-19th century, medicinal herbs as well as their respective preparations had almost 33.1% share in Chinese pharmaceutical market. On annual basis it was estimated that herbal medicine-based market in Malaysia was found to be 1 billion MYR.

## 1.14 WHO'S POLICY ON HERBAL MEDICINES

The WHO aware the other nations about the value of medicinal as well as aromatic plants and their respective traditional preparations. In the late-19th century, World Health Assembly accepted a decision over the regulations as well as managing medicinal plants. This has realized the value of medicinal plants in global healthcare system. World Health Assembly also made criteria for assessing safety and efficacy of herbal products and published the same information. Subsequently from 1987 to 1989, few more declarations have been made in context to the medicinal plant's identification, assessment, processing, cultivation as well as collection, consumption, and guidelines to regulate medicinal plant-based products. In the early phase of 20th century, WHO made some guidelines for herbal products related with their regulation in various countries (Tables 1.3 and 1.4).

**TABLE 1.3**

**WHO Reports—2012 of Herbal Regulation in Different Nations**

| Countries /Registered Products/Revenue | Manufacturing /Medi Claims | NP, NO, Pharmacopoeia | Regulated As |
|---|---|---|---|
| France (787 medicines are registered) | French Medicines Agency (medicine claim may be made about them just like pharmaceuticals), manufactured as like pharmaceuticals | Not having any regulation, policy or guidelines, agenda, headquarters, at country level and no pharmacopoeia at national level present | Herbal medicines are regulated as OTC products |
| Denmark (170 registered products) | Medicine claim may be made about them just like pharmaceuticals, manufactured as like pharmaceuticals | Not having any regulation, policy or guidelines, agenda, headquarters, at country level and no pharmacopoeia at national level present | Herbal medicines are regulated as OTC products |
| Germany (3500 herbal medicines are registered) 2002: US$2.432 | Eudralex (European union rules relating to medicinal products), manufactured as like pharmaceuticals | National policy is present, German and European pharmacopoeia is present in binding, no national research institute | Herbal medicines are legalized as like prescription medicine, OTC products and as medicine for self-care |
| Hungary (no registration system) | Manufactured as like pharmaceuticals | National policy is present, no national research institute or office, European pharmacopoeia is used | Herbal medicines are regulated as prescription medicine, OTC products, dietary supplements and as medicine for self-care |
| Israel (no registration system) | No specific law for manufacturing herbal products, they are legally controlled as like pharmaceuticals to meet up the requirement | Developing National policy, national research institute or office, European pharmacopoeia is used, no national pharmacopoeia, instead homeopathic, United States, French pharmacopoeia is used | Herbal medicines are considered as dietary supplements and by law no medicinal claims may be made from them |
| Ireland (no registration system) | No specific law for manufacturing herbal products, they are legally controlled as like pharmaceuticals to meet up the requirement | Developing National policy, national research institute or office, European pharmacopoeia is used, no national pharmacopoeia | Herbal medicines are regulated as prescription medicine, OTC products, dietary supplements and as medicine for self-care |
| Netherlands (no registration system) | No specific law for manufacturing herbal products, they are legally controlled as like pharmaceuticals to meet up the requirement | Does not have National policy, national research, expert committee, institute or office, regulation, European pharmacopoeia is used, no national pharmacopoeia | Herbal medicines are regulated as prescription medicine, OTC products |
| Spain (2277 registered medicines) | No specific law for manufacturing herbal products, they are legally controlled as like pharmaceuticals to meet up the requirement | Does not have National policy, national research, expert committee, institute or office, regulation, royal Spanish pharmacopoeia 2003, national monographs also exist | Herbal medicines are regulated as prescription medicine, OTC products, dietary supplements and as medicine for self-care, health foods |

*(Continued)*

**TABLE 1.3 (Continued)**

**WHO Reports—2012 of Herbal Regulation in Different Nations**

| Countries/Registered Products/Revenue | Manufacturing/Medi Claims | NP, NO, Pharmacopoeia | Regulated As |
|---|---|---|---|
| Sweden (103 approved natural remedies), US$130 million | Herbal medicines are called natural remedy, for manufacturing herbal medicines are regulated as pharmaceuticals to meet the compliance | Does not have National policy, national research, expert committee, institute or office, regulation, National pharmacopoeia exists but they adhere to European pharmacopoeia | Regulated as OTC products for self-medication, only for certain health signs |
| Switzerland (1000 registered herbal medicines), 2000: US$117 million | For manufacturing herbal medicines are regulated as pharmaceuticals to meet the compliance | The Swiss agency for therapeutic products (national office), Does not have National policy, national research, expert committee, regulation | Herbal products having medicinal value are regulated as prescription and OTC products |
| Great Britain (300–500 licenced medicinal products), 2000: US$123 million | Licensed manufacturing herbal medicines are regulated as pharmaceuticals to meet the compliance | Having, programme, expert team but no national office. British pharmacopoeia, 2002, 124 national monographs | Herbal medicines are regulated under act medicines act 1968, category 1: licensed herbal medicines (Medicinal claims are permitted); category 2: exempt from the licensing requirement |
| Canada (US$715 million) | National policy, programme, regulations are available for traditional or complementary and alternative medicines but no state regulations to some health fields | Ten thousand herbal preparations were recorded, Collection of pharmaceuticals as well as departments, Canadian drug source for health specialists, Collection of non-prescription products, WHO monographs, United States pharmacopoeia, Pharmacopoeia of the People's Republic of China, PDR for herbal medicines, Herbal medicines, Expanded Commission E monographs, ESCOP monographs, British herbal pharmacopoeia, British herbal compendium | Herbal preparations are officially controlled as self-medication, over the counter medicines, natural health products as well as dietary supplements |

Based on these outcomes (mentioned in Tables 1.3 and 1.4), WHO's strategy on herbal products which are having medicinal value is mentioned below:

- The WHO is familiar with value of herbal medicines and their role in healthcare. Such medicines are important, easily available, inexpensive, and their proper utilization must be promoted.
- To encourage the appropriate utilization of plants as well as herbal medicines a complete plan for their identification, assessment of efficacy, and safety profile, formulation must be assessed.
- It is essential to make a methodical list and evaluation of crude drugs based on the clinical investigation; to introduce measures on the regulation of herbal medicines to ensure quality control of herbal products by using modern techniques, applying suitable standards and

## TABLE 1.4
## WHO Reports—2012 of Herbal Regulation in Different Nations

| Countries/ Revenue | NO, EC, NP | Regulation/Marketing | Pharmacopoeia, Registered Products |
|---|---|---|---|
| India (US$149 million) | NO (Department of Medicine and Homeopathy), EC 1962, NP 1964, | Herbal medicines are sold in pharmacies as prescription and over the counter medicines, Manufacturing regulations adhere to pharmacopoeias as well as monographs | Ayurvedic pharmacopoeia of India and the Unani pharmacopoeia of India. 4246 registered herbal medicines, US$149 million, Ayurveda (315 herbal medicines), the Unani (244 herbal medicines), and the Siddha (98 herbal medicines) |
| Myanmar | NPL 1993, NO&EC (Department of Traditional Medicine, 1989, Ministry of Health 1997) | Governing conditions for herbal preparations are restricted to GMP regulations; they are marketed in as OTC medicines | Monograph of Myanmar medicinal plants, 2000; 3678 registered traditional medicines |
| Sri Lanka | NP, 1982. NO (Department of Ayurveda in the Ministry of Health 1961), EC 1962 | Governing conditions for production consist of stick to data available in the pharmacopoeia and monographs Herbal preparation without regulation (do not have any regulatory status) are marketed | Compendium of medicinal plants contains 100 nation al monographs, 200, no state registering scheme |
| Thailand (US$16.7 million) | | Enrolled herbal medicines separated into OTC and prescription medicines | 2003, the Thai herbal pharmacopoeia, two thousand traditional medicines recorded in Thailand |
| Nepal | NPL (National Ayurveda Health Policy)1996, NP 1997, NO (Department of Ayurveda, Ministry of Health 1981), EC, 2001 | Herbal medicines are regulated as prescription and over the counter medicines, Regulatory requirements for herbal medicines as like typical pharmaceuticals | Registration scheme is present in Nepal; however, country has no pharmacopoeias or monographs. |
| Pakistan (US$87 million) | NP 1965, NO 1965, EC 2001. | Herbal preparations are available in pharmacies as OTC drugs | Tibbi pharmacopoeia (1967); Monographs of Unani medicines (Vol. 1), There is no registration system |

good manufacturing practices; and to include herbal medicines in the national standard or pharmacopoeia.

- Also, various plants have medicinal as well as aromatic characters are extinct and some are on the verge of extinction, WHO endorses the call for international cooperation and coordination to establish programmes for the conservation of medicinal plants, to ensure that adequate quantities are available for future generations.

## 1.15 HERBAL PREPARATIONS: MERITS AS WELL AS DEMERITS

There are various merits as well as demerits of herbal drugs. Consultation with health practitioner is required before its utilization for any health conditions.

### 1.15.1 MERITS

There are various benefits associated with using herbal drugs in contrast with other pharma-based drugs. Some of major merits are listed below:

- Less toxicity: Most of the traditional preparations are well accepted by the patient, with less case of toxicity then synthetic drugs and thus can be used safely.
- Therapeutic efficiency: These natural products-based medicines generally doesn't cause any toxicity, side effects or adverse drug reaction like synthetic drugs.
- Inexpensive: Majority of the crude drugs are inexpensive then the synthetic drugs as big infrastructure for testing their efficacy as well as safety is not required like synthetic drugs.
- Easy accessibility: Crude drugs are easily accessible as they are widely available even in the remote areas. Also, prescription or clinic visit is generally not required for such medicines. Even majority of the plants can be cultivated at domestic level.

### 1.15.2 DEMERITS

Even the plant-based products also suffer from various shortcomings as mentioned below:

- Cannot be used in all circumstances: Synthetic drugs are efficient in treating sudden and serious ailments or injuries such as cardiac arrest or asthmatic attack which cannot be treated by herbal treatment; thus, immediate assistance can always be given by lifesaving synthetic drugs with other support measures provided by the hospitals.
- No specifications: Majority of the products as well as raw material doesn't not have any specifications related with storage conditions, dose, mode of administration. Thus, there are always chances of spoilage of drug as well as overdosing which can significantly impact safety as well therapeutic profile of herbal drugs.
- Presence of other plant material: Toxic and poisonous adulterants, substituents, or any similar looking plant material admixture with the parent crude drug can result into serious consequence.
- Herb-based interactions: Herb-herb interactions as herb-synthetic drug interaction may have serious consequences. Some of the phytochemicals showed synergistic, additive, agonistic, and antagonistic effects which can lead to serious consequences thus its always advisable take the physician recommendation before using it in combination with others.
- Herbal products are not regulated: Majority of the nations have different regulations for herbal medicine and some nations failed in establishing regulations for herbal products. It was found that majority of the nations doesn't have strong regulatory system to assess the quality, purity, efficacy, and safety of these medicines. Thus, there is always a chance

of buying substandard, adulterated, spurious, or inferior herbal products by the end consumer. Also, there is always a considerable difference between two batches produced by different manufacturers. Also, batch to batch variation also common among products produced by the same manufacturer. Due to this variation it's difficult to calculate the exact dose of herbs.

## 1.16   CAUSES FOR SUDDEN GROWTH IN HERBAL MEDICINES

From the ancient time crude drugs utilization have been increased exponentially among various nations. This sudden increase in natural products utilizations has boosted herbal drug market with more revenue as well as trade in herbal medicine [43]. Various parameters that affect the growth of herbal drug market at global level are listed below [44]:

- Faith of peoples on herbal medicine since long time
- Toxicity and side effects caused by the synthetic medicine
- Some conventional therapies, where treatment paradigm is quite slow, psychologically patient finds solace in herbals
- Use herbal medicines as these medicines doesn't require prescription most of the time, thus self-medication is easy
- Now majority of the industries are standardizing and validating crude drugs based on their purity, quality, safety, and efficacy
- Herbal drugs are inexpensive and cultivated easily
- These herbal medicines have been also recommended when modern medicine has resulted into undesirable therapeutic outcome or showed less effectiveness [45]
- Complementary medicine has now come up with much better safety profile due to the strong regulations adopted by nations to maintain the quality of herbal medicine [46].

## 1.17   ASSESSMENT OF THE HERBAL MEDICINE BASED ON THEIR EFFICIENCY AS WELL AS SAFETY PROFILE

In 2004, the US National Center for Complementary and Alternative Medicine of the National Institutes of Health started providing support to clinical studies to assess the efficacy as well as safety profile of herbal medicines exploring the effectiveness of herbal medicine. For clinical testing of herbal products standards those are used for the assessment of pharmaceutical products has been applied. Such clinically tested herbal products have been patented by pharmaceutical firms and marketed these products at a considerable benefit. In vivo as well clinical testing results showed positive as well as negative findings. Quality of the clinical trials conducted for herbal products was satisfied; however, some trials also were not done properly [47–49]. It's quite challenging to establish the exact dose for herbal products especially as per the criteria's such as body weight, etc. Thus, evaluating the most effective and safe dose as per the standards is quite challenging [50, 51]. Thus, this could raise several concerns such as:

- Herbal products deprived of standardization as well as quality.
- Lack of clinical trials data of the herbal products.
- Dosage variation among the herbal products.
- Herb-herb interaction or herb-modern medicine interaction.
- Unsatisfactory randomization in nearly all investigations.
- Availability of the subjects for clinical studies of herbal products.
- Challenge in determining proper placebos due to the flavour, smell, etc.
- Large differences in the period of treatments based on herbal medicines.

## 1.18 MOST IMPORTANT CHALLENGES ASSOCIATED WITH CLINICAL STUDIES OF HERBAL MEDICINES

- Ayurvedic HMPs (herbal medicinal products) have been considered as dietary supplements, they are regulated under the Dietary Supplement Health and Education Act, which does not require proof of efficacy as well as safety, thus prevents the stringent quality assessment criteria of Drugs and Cosmetics Act 1940.
- Recently Ayurvedic HMPs has been reported for toxicity due to the presence of heavy metals such as lead which can lead to deadly child encephalopathy, delay in foetal developmental, long-lasting seizures, hearing loss, facial paralysis at the time of the birth. Since the late-19th century, almost sixty-five cases of toxicity due to heavy metals associated with Ayurvedic HMPs among grownups as well as kids have been described in the United States and overseas [52].
- Implementing clinical studied over herbal medicine is very challenging because it's very difficult to arrange groups (control) with the same colour, taste as well as odour. Additionally, placebo utilization for herbal drug assessment also faces concerns such as herbal drugs might demonstrate its potent smell, characteristic taste and thus cannot be correctly mimicked while creating a placebo [53].
- Combined extracts-based formulation also faced so many challenges such as standardization of active compounds in the formulation.
- Due to multicomponent nature of Ayurvedic formulation its quite challenging to assess active compounds in body fluids during the pharmacokinetic analysis (phase-I trial) [54].
- Storage conditions of herbal extracts are not the same and no standard conditions have been established so far, due to the change in storage conditions therapeutic efficacy as well as bioavailability of the herbal formulation are considerably affected. Improper storage can always result into microbial contamination which can ultimately result into the drug spoilage resulting in batch-to-batch variation, as well as poor therapeutic outcome and safety profile [55].
- Importance of GMP in assessment of herbal drugs are typically not well recognized thus Ayurvedic preparations always under the risk of adulteration or substitution.
- Variation in dose treatment can also cause various complications such as issue like over or under dose which can always result into undesirable clinical outcome. Dose can be evaluated based on whether the drug has been used in raw form or extract [55].
- Composition of herbal medicine (whether compound or single), dose, as well as the duration of an overall treatment must be designed in such a way so that it can ensembles the specific psychosomatic nature (Prakriti) of the individual. It is well known in Ayurveda that the "*Prakriti*" determines the effectiveness of herbal medicines which influences response to drugs and a major factor to consider as an inclusion/exclusion criterion in clinical studies. So, for assessing the herbal medicine it's important to assess inclusion as well as exclusion measures. Due to such approach's clinical outcome of herbal medicine always varies [56].
- Various traditional practices have been used for the administration which may or may not be standardized. Additionally, no single formulation has been used which again creates challenge in standardizing medicines.
- Another challenge is under the influence of various external or internal factors chemical composition of plant always varies which can lead to change in therapeutic activity and cause difficulty in standardization of these preparations. Also, it's difficult to get the reproducible results from such preparations [57].
- Additional reasons such as the reality that combined strategy of various herbal therapy does not distinguish the illness from the patient.
- In addition, Ayurveda suggest various involvements at various phases of illness in the similar patient presenting a new difficulty to manage variation during clinical studies.

- Herb-diet interaction cannot be ignored and not studied deeply during the clinical studies.
- Since most crude drugs doesn't cause significant toxicity but still various reports associated with toxicity (to foetus and embryo), and cancer-causing effects cannot be ignored. Recently, embryotoxic effects of *"Pippalyadi vati"* have been evidenced [58, 59].
- Clinical studies over herbal products especially randomized clinical studies require large number of subjects to be involved for longer time and apart from that it requires costly healthcare its quite challenging to get this fund for the assessment of herbal products [60].
- Another challenge in performing the clinical trial is that some drugs have been reported for their poisonous effects, thus in such case clinical trial is not possible [61]
- It's very difficult to track the interactions between dietary components, main course drugs as well as herbal medicine. Some of the common herb-drug interactions includes (a) warfarin with Ginkgo or garlic can cause severe bleeding (b) antidepressants vs *Panax ginseng* can cause mania episodes among patients suffering from depression [62].
- Major concerns related with quality as well as safety profile of herbal products due the presence of certain toxic elements, contaminants, and adulterants.
- One more challenge is associated with fitness of randomized clinical studies for Ayurvedic medicine since this medicine is usually designed for everyone, thus it is unsuitable to give same treatment to all in randomized study [63].
- Even some traditional drugs have been clinically investigated properly but still well-controlled double-blind clinical study-based data to assure their efficacy as well as safety have been lacking [64, 65].

## 1.19   CLINICAL STUDIES ON AYURVEDIC PREPARATIONS

As its quite challenging to patent plants and their extracts (due to ambiguity related with mechanism of action of multicomponent), limited studies have been performed over the plant-based products. In America, it's difficult to assess the plant-based products as per the FDA guidelines as it requires more expenditure and well as time. For new synthetic drug it almost takes 15 years with an expense of US$500 in the United States to meet the requirement of FDA. Due to lack of funds it's difficult for the any private firm to conduct clinical trials on herbal drugs in an order to meet regulatory guidelines linked with safety as well as efficacy of herbal drugs [66].

## 1.20   HERBAL DRUG DEVELOPMENT

Several reports showed that there is a high demand of natural products, nutraceuticals, OTC drugs as well as dietary supplements in different countries. Plant-based products also associated with high risk for the presence of toxic or poisonous compounds, microbial spoilage as well as heavy metals, due to the adulteration, substitution or any other malpractices, or improper harvesting or cultivation procedure [67]. Thus, cultivating such superior cultivars with high medicinal value in a toxic environment impact finished products and ultimately worsen the clinical outcome of the finished product. Such concerns can be regulated by using standard operating procedures to produce safe and effective finished product. Good practices must be adopted while manufacturing these products to ensure their respective safety profile [68]. Herbal products validation, standardization as well as quality control involves the identification, phytochemical analysis, pharmacological investigation followed by the clinical evaluation of the finished products. Standardization as well as quality control of Ayurvedic medicine require more clarity regarding the strict guidelines or framework for standardization made for herbal products. As the standardization and quality assessment of pharmaceutical products (as per the standards of Pharmacopoeia) is completely different than herbal products' standardization and quality assessment, standardization protocols made based on active markers must be ease out as more understanding is required to know the multicomponent nature of natural products as well as their interactions with each other to understand their therapeutic efficacy. Recently, following official bodies encourage and regulate quality control as well as standardization of herbal medicine:

- Department of Indian system of medicine
- European agency for the evaluation of medicinal products
- European pharmacopoeia commission
- European scientific cooperation of phytomedicine
- US agency for healthcare policy and research
- WHO

For herbal products rationalized strategies for standardization as well as quality control are compulsory. For validating herbal products safety profile generally preclinical studies must be required to establish safety as well as toxicity profile. To understand the mechanism of action, pharmacokinetics as well as pharmacodynamics profile, molecular targets of lead compounds as well as overall therapeutic efficacy suitable animal models (both in vivo and in vitro) must be used. To validate the traditional claims of herbal products clinical evaluation is essential to establish exact therapeutic as well as safety profile.

## 1.21 REPUTATION OF PLANT-BASED PRODUCTS AS NEW DRUGS

It has been known that almost 25% of approved synthetic drugs which have been till now are derived from plant origin. This clearly suggest that plant source acts as major source for the synthetic drugs. Among the most dominating as well as potent class of secondary metabolites, alkaloids contributed various drugs to modern medicine system such as:

- Vinblastine/vincristine: antineoplastics
- Quinine: antiparasitics
- Opium alkaloids: analgesics
- Galantamine: anticholinesterases
- Atropine: anticholinergics

Nevertheless, not even alkaloids even other potent class such as terpenoids, glycosides, tannins, flavonoids as well as steroids plays an important role in the treatment of various ailments. Some other important examples such as:

- Cardenolides: Na+/K+ pump-inhibiters
- Paclitaxel: anticancer
- Artemisinin: antimalarial
- Triptolide: anti-inflammatory

*Hemidesmus indicus* (methanolic fraction) one of the lifesaving medicinal plants which have been used in tribal region since ages for its ability to neutralize poisonous effects of snake venom. This plant contains certain active chemical components those are having ability to antagonize poisonous effect of viper and cobra venom, reduce oxidative stress, and potentiate antiserum action in animal's models. Thus, more understanding related with mechanism of action, molecular targets as well pharmacokinetic and pharmacodynamics is required to discover the active chemical component responsible for the efficacy. On other side there are various reports suggested that toxicity caused by the plants-based products impact not only people faith over the natural products but also raised a question about the measures to assess their safety profile. Since extract has not any defined composition it's important to establish clinical profile of the plant by using suitable animal models or by involving human subjects. Its worthy to compare the efficacy, safety, stability profile of synthetic drugs vs herbal medicine to see whether it's required to have the drugs from plants or synthetic origin. Because at the same time overutilization of plants can lead to the extinction of these important cultivars. Objectives of natural products science is to isolate, identify, characterize, and screen therapeutics active components to order to establish their complete phytochemical profile and equate them with therapeutic profile by achieving following set of goals:

- To initially isolate phytochemicals by using various chromatographic procedures such as digitoxin, digoxin, morphine, vinblastine, vincristine, taxol, reserpine, etc.
- To characterize and identify these phytochemicals by using spectroscopic as well as chromatographic methods.
- To evaluate their physical as well as chemical properties.
- To assess their therapeutic profile in comparison with the known standards.
- To develop the synthetic analogues to further improve its pharmacokinetic as well as pharmacodynamics, therapeutic efficacy as well as safety profile.
- To identify the mechanism of action (molecular targets, molecular pathways) of phytochemicals present in the complex extract both at genetic and cellular level.
- To study their interaction with dietary components, other herbs as well as synthetic drugs.
- To formulate the phytochemicals by using an advance drug delivery system.

### 1.21.1  PLANT-DERIVED ETHNOTHERAPEUTICS

Exploration of medicinal features of folklore herbal drugs is one of the extraordinary contributions of Indian researchers. Various traditional formulations (from Ayurveda as well as Siddha) which have been used since ages have been further assessed in terms of standardizing their quality, efficacy as well as safety profile. Various Indian researchers are now focusing on Ayurvedic medicine integrative medicine, as well as evidence-based medicine to explore their therapeutic potential. India biodiversity contain various medicinal as well as aromatic plants, out of which some of them have not been explored yet. Thus, it's important to establish the phytochemical, therapeutic as well as safety profile of existing medicinal plants as well their respective preparations. Other parameters such as dosage, mode of administration, pharmacokinetic as well as pharmacodynamics profile must be established in an order to understand their efficacy against synthetic drugs.

### 1.21.2  SUSTAINABLE UTILIZATION AS WELL AS CONSERVATION OF MEDICINAL PLANTS

To attain sustainable utilization as well as conservation of medicinal and aromatic plants, it's important to understand systematic conservation and large-scale cultivation of the respective plants. Various agro-based or plant tissue culture-based techniques have been used to maintain the sustainable supply of raw material and conserve the plants in their respective habitat. However, due to the over demand of certain elite varieties of medicinal as well as aromatic plants, these plants are always overexploited, resulting in the extinction of plant from their natural habitat. Even the pharmaceutical companies are also dependent on the supply of raw material to produce herbal drug-based formulation and to isolate active component which cannot be synthesized economically.

### 1.21.3  QUALITY CONTROL TECHNIQUES

To achieve the phytochemical uniformity among the plants, it's important to develop the genetically modified elite variety with high level of active components and showed phytochemical uniformity among various plants. However, this can lead to the development of similar (mono-type) plants and restricted variation which could be considerable drawback of this approach. Storage as well processing conditions of the raw material could have significant impact over the phytochemical profile thus its important use the suitable processing as well as storage conditions to prevent the production of finished product with compromised quality. Thus, it's important to establish standard protocol to develop good agricultural practices including development techniques, collecting procedures, effective use of pesticides, fertilizers, and manures. Different challenges faced by the herbal preparations:

- Herbal preparations are the complex mixture with unknown composition, thus also called multicomponent system. This complexity interferes with evaluation of efficacy and safety of herbal preparation.
- It's difficult to identify active component or lead compound responsible for therapeutic activity.
- It was also quite challenging to validate the procedure for isolation, identification as well as characterization of active components.
- Variable phytochemistry of each plant due to the variation in the environmental factors causes variation in all herbal preparation.
- Storage, transportation, processing, and extraction conditions have significant impact on phytochemistry and thus impact the therapeutic profile.
- It's important for all herbal products to follow the standard regulations as well guidelines related to quality assessment, standardization as well as clinical studies. From the step of collection and identification to phytochemical investigation as well as biological as well as safety profile assessment, it's important to compare the chemical profile of finished products with parent raw material. Among them are proper botanical identification, phytochemical screening, and standardization. It's also important to differentiate the herbal extracts with standardized extracts (constant levels of specified compounds), in order to understand the difference between the quality of both.
- For systematic purposes as well as standardization especially when active chemical component is unidentified, marker substance must be recognized. These marker compounds are important for quality assessment of the herbal products.

Quality control parameters for new herbal medicine

- Prescription and its basis
- Physical chemical properties of herbal preparation
- Procedure used to prepare that formulation
- Draft comprising quality standards for all the ingredients
- Stability profile of the formulation or the active ingredient
- Quality assessment as well as regulations for the clinical studies
- Full specifications of the product over the label present over the packaging material
- Clinical, safety profile, pharmacodynamics, pharmacokinetic as well as therapeutic efficacy profile data.

## 1.21.4 Clinical Approach

Clinical studies over the medicinal plants and their respective products have received momentum recently; however, there are various ethical, technical, and logistic challenges involved in completing various clinical studies. Still the less clinical data of the plants and as well as their respective preparations is available. Generally crude drug powders have been utilized for clinical studies without involvement of biostatisticians to authenticate the therapeutic uses of plant-based extracts. To scientifically validate herbal products following procedure can be used:

- It's important to establish phytochemical, preclinical, clinical, safety as well as stability profile of the herbal preparations.
- Active component must be isolated and characterized to establish phytochemical profile.
- Requirement of high throughput bioactivity screening methods and the active fractions are to be explored phytochemically.
- To establish mechanism of action, identify the molecular targets as well as pathways it's important to assess the samples by using suitable pharmacological assays.
- For pharmacological assessment traditional knowledge or phytochemical profile can be used to select the suitable model to assess the pharmacological activity.

## 1.22   MARKET SCENARIO OF HERBAL PRODUCTS

Plants are considered as the main source for therapeutic compounds. Among all nations China as well as India uses the medicinal as well as aromatic plants extensively. So far more than five thousand medicinal plants have been explored in indigenous system of medicine (TCM) of China and almost seven thousand have been explored in Indian traditional system of medicine. As per the data collected from the Export-Import Bank, medicinal plants presented market with the growth of 7% per annum. In the international herbal drug market, stake of China is US$6 billion while India herbal market is US$1 billion. Vast export of raw crude drug material from India to other nations such as America, Japan, France, Switzerland, and Germany have been observed. Crude drugs like senna tora, plantago, ephedra, dioscorea, cinchona, digitalis, belladonna, aloe, acorus, and aconite have been exported at international level from India. Compromised or substandard quality of crude drugs due to malpractices, lack of safety, stability as well as therapeutic profile, improper regulations, and lack of standardization as well as quality assessment of the crude drugs effects the trade (Table 1.5). Phytomedicine-based herbal market as well as their trade in US$ is mentioned in Table 1.5. Medicinal plants as well as its parts have been shipped from India. It has been also noticed that several pharmaceutical drugs have been derived from crude drugs. Even crude drugs have been considered as source for various synthetic analogues

## TABLE 1.5
## Total Market Size Herbal Medicine and Their Sale in US$

| No. | Country | Years | Drug Sales in US$(Billion) |
|---|---|---|---|
| 1 | Europe | 1991 | 6.0 |
| | Germany | | 3.0 |
| | France | | 1.6 |
| | Italy | | 0.6 |
| | Others | | 0.8 |
| 2 | Europe | 1996 | 10.0 |
| 3 | United States | 1996 | 4.0 |
| 4 | India | 1996 | 1.0 |
| 5 | Other countries | 1996 | 5.0 |
| 6 | All countries | 1998 | 30.0–60.0 |

with minimum or no side effects. Various phytochemicals such as vincristine, vinblastine, taxol, podophyllotoxin, pilocarpine, morphine, gitoxigenin, ephedrine digoxigenin, digitoxigenin, curcumin, codeine, capsaicin, camptothecin, atropine, aspirin, artemisinin, and allicin have been isolated. Even certain crude drugs have been used in raw form after processing. At the end the exploration of phytochemical profile, therapeutic potential, safety, pharmacodynamic as well as pharmacokinetic profile and stability of herbal products or phytochemicals are of prime concern. Recently, almost 121 medicinal plants have been explored out of which none of them have been utilized to produce synthetic drugs.

## 1.22.1 HERBAL MEDICINE MARKET

In the end of 19th century, European herbal medicine market was found to be US$6 billion and hopefully it would be more in the 21st century and it seems exponentially increased further. Herbal preparations sold in France as well as Germany as prescription drugs which is protected by national health insurance. In the late-19th century, US herbal drug market was found to be US$4 billion and hopefully it has been increased further in the 21st century. Overall herbal drug market is found to be between US$30 and US$60 billion. Market size of US$1 billion has been estimated in India out of which US$80 million accountable to export of herbal extracts. Among the developed nations most popular herbals preparations are demonstrated in Figure 1.3. Overall sales of these herbal drugs were found to be more than 50% of the herbal medicine market. These crude drugs, including ginseng, saw palmetto, garlic, and gingko, have been standardized and studied for the mechanism of action as well as clinical studies. Among developed nations, Germany published high number of monographs mainly on therapeutic merits of more than 300 medicinal plants. Other nation such as China has made systematic data over almost 800 medicinal plants medicinal plants. Considering the growth in herbal market as well as great development in the market of nutraceuticals India has started exporting crude drugs. India evidenced high sale of *Aloe*, *Garlic* and *Panax* sp. India produce high level of crude drugs *Plantago ovata*, *Cassia senna*, and *Ricinus communis* and thus exported in high amount. However, Indian market is not only known for crude drugs, but also for the export of extracts and phytochemicals and finished preparations. Apart from these plants India export twenty more plants including *Azadirachta indica*, *Glycyrrhiza glabra*, *Commiphora mukul*, *Aloe barbadensis*, and *P. ovata* with net worth of US$8 million. There are almost 52 medicinal plants which have been used in 52–141 herbal formulations. Triphala, polyherbal formulation have been used for various decades as an active ingredient of 219 formulations. Various plant tissue culture-based approaches used to conserve as well to improve the genetic makeup of the plant in order to get the high yield of phytochemicals. Over exploitation of medicinal plants resulted in their extinction from the nature [69–97]. Various in vitro propagation methods have been used to prevent their extinction as well as such techniques have been used to improve their secondary metabolite synthesis in the medicinal plants. Crude drug market includes various forms of crude drugs (organized as well as unorganized) [69–97]. Essential oils are one the emerging class of natural products which is dominating in herbal drug market because of their considerable therapeutic benefits and also act as potential insecticide as well as psychological benefits. However, owing to the poor pharmacokinetic as well as pharmacodynamic properties of certain phytochemicals various polymers have been used in different drug delivery system to achieve the optimum therapeutic effects. Nanotechnology-based drug delivery system has been used to achieve the optimum therapeutic effects [69–97]. There are several crude drugs commonly available in the Indian market (Figures 1.5–1.8). Majority of these crude drugs are extensively utilized in the Indian system of medicine and widely available in India. Details of these crude drugs are mentioned in Table 1.6.

**FIGURE 1.5**    Traditional crude drugs of India (1–16).

**FIGURE 1.6**    Traditional crude drugs of India contd (17–32).

**FIGURE 1.7** Traditional crude drugs of India contd (33–48).

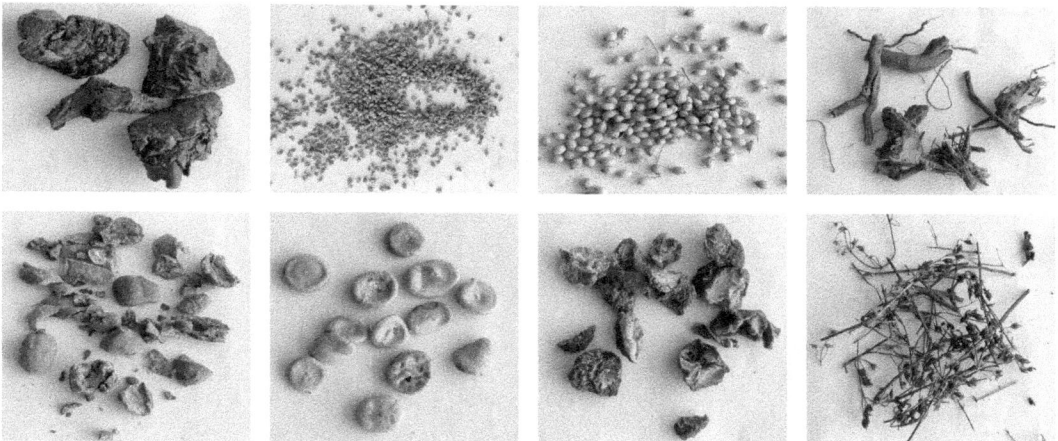

**FIGURE 1.8** Traditional crude drugs of India contd (49–56).

**TABLE 1.6**

**List of Traditional Drugs Widely Available in India**

| Sr. No. | Common Name | Biological Source | Family | Uses |
|---|---|---|---|---|
| 1 | Guggal | *Commiphora wightii* | Burseraceae | Arthritis |
| 2 | Ashoka | *Saraca indica* | Leguminosae | Dysmenorrhea and menorrhagia |
| 3 | | Red sandalwood | | |
| 4 | Sarpgandha | *Rauvolfia serpentina* | Apocynaceae | In the treatment of high blood pressure |
| 5 | Bhringraj | *Eclipta prostrata* | Compositae | Promoting hair growth, treating dandruff |
| 6 | Coffee beans | *Coffea arabica* | Rubiaceae | CNS stimulant |
| 7 | Shatavari | *Asparagus racemosus* | Liliaceae | Used to promote fertility |
| 8 | Guar gum | *Cyamopsis tetragonoloba* | Leguminosae | Gelling agent and emulsion stabilizer |
| 9 | Bhilava | *Semecarpus anacardium* | Anacardiaceae | To treat skin diseases |
| 10 | Tulsi | *Ocimum sanctum* | Lamiaceae | Antimicrobial |
| 11 | Nutmeg | *Myristica fragrans* | Myristicaceae | For stomach and kidney disorders |
| 12 | Shikakai | *Acacia concinna* | Fabaceae | Hair loss |
| 13 | Curry leaves | *Murraya koenigii* | Rutaceae | Antioxidants |
| 14 | Turmeric | *Curcuma longa* | Zingiberaceae | Anti-inflammatory |
| 15 | Cumin | *Cuminum cyminum* | Apiaceae | Carminative |
| 16 | Methi | *Trigonella foenum graecum* | Leguminosae | To treat digestive problems |
| 17 | Neem | *Azadirachta indica* | Meliaceae | Antimicrobial |
| 18 | Reetha | *Sapindus mukorossi* | Sapindaceae | Mucolytic agent |
| 19 | Fennel | *Foeniculum vulgare* | Umbelliferae | Carminative |
| 20 | Ashoka | *S. indica* | Leguminosae | Dysmenorrhea and menorrhagia |
| 21 | Punarnava | *Boerhaavia diffusa* | Nyctaginaceae | To rejuvenate the whole body |
| 22 | Tejpata | *Cinnamomum tamala* | Lauraceae | Respiratory ailments |
| 23 | Ginger rhizome | *Zingiber officinale* | Zingiberaceae | Throat infection |
| 24 | Picrorhiza (Kutki) | *Picrorhiza kurroa* | Scrophulariaceae | To treat various liver and upper respiratory tract infections |
| 25 | Mulethi | *Glycyrrhiza glabra* | Leguminosae | Throat infections |
| 26 | Adusa (Vasaka) | *Adhatoda vasica* | Acanthaceae | Cough, respiratory infections |
| 27 | Aloin | *Aloe vera* | Liliaceae | Anti-inflammatory and antioxidant |
| 28 | Sandalwood | *Santalum album* | Santalaceae | In cosmetics, perfumes and soaps |
| 29 | Hyoscyamus niger | *Hyoscyamus niger* | Solanaceae | Toothache, asthma, cough |
| 30 | Isapghol | *Plantago ovata* | Plantaginaceae | Laxative |
| 31 | Garlic (Kashmiri) | *Allium sativum* | Amaryllidaceae | Cardiac complications |

*(Continued)*

**TABLE 1.6** *(Continued)*
**List of Traditional Drugs Widely Available in India**

| Sr. No. | Common Name | Biological Source | Family | Uses |
|---|---|---|---|---|
| 32 | Orange peel | *Citrus aurantium* | Rutaceae | To treat various skin problems |
| 33 | Shankhpushpi | *Convolvulus pluricaulis* | Convolvulaceae | Brain tonic |
| 34 | Gum acacia | *Acacia arabica* | Leguminoseae | Wound healing |
| 35 | Gokhru (bada) | *Pedalium murex* | Pedaliaceae | Diuretic and anti-inflammatory effects |
| 36 | Sugar badam | *Prunus dulcis* | Rosaceae | Mild laxative |
| 37 | Colchicum | *Colchicum autumnale* | Liliaceae | Gout suppressant |
| 38 | Clove | *Eugenia caryophyllus* | Myrtaceae | Carminative, aromatic, and as a flavouring agent |
| 39 | Black pepper | *Piper nigrum* | Piperaceae | Spice and condiments |
|  | Commiphora | *Commiphora mukul* | Burseraceae | Anti-inflammatory |
|  | Gokhru (Chota) | *Tribulus terrestris* | Zygophyllaceae | For liver disease (hepatitis), inflammation, joint pain |
| 40 | Ashwagandha | *Withania somnifera* | Solanaceae | Phytosteroid |
| 41 | Gond Katira | *Astragalus gummifer* | Leguminosae | Treatment of diabetes |
| 42 | Senna | *Cassia acutifolia* | Leguminosae | To treat constipation |
| 43 | Gurmar | *Gymnema sylvestre* | Apocynaceae | Fight sugar cravings and lower high blood sugar levels |
| 44 | Arjuna bark | *Terminalia arjuna* | Combretaceae | Uses: hypertension, congestive heart failure |
| 45 | Vidarikand | *Pueraria tuberosa* | Fabaceae | Aphrodisiac, cardio-tonic, galactogogue, and diuretic |
| 46 | Harad | *Retz. Terminalia chebula* belonging to Family-increase appetite | Combretaceae | Digestive aid, liver stimulant, stomachic, gastrointestinal prokinetic agent, and mild laxative |
| 47 | Cinnamon | *Cinnamomum zeylanicum* | Lauraceae | Throat infections |
| 48 | Gudhal | *Hibiscus rosa* | Malvaceae | Treatment of excessive and painful menstruation, cystitis, bronchial catarrh, coughs and to promote hair growth |
| 49 | Rhubarb (Revand chini) | *Rheum emodi* | Polygonaceae | Laxative, tonic, diuretic |
| 50 | Ajowan (Carom seeds) | *Trachyspermum ammi* | Apiaceae | To treat GIT complications |
| 51 | Coriander | *Coriandrum sativum* | Apiaceae | Anti-microbial, anti-oxidant, anti-diabetic, anxiolytic, anti-epileptic, anti-depressant, anti-mutagenic, anti-inflammatory, anti-dyslipidemic, anti-hypertensive |
| 52 | Chitrak | *Plumbago zeylanica* | Plumbaginaceae | In the treatment of menstrual disorders, viral warts, and chronic diseases of nervous system |
| 53 | Bahera | *Terminalia bellirica* | Combretaceae | Antimicrobial and anti-allergic |
| 54 | Nux vomica | *Strychnos nux-vomica* | Loganiaceae | Anticancer |
| 55 | Amla or Indian goose berry | *Emblica officinalis Gaerth (Phyllanthus emblica Linn.)* | Euphorbiaceae | To promote longevity, enhance digestion, treat constipation, reduce fever and cough, alleviate asthma, strengthen the heart, benefit the eyes, stimulate hair growth, enliven the body, and enhance intellect |
| 56 | Chirata | *Swertia chirata* | Gentianaceae | In the management of weight loss |

## REFERENCES

1. Firenzuoli F, Gori L. Herbal medicine today: clinical and research issues. *Evid Based Complement Alternat Med.* 2007;4:37–40.
2. Rawls R. Europe's strong herbal brew: Chemical and biological research, mostly from Europe, supports the growing respectability of herbal medicines in US *Chem Eng. N.* 1996:53–60
3. Biotech Consortium India Ltd. Sectoral Study on Indian Medicinal Plants: Status, Perspective and Strategy for Growth. Biotech Consortium India Ltd, New Delhi, 1996.
4. Brower V. Nutraceuticals: poised for a healthy slice of the healthcare market? *Nat Biotechnol.* 1998; 16:728–731.
5. Ampofo AJ, Andoh A, Tetteh W, Bello M. Microbiological profile of some Ghanaian herbal preparations-safety issues and implications for the health professions. *Open J Med Microbiol.* 2012;2:121–130.
6. Mosihuzzaman M, Choudhary MI. Protocols on safety, efficacy, standardization, and documentation of herbal medicine. *Pure Appl Chem.* 2008;80(10):2195–2230.
7. World Health Organization (WHO). Regional strategy for traditional medicine in western Pacific. World Health Organization, Western Pacific, Manila, 2002.
8. Evans M. A Guide to Herbal Remedies. Orient Paperbacks. 1994:40–57.
9. Kamboj VP. Herbal medicine. *Curr Sci.* 2000;78:35–39.
10. Gupta LM, Raina R. Side effects of some medicinal plants. *Curr Sci.* 1998;75:897–900.
11. Winslow LC, Kroll Winslow LC, Kroll DJ. Herbs as medicine. *Arch Intern Med.* 1998;158:2192–2199.
12. Vickers A, Zollman C. ABC of complementary medicine: herbal medicine. *Br. Med. J.* 1999; 319:1050–1053.
13. Tirtha SSS. Overview of Ayurveda. In: The Ayurveda Encyclopedia: Natural Secrets to Healing, Prevention and Longevity (eds. Amrit Kaur Khalsa and Rob Paon). Satyaguru Publications, 1998:3–11.
14. Tyler VE, Phytomedicine: back to the future. *J Nat Prod.* 1999;62:1589–1592.
15. Miller LG. Herbal. Medicinal: selected clinical considerations focusing on known or potential drug-herb interactions. *Arch Intern Med.* 1998;158:2200–2211.
16. Kamboj VP. Herbal medicine. *Curr Sci.* 2000;78(1):35–39.
17. Krishnan R. *Indian Drug Manuf Assoc Bull.* 1998;13:318–320.
18. Kokate CK, Purohit AP, Gokhale SB. Pharmacognosy. In Nirali Prakashan (30th Ed.) 2005:117–118.
19. Kalia AN. Textbook of Industrial Pharmacognosy. Oscar Publishers, India. 2005:27–47.
20. Kamboj VP. Herbal medicine. *Curr Sci.* 2000;78(1):35–39.
21. Anonymous. *Sectoral Study an Indian Medicinal Plants-Status, Perspective and Strategy for Growth.* Biotech Consortium India Ltd., New Delhi, 1996:28–41.
22. Inamdar N, Edalat S, Kotwal VB, Pawar S. Herbal drugs in milieu of modern drugs. *Int J Green Pharm.* 2008;2(1):2–8.
23. Borris J. Natural products research perspectives from a major pharmaceutical company. *J Ethnopharmcol.* 1996;51:29.
24. Kamboj VP. Herbal medicine. *Curr Sci.* 2000;78(1):35–39.
25. World Health Organization (WHO). General guidelines for methodologies on research and evaluation of traditional medicines. 2001.
26. Kokate CK, Purohit AP, Gokhale SB. Pharmacognosy. Nirali Prakashan (30th Edn.) 2005:118–119.
27. Gadre AY, Uchi DA, Rege NN, Daha SA. Nuclear variations in HPTLC fingerprint patterns of marketed oil formulations of *Celastrus paniculatus. Ind J Pharmacol.* 2001;33:124–145.
28. Sapna, S, Ravi TK. Approaches towards development and promotion of herbal drugs. *Pharm Rev.* 2007;1(1):180–184.
29. Hou JP. The development of Chinese herbal medicine and the Pen-ts'ao. *Comp Med East West.* 1977 Summer;5(2):117–122. doi: 10.1142/s0147291777000192
30. Garodia P, Ichikawa H, Malani N, Sethi G, Aggarwal BB. From ancient medicine to modern medicine: ayurvedic concepts of health and their role in inflammation and cancer. *J Soc Integr Oncol.* 2007;5:25–37.
31. Alschuler L, Benjamin SA, Duke JA. Herbal medicine—what works, what is safe. *Patient Care.* 1997;31:48–103.
32. Brevoort P. The booming US botanical market. A new overview. *Herbal Gram.* 1998;44:33–44.
33. Cragg GM, Newmann DJ, Snader KM. Natural product in drug discovery and development. *J Nat Prod.* 1997:60:52–60.
34. Dwyer J, Rattray D. Anonymous. Plant, people and medicine. In: Magic and Medicine of Plant. Reader's Digest General. 1993:48–73.

35. Harvey AL. Medicines from nature: are natural product still relevant to drug discovery? *Trends Pharmacol Sci.* 1999;20:196–198.
36. Kamboj VP. Herbal medicine. *Curr Sci* 2000;78:35–39.
37. Murray MT, Pizzorno JE Jr. Botanical medicine – a modern perspective. In: Textbook of Natural Medicine Vol 1 (eds. J.E. Jr Pizzorno, M.T. Murray) Churchill Livingstone, London, 2000:267–279.
38. Lele RD. Ayurveda (Ancient Indian System of Medicine) and modern molecular medicine. *J Assoc Phys.* 1999;47:625–628.
39. Jain SK, DeFilipps RA. Medicinal Plants of India. Reference Publication, Inc. 1991:38–42. https://agris.fao.org/agris-search/search.do?recordID=US9314175
40. Barrett B, Shadick K, Schilling R, Spencer L, et al. Hmong/medicine interactions: improving cross-cultural health care. *Fam Med.* 1998;30(3):179–184.
41. Pal SK. Complementary and alternative medicine: an overview. *Curr Sci.* 2002;82:518–524.
42. Shiapush M. Postmodern values, dissatisfaction with conventional medicine and popularity of alternative therapies. *J Soc.* 1998;34:58–70.
43. MacLennan AH, Wilson DH, Taylor AW. Prevalence and cost of alternative medicine in Australia. *Lancet* 1996;347:569–573.
44. Tapsell LC, Hemphill I, Cobiac L, Patch CS, Sullivan DR, Fenech M, Roodenrys S, Keogh JB, Clifton PM, Williams PG, Fazio VA, Inge KE. Health benefits of herbs and spices: the past, the present, the future. *Med J Aust.* 2006;185:S4–24.
45. Armstrong NC, Ernst E. The treatment of eczema with Chinese herbs: a systematic review of randomized clinical trials. *Br J Clin Pharmacol.* 1999;48:262–264.
46. Eisenberg DM, Kessler RC, Foster C, Norlock FE, Calkins DR, Delbanco TL. Unconventional medicine in the United States. *New Engl J Med.* 1993;328:246–252.
47. Robbers JE, Tyler VE. Tyler's Herbs of Choice: The Therapeutic Use of Phytomedicinals. The Haworth Herbal Press, Inc., New York, NY, 1999:124–158.
48. Pittler MH, Abbot NC, Harkness EF, Ernst E. Location bias in controlled clinical trials of complementary/alternative therapies. *Int J Epidemiol.* 2000;53:485–489.
49. Linde K, Jonas WB, Melchart D, Willich S. The methodological quality of randomized controlled trials of homeopathy, herbal medicines and acupuncture. *Int J Epidemiol.* 2001;30:526–531.
50. Srinivasan K. Spices as influencers of body metabolism: an overview of three decades of research. *Food Res Int.* 2005;38:77–86.
51. Robert BS, Stefanos K, Janet P, Michael JB, David ME, Roger BD, Russell SP. Heavy metal content of ayurvedic herbal medicine products. *JAMA.* 2004;292:2868–2873.
52. Rothman KJ, Michels KB. The continuing unethical use of placebo controls. *N Engl J Med.* 1994;331:394–398.
53. Bhattaram VA, Graefe U, Kohlert C, Veit M, Derendorf H. Pharmacokinetics and bioavailability of herbal medicinal products. *Phytomedicine.* 2002;3:1–33.
54. Thatte UM. Clinical research with ayurvedic medicines. *Pharma Times.* 2005;37:9–10.
55. Dahanukar SA, Thatte UM. Current status of ayurveda in phytomedicine. *Phytomedicine.* 1997;4:359–368.
56. Bauer R, Tittel G. Quality assessment of herbal preparations as a precondition of pharmacological and clinical studies. *Phytomedicine.* 1996;2:193–198.
57. Roy C, Chandrashekharan M, Mishra S. Embryotoxicity and teratogenicity studies of an ayurvedic contraceptives-pippalyadi vati. *J Ethnopharmacol.* 2001;74:189–193.
58. Chan K. Some aspects of toxic contaminants in herbal medicines. *Chemosphere.* 2003;52:1361–1371.
59. Shukla NK, Narang R, Nair NGK, Radhakrishna S, Satyavati G. Multicentric randomized controlled clinical trial of Kshaarasootra (Ayurvedic medicated thread) in the management of fistula-in-ano. *Indian J Med Res.* 1991;94:177–185.
60. Moore C, Adler R. Herbal vitamins: lead toxicity and developmental delay. *Am Acad Pediatrics.* 2000;106:600–602.
61. Fugh-Berman A. Herb-drug interactions. *Lancet* 2000;355:134–138.
62. Kristofferson S. Uptake of alternative medicine in Australia. *Lancet* 1996;347:569–573.
63. Ernst E, Rand JI, Stevinson C. Complementary therapies for depression – an overview. *Arch Gen Psychiat.* 1998;55:1026–1032.
64. Harrer G, Schimidt U, Kuhn U, Biller A. Comparison of equivalent between the St. John's Wort extract LoHyp-57 and fluoxetine. *Arzneimittel-Forschung.* 1999;49:289–296.
65. Vaidya RA, Vaidya ADB, Patwardhan B, Tillu G, Rao Y. Ayurvedic pharmacoepidemiology: a proposed new discipline. *J Assoc Physicians India* 2003;51:528.

66. Grunwald J. The European phytomedicines market: figures, trends, analysis. *Herbal Gram* 1995;34:60–65.

67. Fisher P, Ward A. Complementary medicine in Europe. *BMJ.* 1994;309(6947):107–111. doi: 10.1136/bmj.309.6947.107

68. Straus S. Herbal medicines: what's in the bottle? *New Engl J Med* 2002;347:1997–1998.

69. Bhatia S. Stem cell culture. In: Introduction to Pharmaceutical Biotechnology, Volume 3: Animal Tissue Culture and Biopharmaceuticals. IOP Publishing Ltd, Bristol, UK; 2019;3:1–24.

70. Bhatia S. Organ culture. In: Introduction to Pharmaceutical Biotechnology, Volume 3: Introduction to Animal Tissue Culture Science. IOP Publishing Ltd, Bristol, UK; 2019;3:1–28.

71. Bhatia S. Animal tissue culture facilities. In: Introduction to Pharmaceutical Biotechnology, Volume 3: Animal Tissue Culture and Biopharmaceuticals. IOP Publishing Ltd, Bristol, UK; 2019;3:1–32.

72. Bhatia S. Characterization of cultured cells. In: Introduction to Pharmaceutical Biotechnology, Volume 3: Animal Tissue Culture and Biopharmaceuticals. IOP Publishing Ltd, Bristol, UK; 2019:3;1–47.

73. Bhatia S. Introduction to genomics. In: Introduction to Pharmaceutical Biotechnology, Enzymes, Proteins and Bioinformatics. IOP Publishing Ltd, Bristol, UK; 2018;2:1–39.

74. Bhatia S. Bioinformatics. In: Introduction to Pharmaceutical Biotechnology, Enzymes, Proteins and Bioinformatics. IOP Publishing Ltd, Bristol, UK; 2018;3:1–16.

75. Bhatia S. Protein and enzyme engineering. In: Introduction to Pharmaceutical Biotechnology, Enzymes, Proteins and Bioinformatics. IOP Publishing Ltd, Bristol, UK; 2018;2:1–15.

76. Bhatia S. Industrial enzymes and their applications. In: Introduction to Pharmaceutical Biotechnology, Enzymes, Proteins and Bioinformatics. IOP Publishing Ltd, Bristol, UK; 2018;2:21.

77. Bhatia S. Introduction to enzymes and their applications. In: Introduction to Pharmaceutical Biotechnology, Enzymes, Proteins and Bioinformatics. IOP Publishing Ltd, Bristol, UK; 2018;2:1–29.

78. Bhatia S. Biotransformation and enzymes. In: Introduction to Pharmaceutical Biotechnology, Enzymes, Proteins and Bioinformatics. IOP Publishing Ltd, Bristol, UK; 2018;3:1–13.

79. Bhatia S. Modern DNA science and its applications. In: Introduction to Pharmaceutical Biotechnology, Volume 1: Basic Techniques and Concepts. IOP Publishing Ltd, Bristol, UK; 2018;1(3):1–70.

80. Bhatia S. Introduction to genetic engineering. In: Introduction to Pharmaceutical Biotechnology, Volume 1: Basic Techniques and Concepts. IOP Publishing Ltd, Bristol, UK; 2018;1(3):1–63.

81. Bhatia S. Applications of stem cells in disease and gene therapy. In: Introduction to Pharmaceutical Biotechnology, Volume 1: Basic Techniques and Concepts. IOP Publishing Ltd, Bristol, UK; 2018;1:1–40.

82. Bhatia S. Transgenic animals in biotechnology. In: Introduction to Pharmaceutical Biotechnology, Volume 1: Basic Techniques and Concepts. IOP Publishing Ltd, Bristol, UK; 2018;1:1–67.

83. Bhatia S. History and scope of plant biotechnology. In: Modern Applications of Plant Biotechnology in Pharmaceutical Sciences. Academic Press, 2015:1–30.

84. Bhatia S. Plant tissue culture. In: Modern Applications of Plant Biotechnology in Pharmaceutical Sciences. Academic Press, 2015: 31–107.

85. Bhatia S. Laboratory organization. In: Modern Applications of Plant Biotechnology in Pharmaceutical Sciences. Academic Press, 2015:109–120.

86. Bhatia S. Concepts and techniques of plant tissue culture science. In: Modern Applications of Plant Biotechnology in Pharmaceutical Sciences. Academic Press, 2015:121–156.

87. Bhatia S. Application of plant biotechnology. In: Modern Applications of Plant Biotechnology in Pharmaceutical Sciences. Academic Press, 2015:157–207.

88. Bhatia S. Somatic embryogenesis and organogenesis. In: Modern Applications of Plant Biotechnology in Pharmaceutical Sciences. Academic Press, 2015:209–230.

89. Bhatia S. Classical and nonclassical techniques for secondary metabolite production in plant cell culture. In: Modern Applications of Plant Biotechnology in Pharmaceutical Sciences. Academic Press, 2015:231–291.

90. Bhatia S. Plant-based biotechnological products with their production host, modes of delivery systems, and stability testing. In: Modern Applications of Plant Biotechnology in Pharmaceutical Sciences. Academic Press, 2015:293–331.

91. Bhatia S. Edible vaccines. In: Modern Applications of Plant Biotechnology in Pharmaceutical Sciences. Academic Press, 2015:333–343.

92. Bhatia S. Microenvironmentation in micropropagation. In: Modern Applications of Plant Biotechnology in Pharmaceutical Sciences. Academic Press, 2015:345–360.

93. Bhatia S. Micropropagation. In: Modern Applications of Plant Biotechnology in Pharmaceutical Sciences. Academic Press, 2015:361–368.

94. Bhatia S. Laws in plant biotechnology. In: Modern Applications of Plant Biotechnology in Pharmaceutical Sciences. Academic Press, 2015:369–391.

95. Bhatia S. Technical glitches in micropropagation. In: Modern Applications of Plant Biotechnology in Pharmaceutical Sciences. Academic Press, 2015:393–404.

96. Bhatia S. Plant tissue culture-based industries. In: Modern Applications of Plant Biotechnology in Pharmaceutical Sciences. Academic Press, 2015:405–417.

97. Bhatia S, Al-Harrasi A, Kumar A, Behl T, Sehgal A, Singh S, Sharma N, Anwer MK, Kaushik D, Mittal V, Chigurupati S, Sharma PB, Aleya L, Vargas-de-la-Cruz C, Kabir MT. Anti-migraine activity of freeze-dried latex obtained from Calotropis gigantea Linn. *Environ Sci Pollut Res Int.* 2022 Apr;29(18):27460–27478. doi: 10.1007/s11356-021-17810-x

# 2 Introduction to Phytopharmaceuticals

*Ahmed Al-Harrasi*
Natural and Medical Sciences Research Center, University of Nizwa

*Saurabh Bhatia*
Natural and Medical Sciences Research Center, University of Nizwa
School of Health Sciences, University of Petroleum and Energy Studies

*Mohammed F. Aldawsari*
Prince Sattam Bin Abdulaziz University

*Deepak Kaushik*
M. D. University

## 2.1 INTRODUCTION

Phytochemicals are secondary metabolites which are synthesized by the plants through different biosynthetic pathways when exposed to the different stress factors present in the environment. These stress factors could be the reason behind change in the environmental conditions as well as pathogens or microbes that trigger synthesis of secondary metabolites in order to control its defence mechanisms. Thus, these stress compounds are named as elicitors because of their capability to elicit or trigger the synthesis of phytochemicals. Elicitors are broadly classified as biotic and abiotic elicitors. These elicitors impact the rate of biosynthesis via influencing expression of certain key enzymes involved in respective biosynthetic pathway. Biosynthetic pathways which are known as bio-machines to produce secondary metabolites must be deeply understood by various radioisotopic tracer techniques for the efficient production of secondary metabolites. These biosynthetic pathways get stimulated against pathogen(s) or in any other stress condition, which may lead to conversion of primary metabolites into secondary metabolites such as ethanol, lactic acid, and certain amino acids. The entire terpenoid-based compounds such as primary metabolites as well as over 25,000 secondary metabolites are obtained from the isopentenyl diphosphate (IPP). So far 12,000 known alkaloidal compounds have been explored, and these alkalis like substances are synthesized mainly from amino acids. Similarly, almost 8,000 phenol-based components explored so far are derived by either the shikimic acid pathway or malonate/acetate pathway [1]. Journey of secondary metabolites always starts from the precursor(s) by the involvement of different enzymes and their respective genes which are extracted, isolated, purified, and at last characterized to form finished product in form of pure marker compound. Several marker compounds have high importance in pharmaceutical industry mainly as potential therapeutics such as anticancer, antimicrobial, antioxidant, or as pharmaceutical excipients. This chapter is mainly emphasized on production of certain key phytopharmaceuticals such as quinine, Ca-sennosides, diosgenin, glycyrrhizin, hesperidin, andrographolides, curcumin, podophyllotoxin, solasodine, and caffeine.

DOI: 10.1201/9781003274124-2

## 2.2   CATEGORIZATION OF PLANT-BASED PHYTOCHEMICALS

In general, plant-based phytochemicals are classified in to three major classes [2, 3]:

- Phenolic compounds
- Terpenoidal components
- Nitrogenous components

Reports have established that terpenoidal compounds are obtained from the mevalonic pathway from precursor acetyl-CoA, while phenolic components are derived from the shikimic acid or the mevalonic acid pathway. Nitrogen-based organic phytochemicals including alkaloidal compounds are derived from amino acids (aliphatic) obtained from the tricarboxylic acid cycle or derived from the shikimic acid pathway (via aromatic acids), as well as mevalonic acid (by acetyl-CoA). Such three potential categories are used in synthesizing pharmacologically active compounds [4].

### 2.2.1   PHENOLIC COMPOUNDS

Phenols are generated in plants as secondary metabolites that include several groups of phytochemicals synthesized from the shikimate-phenylpropanoids-flavonoids pathways [3]. Plants generate these components for its development and protection against pathogenic organisms [5]. The word "phenolic" or "polyphenolic" could be accurately explained as a chemical substance having aromatic ring-based hydroxyl groups with or without other functional groups to possess oestrogenic properties [4, 6]. Phenolic compounds can be generally classified into non-soluble as well as soluble compounds [3]. Alternatively, the complexes, e.g., furanocoumarins, lignin, flavonoids, isoflavonoids, and tannins, are incorporated in phenols. Several investigators reported that biotic as well as abiotic factors can trigger the synthesis of phenolics, for example, increase in phenolics has been found in blueberries as well as silver birch. It was found that natural exposure of red clover leaves to ozone increases phenolics level [7, 8]. Activity of some pathogenic microorganisms reduced in a serious coffee pest plant due to the release of phenolic compounds which offers more resistant to their leaves against pathogenicity caused by microbes [9]. Synthesis of lignin increased when plant was found to be infected with fungus. Lignin falls under the category of phenolic glycoside which is diversely present in plants. Overproduction as well as expression of components like salicylic acid, catechol, protocatechuic acid, benzaldehyde, and others have been observed after exposure to fungi [2, 10]. Increase in the level of chlorogenic acid has been reported in potato tubers when exposed to *Streptomyces scabies*, *Phytophthora infestans*, and *Verticillium albo-atrum* [3]. It was also found that phytochemicals especially phenolics production triggered under the influence of pathogenic microorganisms has its serious implication over its development such as benzaldehyde release impacted the germination of *Monilia fructicola* as well as *Botrytis cinerea*. Phenolic compounds such as kaempferol, naringenin, dihydroquercetin, and quercetin showed antifungal effect against rice pathogenic fungal strains *Rhizoctonia solani* as well as *Pyricularia oryzae* [11]. Flavonoids forms an important class of polyphenolic compounds. Quercetin, isorhamnetin as well as kaempferol (flavonols) diversely appear in broccoli, onions, endives as well as leeks while chrysoeriol, luteolin as well as apigenin (flavones) abundantly appear in celery, parsley, and thyme [12, 13]. Whereas other vegetables like pak choi as well as lettuce contain high level of phenolic including such as caffeic acids and gallic; spinach, parsley, onions contain cinnamic as well as vanillic acids and garlic, tomatoes, and carrots contain coumaric acid [13].

### 2.2.2   NITROGEN-CONTAINING COMPOUNDS

Nitrogen-containing compounds mainly include alkaloids containing (alkali like organic nitrogen (derived from an amino acid) in a heterocyclic ring. However, when compounds contain nitrogen in aliphatic ring those are called protoalkaloids.

- Pyridine-piperidine: tobacco, areca, and lobelia.
- Tropane: belladonna, hyoscyamus, datura, duboisia, coca, and withania.
- Quinoline and isoquinoline: cinchona, ipecac, opium.
- Indole: ergot, rauwolfia, catharanthus, nux-vomica, and physostigma.
- Imidazoles: pilocarpus.
- Steroidal: vertrum, kuruchi.
- Alkaloidal amines: ephedra and colchicum.
- Glycoalkaloids: solanum.
- Purines: coffee, tea, and cola.

Classification of the alkaloidal compounds is generally done based on nitrogen-based ring system as well as on their biosynthetic sources such as amino acids, cyanogenic glycosides, glucosinolates amines as well as alcamides [14–16]. Alkaloids are the physiologically active compounds because of their considerable effect over the biological system (Figure 2.1). Usually, these compounds are divided based on their precursors such as [16, 17]:

- Pyridine: nicotine and anabasine
- Tropane: atropine, cocaine, scopolamine
- Isoquinoline: codeine, morphine
- Purine: caffeine, theobromine, theophylline
- Steroidal: solanine (veratrum and solanum alkaloids)

Most of the alkaloidal compounds are bitter and toxic, and synthesized by the plants to protect against microbial as well as herbivores attack. These organic compounds have been also reported as UV protective agents [4]. Certain alkaloidal compounds showed central nervous system (CNS) stimulant effects such as ephedrine, caffeine however some possess toxic effects on the nerve system such as coniine from *Conium maculatum* and some has toxic action on Na+-K+-ATPase such as sanguinarine [16, 18]. Nicotine has been reported for its insecticidal effects such as inhibitory effects on cigarette beetle as

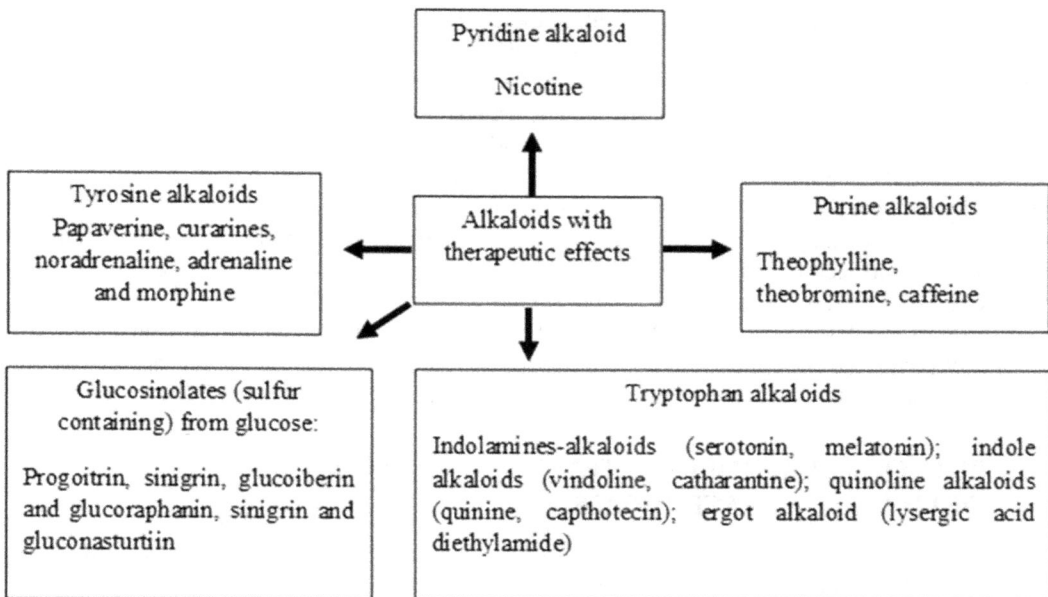

**FIGURE 2.1** Alkaloids with therapeutic effects.

well as tobacco hornworm and showed toxicity against herbivores via binding to acetylcholine receptors [4]. Similarly, caffeine has been also reported for its property to provide protection to plants against herbivores as well as other pathogens and stimulate/inhibit release of secondary metabolites of surrounding plants [19, 20]. In the growing phytopharmaceutical industry, alkaloids have an important place such as reserpine from *Rauvolfia serpentina*, Quinine from cinchona tree, Vincristine/vinblastine from *Catharanthus roseus* and rauwolscine from *Rauwolfia canescens*, etc. [4]. There are certain alkaloids that draw a clear relationship (phytochemotaxonomical) between plant family and its chemical components such as tropane alkaloids such as atropine as well as cocaine. Some of the important alkaloids those are having considerable therapeutic effects are illustrated below [15, 17, 21, 22].

Purine alkaloids have been reported for their improved CNS stimulation with reduced sleep, improved attentiveness via affecting CNS activity by inhibiting the synthesis of adenosine (A1, A2) receptors. Physostigmine from *Physostigma venenosum* has been evidenced for its effects on inhibiting neurotransmitter-destroying enzymes, acetylcholinesterase inhibitors, as well as its effects on neuromuscular intersection [21]. Glucosinolate alkaloids have been reported for their there are insecticidal as well as pesticidal effects via acting as allelochemical [21]. This occurs after the hydrolysis of glucosinolates by enzyme myrosinase to form thiocyanate as well as isothiocyanate, with free glucose [15].

### 2.2.3 TERPENES

Terpenes are non-polar organic components which include isoprene units. These organic compounds are synthesized by mevalonate as well as deoxy-D-xylulose pathways [23]. Based on number of isoprene units, terpenoids are classified into [15]:

- Hemiterpene
- Monoterpenes
- Sesquiterpenes
- Diterpenes
- Triterpenes
- Tetraterpenes

Terpenes and terpenoids are the basic components present in different types of plant essential oils. In the presence of a pathogenic environment, marked increase in the synthesis of these compounds has been noticed. Apart from protecting plants from herbivores, insects, and microbial pathogens, terpenes possess an array of therapeutic effects, thus gaining immense importance in phytopharmaceutical industry [24]. Menthol is a monoterpenoidal organic compound known for its antiherbivores as well as various therapeutic effects [4]. In phytopharmaceutical industry, the word "traditional use" is extensively utilized in informal language but has also immense importance for registration procedures of natural products. It is thus necessary to discover its real meaning from the viewpoint of social and pharmaceutical history. It is revealed that a tradition is founded by broadcast of knowledge and techniques over at least three decades, while the life span of a generation may vary extensively. Variations and adaptation to scientific advancements do not at all circumvent but even represent the development of traditions. Past research offers important information about value and risk of phytopharmaceuticals. Criteria of traditional use could balance but not replace contemporary data-based methods; nevertheless, they always need to be applied historically sound [24].

## 2.3 ROLE OF BIOSYNTHETIC PATHWAYS IN THE PRODUCTION OF PHYTOPHARMACEUTICALS

Biosynthetic pathways play an important role in production of phytopharmaceuticals in the form of secondary metabolites. There are several ways to elicit secondary metabolites synthesis such as elicitation, precursor feeding, cell culture, suspension culture, metabolic engineering, cell immobilization, scale-up techniques, and hairy root culture (Figure 2.2). These techniques offer scope for

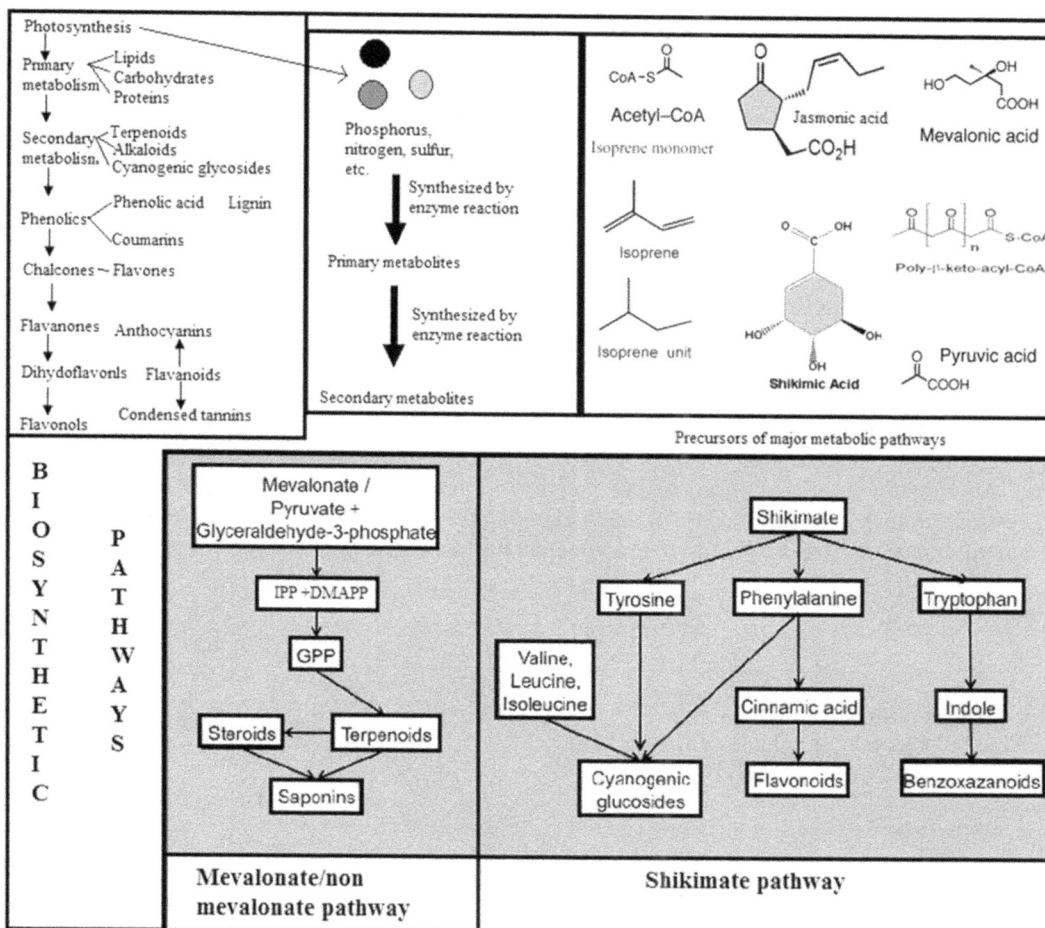

**FIGURE 2.2** Role of biosynthetic pathways in the production of phytopharmaceuticals.

phytopharmaceuticals industry at larger scale. In this chapter, we will emphasize on certain phytopharmaceuticals such as quinine, Ca-sennosides, diosgenin, glycyrrhizin, hesperidine, andrographolides, curcumin, podophyllotoxin, solasodine, and caffeine, which are important from commercial as well as therapeutic point of view [25–33].

## 2.4 ROLE OF EXTRACTION PROCESS IN PRODUCTION OF PHYTOPHARMACEUTICALS

So from above its clear that plant contains different chemical compounds in the form of secondary metabolites which posses' significant therapeutic activity at a particular concentration. These therapeutic compounds can be isolated and utilized in the prevention of disease. So far various methods have been explored to extract these components [34].

## 2.5 ISOLATION AS WELL AS PURIFICATION OF PHYTOPHARMACEUTICALS

Secondary metabolites offer wide range of biological properties. This therapeutic potential of secondary metabolites can be further utilized to treat various ailments. Secondary metabolites can be extracted out of plant tissue system via various conventional and advance methods. However, complex composition of extract always resists the qualitative and quantitative assessment of targeted compound. Thus, targeted chemical compound must be isolated and purified to derive pure or active fraction of respective chemical compound. In this context, bioassay-guided fraction can help in the identification of lead compound responsible for biological activity. In contrast to extraction methods, separation or the isolation procedures are not common for all phytochemicals. Procedures involved in separation are sometimes selected on the basis of physical or chemical properties of phytochemicals such as molecular size, ionic strength, partition coefficient, and adsorption properties. Separation methods such as chromatography, especially column chromatography, are the most common procedures to obtain pure phytochemicals from an extract [35].

Due to high requirement of pure marker compounds in research and development, formulation and development as well as in pharmaceutical and cosmetic industries, isolation and purification of secondary metabolites from medicinal and aromatic plants have recently become more advance and robust. These robust procedures based on isolation, separation, and purification of bioactive compounds also help in the biological assessment of phytochemicals (Figure 2.3). In vitro procedures are more suitable as animal-based trials are usually time consuming, expensive, and need ethical clearance. Nature (structural configuration) and quantity of biomarker compound in the extract determine the overall biological activity; however, cultivation, processing, and collection of plant material significantly impact the synthesis of secondary metabolites. It's important to establish a library of pure

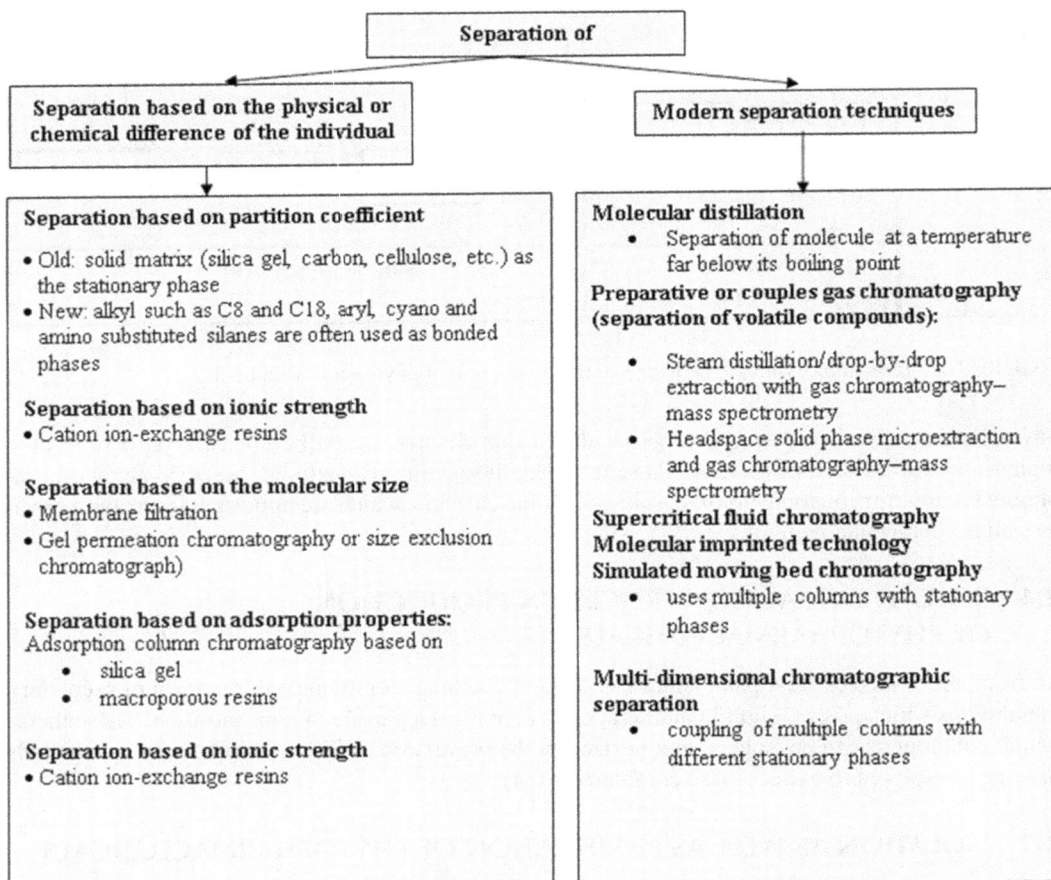

**FIGURE 2.3** Separation of phytochemicals based on physical and chemical properties of natural products and modernization in separation of phytochemicals.

biomarker compounds (purity > 90%) (https://www.nccih.nih.gov/grants/natural-product-libraries) in order to understand the possibility of their synthesis in lab, to quantify their presence in plants, for qualitative analysis as well as the characterization of the compounds and for formulation and development of phytochemicals. After collection and processing of plant material, it can be subjected to extraction by using suitable method as well as the solvent. Once extract is derived, separation procedure (chromatographic methods) can be used to separate as well as further purify the bioactive compounds. Pure compound can be further characterized using suitable spectroscopic techniques [36–47].

## 2.6   PURIFICATION OF THE BIOACTIVE MOLECULE

Purification of phytochemicals is usually performed by using thin-layer and column chromatographic procedures. Extract is complex mixture of various phytochemicals and this complexity of presence of multiple components in one system makes separation more challenging. Thus, for more effective separation, polarity must be increased by using multiple solvents. Different stationary phases including silica, polyamide, cellulose, and alumina have been proposed to run mobile phase (binary, ternary, quaternary). Thin-layer chromatography has constantly been employed to examine the active fractions of compounds by column chromatography [36–47].

## 2.7   STRUCTURAL ELUCIDATION OF THE BIOACTIVE MOLECULES

It's important to elucidate the structure of phytochemical mainly for its identification, for its possible chemical synthesis as well as its derivatives, to study structure-based biological interactions using computational analysis techniques and to establish its database in phytochemical-based library. This can be done by using various spectroscopic techniques. Underlying principle behind these spectroscopic techniques is that using an emission source electromagnetic radiation passed through an organic molecule. These phytochemicals absorb with some of the radiation; however, some radiations passed without interference and reached to the detector [47–61]. Thus, spectrum can be generated via determining the extent of absorption of electromagnetic radiation. This spectrum represents specific structural features such as specific bonds in a molecule. Based on these spectral arrangements, structural features of any phytochemical can be elucidated [36]. Spectral information of any phytochemicals helps in assessing identity of phytochemicals [47–61].

## 2.8   PHYTOPHARMACEUTICALS: COMMERCIALLY IMPORTANT PHYTOPHARMACEUTICALS

Some of the commercially important phytopharmaceuticals which are commonly used in the herbal as well as pharmaceutical industries are mentioned below:

### 2.8.1   QUININE

Quinine with other related alkaloids can constitute up to 16% of the bark weight. Quinine is a key alkaloid which is considered one of the oldest drugs obtained from cinchona bark. Quinine is classified under quinoline derivatives which are well known for their use in the treatment of malarial, fever, pain, and inflammation. Quinoline derivatives including quinine obtained from natural resource, quinidine, and synthetic drug known as chloroquine are well known for their use in the treatment of malaria. Quinine isolation from plant material (cinchona bark) unlocked different opportunities for the large scale production of quinine that was appropriate for the treatment of malaria and other illnesses. Utilization of cinchona bark, to cure recurrent fevers, was first reported in print by an Augustinian monk (1638). Due to the more emphasis on synthetic antimalarial drugs like chloroquine, production of quinine from *Cinchona* genus was decreased; however, due to development of antimalarial drugs resistance malarial parasites, production of quinine has regained the importance

[62]. In spite of its antimalarial effects, quinine is also used in beverage industry as bitter agent, in chemical industry as a catalyst especially in asymmetric organic synthesis, and in chromatography, quinine has role in stationary phases as chiral selector. This bitter bioactive compound is also used as a standard substance in the determination of bitter taste. Quinine is considered standard for assessing bitterness value (threshold of sulphate salt: 0.000008) in gustation and as optical resolution agent. Diastereomer of quinine (i.e. Quinidine) is evidenced for its antiarrhythmic effect [63].

In an international trade as well as the domestic level, the rating of cinchona bark is not done on the basis of its quality as there is no rating scheme, although cataloguing has been done into industrial bark (small fragments) as well as druggist quills. Industrial bark is used to meet the for alkaloid demand while druggist quills have been utilized for the preparation of gelatin. For international trade, state of derivation, inside colour of the bark, moisture content and most importantly quantity of alkaloidal compounds are considered. Different grades of quinine are considered during trade such as:

- Quinine content: % of quinine as anhydrous alkaloid
- SQ(7): quinine sulphate with seven water molecules
- SQ(2): quinine sulphate with two water molecules

Generally, pharmacopoeias suggest a least total alkaloidal level is 5–7% in the bark. As per the quality standards after the collection of the cinchona bark, its first dried to reduce 10% moisture content. In beverage industry, both yellow and red cinchona barks have been used with a limitation that overall alkaloidal content must not go beyond 83 ppm in the final beverage available at the counter. Apart from the *Cinchona* genus, quinine and quinidine alkaloids can also be derived from other medicinal plants such as from the bark of *Remijia pedunculata* (Rubiaceae) [64]. As per pharmacopoeias recommendations, least total alkaloidal level must lie between 5 and 7% in the bark. Commercially cinchona bark is classified as industrial bark for alkaloid extraction and druggist quills for its utilization in gelatin preparation. During commercialization, cinchona bark container or packaging must include total colour of the inner side of the bark, alkaloid content, moisture content, and country of origin. Content of quinine i.e. up to 16% (sometime ranges to 6–10%) varies among different species such as *Cinchona calisaya*, *Cinchona ledgeriana*, *Cinchona officinalis*, *Cinchona robusta*, and *Cinchona succirubra* (Rubiaceae). Alkaloids resemble ammonia [NH3] in chemical they form salts with acids without liberation of water. Due to their basic character, alkaloids react with acids to from salts. The salt insoluble in water is quinine sulphate.

### 2.8.1.1 Isolation of Quinine

Quinine was extracted and isolated from cinchona bark in 1980 by Pierre Joseph Pelletier and Joseph Caventou. Alkaloids are alkali-like substances having basic character and thus react with acids to from water insoluble salts such as quinine sulphate. Powdered cinchona bark material was first initially treated with calcium hydroxide/NaOH to release (precipitate all alkaloids) cinchonine, quinine, cinchonidine, and quinidine (Figure 2.4). Alkali treatment of bark allows the release of alkaloids to enhance their solubility in a non-polar solvent such as benzene. End product received from this process is subjected to Soxhlet extraction using benzene for almost 6 h. Benzene fraction is filtered and eventually treated with 5% sulphuric acid for the chemical conversion all alkaloids in sulphate alkaloids apart from quinine. Later on by adding sodium hydroxide, pH of the solution is maintained at 6.5 (Figure 2.4). Subsequently solution received is kept at reduced temperature for nearly 12 hours, and further by using hot water, quinine was recrystallized. Crystals of neutral sulphated quinine derived from this step is subjected for the treatment with weak sulphuric acid and strong base again to liberate the quinine (Figure 2.4).

### 2.8.1.2 Estimation of Quinine

Galloway et al. 1990 developed simple and rapid method for the estimation of quinine using reversed-phase high-performance liquid chromatography with UV detection (absorbance at 240 nm) [65]. The procedure is sensitive and free from analytical interferences from other drugs, such as

Size reduction process

Cinchona bark → Cinchona powdered bark

Sufficient quantity of 5% NaOH ← Treatment with Ca(OH)₂ or calcium oxide (CaO)

*To liberate/precipitate free alkaloid bases and increase their solubility in organic solvent such as benzene*

Soxhlet extraction (Benzene), 6hrs

Treatment of extract with 5% H₂SO₄ upto 90°C at 20-30 min

*To convert all alkaloids into alkaloidal suphate salts*

Allow to cool at room temperature & adjust pH of aq. extract at 6.5 with dil NaOH

Treatment of alkaloidal sulphate solution with activated charcoal powder (1g per 1L), (to remove coloring agent) boil, shake vigorously and filter.

Quinine

**1st Process**

Quinine (Crystalline form)

↑ Dried at 30-40°C

↑ Washed to remove sodium and ammonium salts

↑ Quinine (Amorphous form)

**2nd Process**

Cinchonine Cinchonidine Quinidine

Alkaline treatment with ammonia (in boiling water)

Filtrate ← Residue of quinine sulphate

Cool the hot filtrate slowly, keep overnight at 2-10°C and again filter. Collect the residue and the filtrate separately.

*Favors crystallization of only quinine at low temperature & rest alkaloids remain soluble at low temperature*

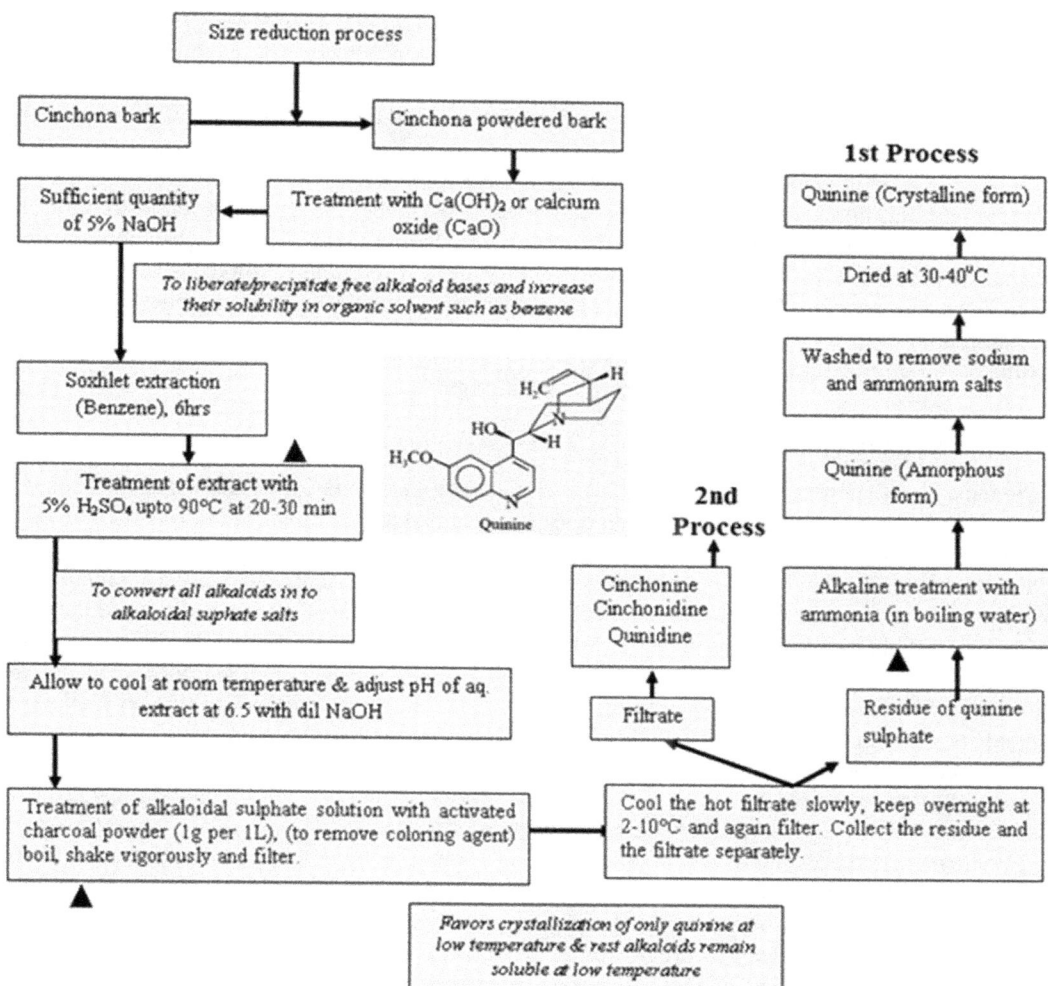

**FIGURE 2.4** Isolation procedures (two) of quinine [starts from left top (cinchona bark) to right side].

quinine diastereoisomer quinidine. A sensitive, specific, and rapid high-performance liquid chromatographic method with ultraviolet detection for therapeutic drug monitoring of the cardioactive compound quinidine is reported by Meineke et al. [66]. A simple, specific, rapid, and sensitive high-performance liquid chromatography (HPLC) method has been developed by Karbwang et al. (1998) to measure plasma level of quinine and quinidine [67]. Similarly, a new simple, selective, and reproducible high-performance liquid chromatographic method (reversed-phase C18 column, using perchlorate ion as the counter ion and ultraviolet detection at 254 nm) for the determination of quinine in plasma, saliva, and urine is reported by Babalola et al. [68].

Apart from HPLC method, quinine can also be measured by fluorimetry and non-aqueous titration. Fluorimetry-based estimation of quinine can be performed by measuring blue fluorescence (366 nm) generated via irradiation of the quinine in dil. H₂SO₄ at 450 nm. Graph between fluorescence vs concentration for standard compound is plotted to measure the fluorescence. Similarly, quinine content in test sample is estimated by measuring its fluorescence and intrapolation from standard graph. HPLC method with fluorescence detection has been reported for the simultaneous estimation of quinine and quinidine in plasma, whole blood, and erythrocytes using ultrasphere C18 reversed-phase column (25 cm × 4.6 mm inside diameter, 5 μm particle size). This reported procedure is simple, accurate, and selective, and requires only a single-step liquid-liquid extraction and uses the structurally similar alkaloid, cinchonine, as the internal standard [69]. TLC-based colorimetric method has been also reported for the determination of quinine and quinidine in the presence of cinchonine, cinchonidine, and other cinchona alkaloids in raw materials, formulations, and biological fluids [70]. A simple non-aqueous titration method has been reported for determining the sulphates of quinine and quinidine.

A simple non-aqueous titration procedure has been reported for estimating the sulphates of quinine and quinidine by use of barium perchlorate. This procedure has been found to be accurate, precise, and suitable for estimation of pure materials and tablets [71]. Further non-aqueous titrimetric procedure has been reported for estimating the diastereomeric sulphates of quinine and quinidine using barium acetate [72]. Volumetric estimation, nephelometric estimations, gravimetric estimation of quinine, Herapath test, thalleioquin, etc. have been also used for the qualitative estimation of quinine in blood and urine [73].

### 2.8.2 EPHEDRINE (1-PHENYL, 1-HYDROXY, 2-METHYL AMINO PROPANE)

Ephedrine (non-heterocyclic, protoalkaloid) is a natural alkaloid derived from almost 45 species (such as *Ephedra sinica*, *Ephedra equisetina*, *Ephedra nebrodensis*, *Ephedra intermedia*, *Ephedra major*, and others) belonging to genus *Ephedra*, a Chinese medicinal plant which has been used in China for several thousand years (3000 BC). The chemical constituents those are accountable for therapeutic effects are the alkaloids ephedrine and pseudoephedrine. Commercially dried ephedra plant material must have not less than 1.25% ephedrine. The natural protoalkaloid alkaloid (ephedrine) was initially derived and characterized by Nagai in 1885. It was then overlooked till it was revived by Chen and Schmidt in beginning of 1920s [74]. Due to the presence of this protoalkaloid (ephedrine), various natural supplements as well as herbal extract have been used in the treatment of asthma treatment, weight loss, and to improve athletic performance. Because of its bronchiodilatory effects, it's used in the management of asthma as well as allergic conditions including hay fever. It also elevates blood pressure via peripheral contraction of arterioles [74].

Ephedrine acts in a similar way as like hormone adrenaline by acting on adrenoceptors (alpha and beta) as well as suppressing norepinephrine reuptake and enhancing the release of norepinephrine from vesicles present in neurons. Epinephrine is a CNS stimulant that causes bronchodilation in the lungs, increases blood pressure as well as heart rate and ultimately give the sensation of high energy. Apart from its CNS stimulant and vasoactive effects, ephedra has been also used in reducing body weight due to its fat burning properties and capability in increasing metabolic rate. Ephedrine-based treatment got more recognition because of its effectiveness in the asthma treatment, mainly due to the fact that it can be administered by oral route unlike adrenaline which was used as standard therapy earlier [75–80]. Journey of Ephedrine started from traditional utilization in China to mainstream medicine and then moved to dark phase of street drugs and nutritional supplements. Over the last ten years, several reports suggested considerable adverse effects and mortalities due to the consumption of ephedra or ephedrine. A review based on 140 studies of adverse effects linked to the administration of ephedra alkaloids that were presented by Haller and Benowitz to the FDA (1997–1999). This report included adverse events experienced by individuals such as cardiovascular (47%) and CNS (18%) complications. One of the most common adverse effects was hypertension; however, other adverse effects such as palpitations, tachycardia, or both stroke and seizures were also reported. Currently ephedra- and ephedrine-related products are banned in several nations, mainly due to the adverse effects and becoming a main source for the synthesis of the addictive stimulant, i.e., methamphetamine or crystal meth; its use can lead to addiction [75–80].

As far as pharmacokinetic parameters are concerned, alkaloids such as ephedrine, synephrine, and caffeine show high absorption via gastrointestinal route, therefore present wide distribution across the body. After oral administration, generally clinical effects are observed in 60 minutes and last from 2 to 6 h [75–80]. In liver, ephedrine is converted in to its metabolites such as norepinephrine, benzoic acid, and hippuric acid by biotransformation reactions such as oxidative deamination, demethylation, aromatic hydroxylation, and conjugation. However, it cannot be easily hydrolyzed in the body by enzyme monoamino oxidase; its onset of action is slow, and thus, it displays extended action in comparison with adrenaline. Ephedrine is supposed to cross the placenta up to a greater extent and distribute into breast milk. Rate of elimination of ephedrine is 3 h when urine pH is acidic (pH 5); however, rate of elimination increased up to 6 h after increase in urinary pH 6.3 [75–80].

### 2.8.2.1 Isolation of Ephedrine

A Japanese chemist, Nagajosi Nagai (1887), initially isolated pure ephedrine from Ma Huang. As per Nagai (1887) method, ephedra powder was treated with sodium carbonate and then subjected to extraction (overnight) by using benzene. Benzene fraction was separated by filtration and subsequently treated with dil. HCl (Figure 2.5) to convert alkaloids into salts. Later on, aqueous acid layer is separated and treated with sodium potassium carbonate. Treatment with alkaline sodium potassium carbonate liberated free alkaloids bases present in the solution. Later on, they were extracted with non-polar organic solvent mainly chloroform. Chloroform layer was subjected to evaporation

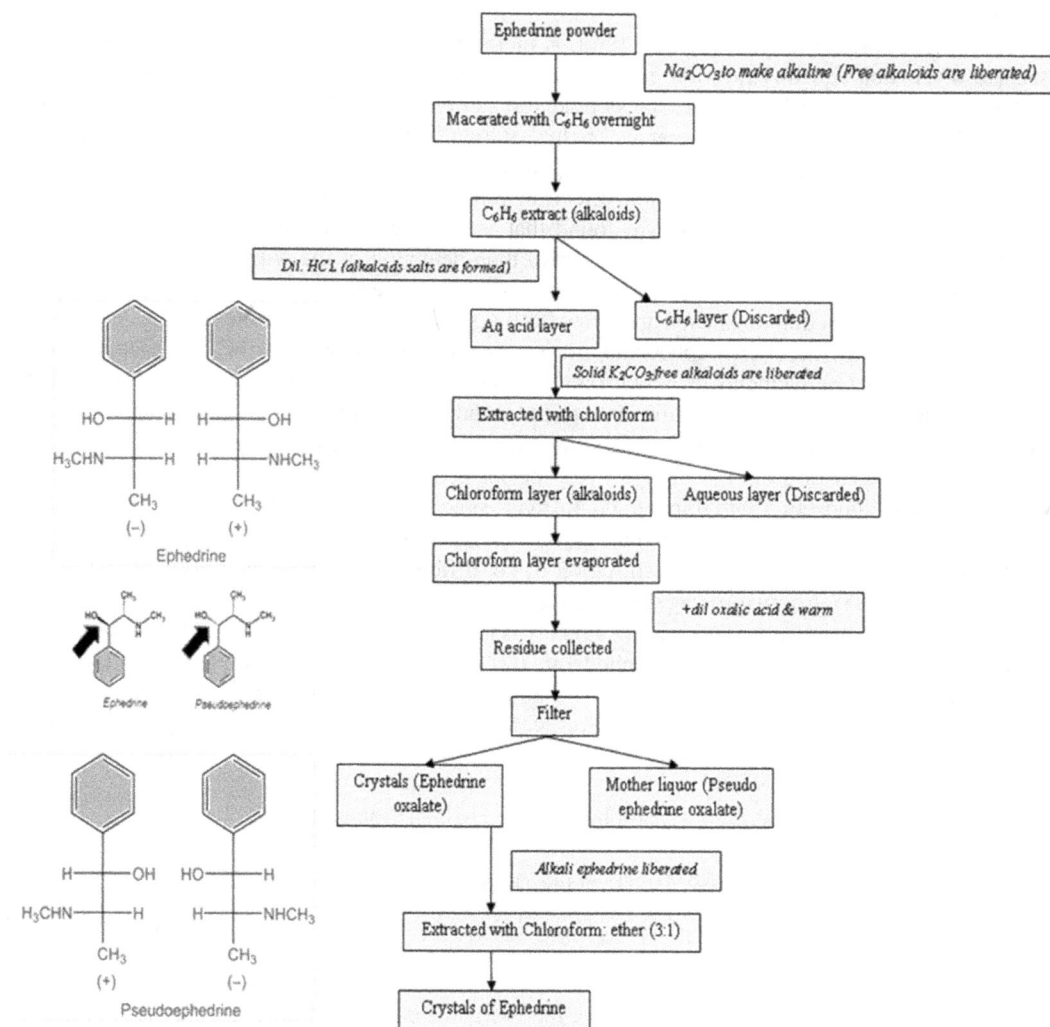

**FIGURE 2.5** Isolation of ephedrine from ephedra powder.

to get the final residue (Figure 2.5). Residual material is treated with oxalic acid solution and heated to obtain alkaloidal oxalates (ephedrine oxalate and pseudo-ephedrine oxalate). These isolated compounds placed in ice cold water for the crystallization. The soluble pseudo-ephedrine oxalate remains in solution; however, ephedrine oxalate crystallizes out. Later on, pure ephedrine is derived by liberating it's from its oxalate salt form via treatment with alkaline solution and then extraction with a combination of chloroform and ether (1:3) (Figure 2.5).

### 2.8.2.2   Identification Tests for Ephedrine

- Aqueous solution of ephedrine is treated with dil. HCl followed by the treatment with 10% $CuSO_4$ solution and 20% NaOH solution. Purple or violet colour initially appears and extracted with ether so as to get the organic layer in purple colour, whereas aqueous layer in blue colour.
- With ninhydrin reagent, alkaline solution of ephedrine presents violet colour.
- With concentrated sulphuric acid and 40% formaldehyde (3–4 drops), ephedrine gives pink to red coloration.

### 2.8.2.3   Estimation of Ephedrine

Ephedrine HCl was earlier estimated by titration method on the basis of copper complex formation [81]. However, non-aqueous titration can also be used to estimate ephedrine in samples. For this method, A is initially used to prepare the sample: test sample (0.17 g) is dissolved in mercuric acetate solution (10 ml), by warming it moderately. To this solution, acetone (50 ml) is added and then method b (non-aqueous titration) is followed by titrating with perchloric acid (0.1 M). Ultimately end point is determined by potentiometric titration. Electron-capture GLC method for estimation of plasma ephedrine level has been reported. It was found that estimating N-pentafluorobenzoyl (derivative of ephedrine) is almost hundred times more sensitive than detection of the N-trifluoroacetyl derivative [82]. In another report, GC-based procedure was used for the estimation of pseudoephedrine and norpseudoephedrine in human plasma and urine [83]. A simple procedure for the estimation of ephedrine alkaloids: ephedrine, pseudoephedrine, norpseudoephedrine, norephedrine, and methylpseudoephedrine in dietary supplements by gas chromatography-mass spectrometry was also reported (using 3,4-methylenedioxy-propylamphetamine as internal standard). This procedure can be used to assess various dietary supplements containing Ma Huang (*E. sinica*) and *Sida cordifolia* plant extracts [84]. Another report based on joint investigation was performed to assess the precision and accuracy of a method using liquid chromatography/tandem mass spectrometry for ephedrine-type alkaloids in plasma as well as urine samples. On the basis of results, it was suggested that this procedure can be adopted Official First Action for the estimation of six different ephedrine-type alkaloids in human urine and plasma [85]. Another report suggested that HPLC-UV method for the estimation of ephedra alkaloids in human urine and plasma is not suggested for adoption as Official First Action [86].

Additionally, to estimate content of ephedrine, test sample must be weighed (3.5 g) properly and subsequently dissolved in ethanol (5 ml). To the solution recovered from last step, 0.1-M hydrochloric acid (50 ml) is added, followed by titration with 0.1-M sodium hydroxide, using methyl red as indicator until a yellow colour is produced.

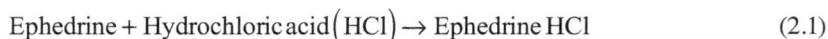

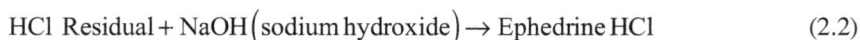

$$\text{Ephedrine} + \text{Hydrochloric acid}(\text{HCl}) \rightarrow \text{Ephedrine HCl} \tag{2.1}$$

$$\text{HCl Residual} + \text{NaOH}(\text{sodium hydroxide}) \rightarrow \text{Ephedrine HCl} \tag{2.2}$$

### 2.8.3   Cardiac Glycosides

As the name suggests, primary site of action of these organic compounds is heart muscles to improve the output of the heart and also enhance the rate of cardiac muscular contractions via acting directly on the sodium-potassium ATPase pump. This powerful action exerted by cardiac glycosides over the heart muscles can be utilized actively to restore the strength of weakened heart and to support its function more actively. However, this mechanism of action comes under the toxic effects of cardiac glycosides. These glycosides act by binding to the α-subunit of the Na+/K+-ATPase pump to inactivate its action [87, 88]. As this pump is responsible for exchange of intracellular sodium ions with extracellular potassium ions, its inactivation by cardiac glycosides results into overall rise in intracellular sodium ions level. Increase in intracellular sodium level cause increases in intracellular calcium levels, resulting into positive inotropic effect of cardiac glycosides. Cardiac glycosides are diversely present in medicinal plants belonging to family such as Scrophulariaceae, Oleaceae, and Liliaceae represented by digitalis. Digitoxin, digoxin, ouabain, and oleandrin have been considered most important cardiac glycosides obtained from medicinal plants such as *Digitalis purpurea* (foxglove), *Digitalis lanata*, *Strophanthus gratus* (ouabain), and *Nerium oleander* (oleander). Other medicinal plants such as *Convallaria majalis* (lily of the valley), *Thevetia peruviana* species (yellow oleander), *Urginea indica* (squill), and *Urginea maritima* also considered important source of cardiac glycosides. As far as the toxicity is concerned, cardiac glycosides are distributed throughout the plants; that's why all parts are considered toxic. However, relatively seeds contain more glycoside than other parts of the plant [87, 88]. Digoxin at therapeutic levels is used to treat congestive heart failure but becomes toxic at high doses. Since these compounds are glycosides, they contain glycone and aglycone parts where aglycone part of these glycosides is the steroidal in nature. Since aglycone is structurally related to steroid hormones, so mainly it's responsible for biological activity. Based on the structural features of aglycone part, it is divided into two categories (cardenolides and bufadienolides) [87, 88]:

- Cardenolides (C23glycosides): These organic compounds contain five-membered α-β-unsaturated lactone at the 17th position (β-orientation). This class includes medicinal plants like digitalis and strophanthus. Cardenolides are commonly found in the Asclepiadaceae as well as Apocynaceae families.
- Bufadienolides (C24 glycosides): These organic compounds contain six-membered doubly unsaturated lactone at 17th position. This class includes medicinal plants like squill. Bufanolides are found in Ranunculaceae as well as Liliaceae families. Genins are steroidal aglycones present in toad venoms in free as well as conjugated (with suberylarginine).

### 2.8.3.1  Isolation Estimation and Identification of *Digitalis purpurea* Glycosides

Cardiac glycosides of *D. purpurea*: It contains a number of glycosides. The primary glycosides are Purpurea glycoside A, Purpurea glycoside B, and glucogitaloxin. Primary glycosides of *D. purpurea* are tetrasides. On enzymatic hydrolysis, primary glycosides can be converted into secondary glycosides which can finally release the aglycone part as mentioned in Figure 2.6.

Digitoxin is the main active chemical component of *D. purpurea*. Digitoxigenin, gitoxigenin, and digitaloxigenin are the aglycones of digitoxin, gitoxin, and gitaloxin, respectively. These are formed by the acid hydrolysis. *D. purpurea* also contains a number of cardenoloides such as biosides and monosides. There are various biotechnological approaches which have been used to modulate to the cardenolide pathway [89–105].

### 2.8.3.2  Isolation of Cardiac Glycosides

Earlier various methods have been introduced for the extraction of cardiac glycosides such as supercritical mixture of carbon dioxide and methanol to extract the cardiac glycoside, digoxin, from the *D. lanata* leaf was employed successfully [106]. For the extraction of cardiac glycosides from *D. purpurea* and *D. lanata*, modified Stas-Otto method can be used. In this procedure, two different steps are available (Figure 2.7). Selection is dependent on the plant material type and whether the plant material is present in fresh or dry form. The first method (method A) involves the extraction of cardiac glycosides from dried material. In dried form, plant material enzymes are already inactivated. Plant material is first processed and then extracted by maceration using with alcohol (70%) for

| *Primary* *glycosides* | | *Secondary* *glycosides* | | *Agycone* | | |
|---|---|---|---|---|---|---|
| P.G.A | $\xrightarrow{EH}$ | Digitoxin and glucose | $\xrightarrow{AH}$ | Digitoxigenin | + | 3 Digitoxose |
| P.G.B | $\xrightarrow{EH}$ | Gitoxin and glucose | $\xrightarrow{AH}$ | Gitoxigenin | + | 3 Digitoxose |
| Glucogitaloxin | $\xrightarrow{EH}$ | Gitaloxin and glucose | $\xrightarrow{AH}$ | Gitaloxigenin | + | 3 Digitoxose |

Enzyme: *Digipuridase*    AH: Acid hydrolysis    EH: Enzymatic hydrolysis
Purpurea glycoside A : PGA  Purpurea glycoside B : PGB

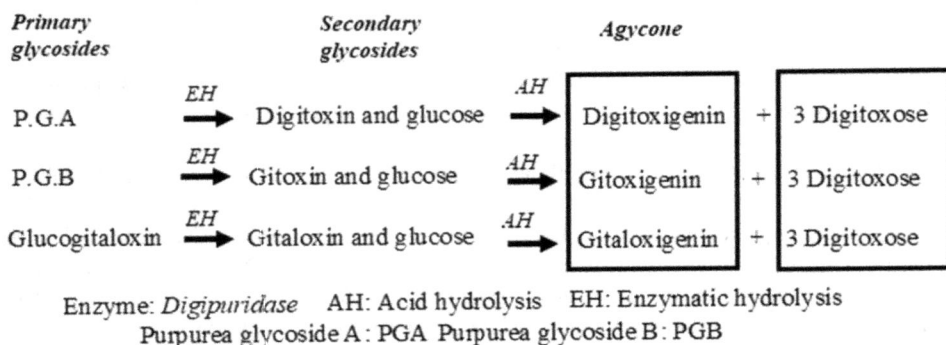

**FIGURE 2.6**   Isolation of PGA and PGB to yield secondary glycosides.

almost 1 h (Figure 2.7). Alcoholic fraction is filtered and filtrate is treated with strong solution of lead subacetate to precipitate pigments, proteins, and alkaloidal substances. After treatment with lead, subacetate supernatant layer is separated by using centrifugation (Figure 2.7). Afterwards excess amount of lead subacetate is precipitated with sodium sulphate anhydrous solution (6.3%). Again supernatant layer is separated by using centrifugation. Supernatant layer is treated with chloroform. At last extract is evaporated to initiate the crystallization of glycosides (Figure 2.7).

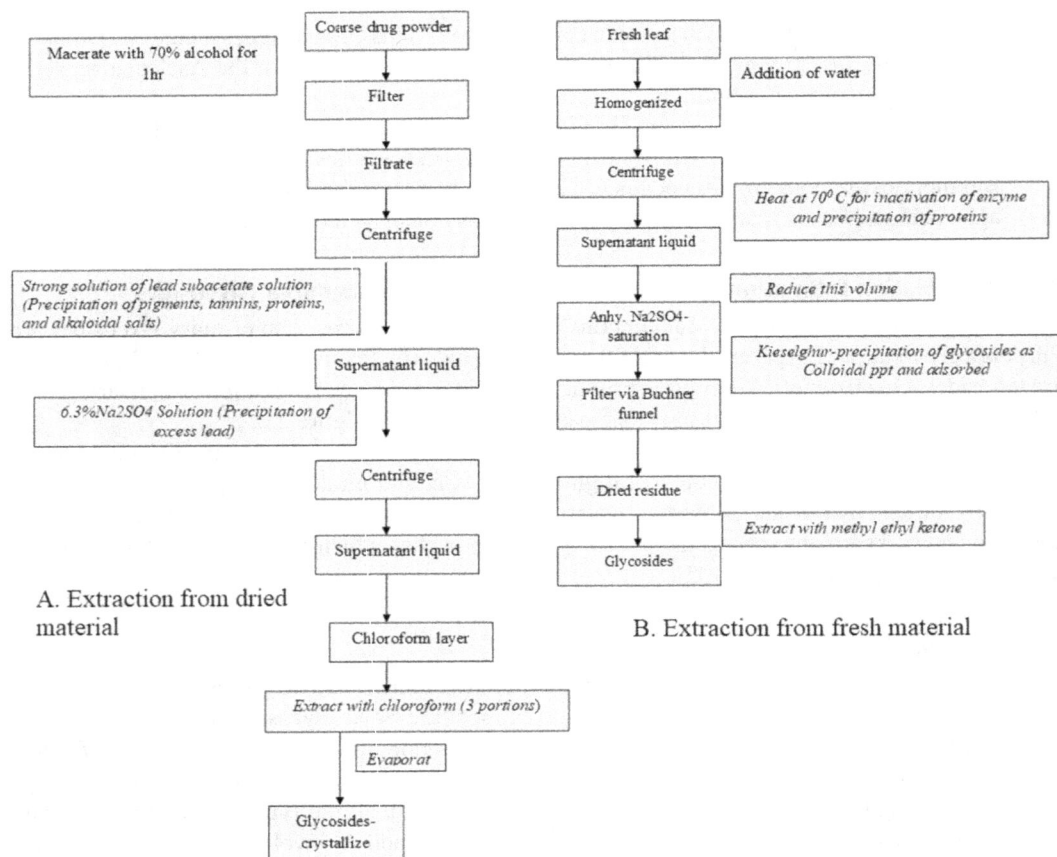

A. Extraction from dried material

B. Extraction from fresh material

**FIGURE 2.7**   Methods for extraction of cardiac glycosides from dried as well as fresh plant materials.

Second method (method A) involves the extraction of cardiac glycosides from fresh material. This method involves prior inactivation of enzymes by heating homogenized plant material at elevated temperature (70°C) to coagulate the proteins. Later on, supernatant layer is separated by using centrifugation (Figure 2.7). Afterwards supernatant layer is treated with anhydrous sodium sulphate to saturate the solution. Eventually glycosides are absorbed on kieselguhr to form colloidal glycoside precipitate. Precipitates formed are collected by using filtration process (Buchner funnel). Ultimately the dried residual material is extracted with methyl ethyl ketone (Figure 2.7).

### 2.8.3.2.1 Isolation of Digitoxin

For the isolation of digitoxin, plant material is first extracted with water at 45°C till 4–5 h. Afterwards, the marc is extracted by maceration with methanol (20%), and methanolic fraction is collected (Figure 2.8). Mix both aqueous and methanolic fractions and hydrolyze the mixture with

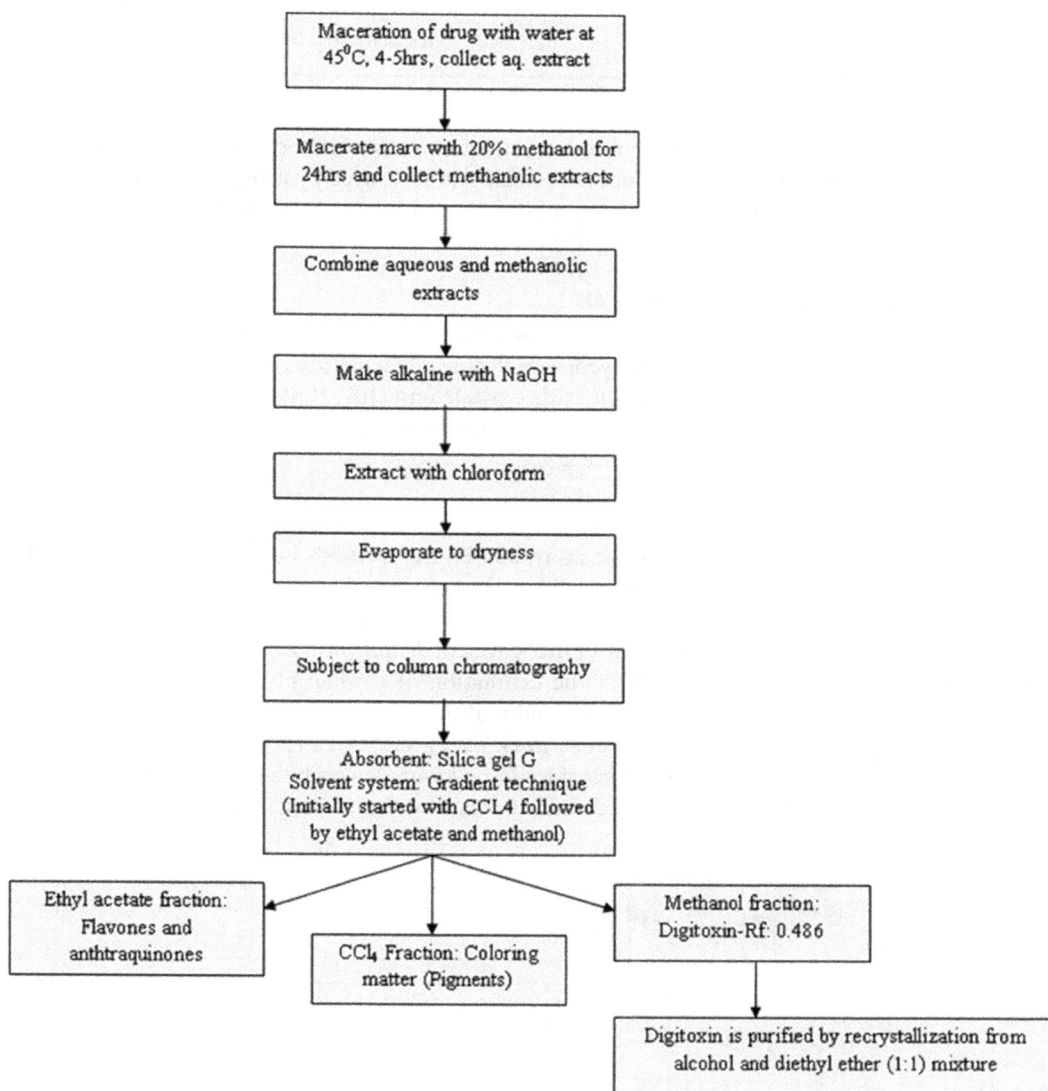

**FIGURE 2.8** Isolation of digitoxin.

## TABLE 2.1
## General Qualitative Chemical Tests for Cardiac Glycosides

| Chemical Tests | Legal Test | Baljet Test | Keddes Test | Raymond Test |
|---|---|---|---|---|
| Procedure | The extract of drug is dissolved in pyridine Sodium nitroprusside solutions added It gives yellow-to-orange colour on making alkaline | Sample gives pink-to-red coloration with sodium picrate | Sample is treated with 3,5-dinitrobenzoic acid in methanol and KOH solution. It gives reddish/bluish violet colour | Sample gives bluish violet coloration with dinitrobenzene and methanolic KOH |
| Inferences | Yellow-to-orange colour | Pink-to-red colouration | Reddish/bluish violet colour | Bluish violet colouration |
| Observation | Presence of cardiac glycosides | Presence of cardiac glycosides | Presence of cardiac glycosides; Bufadienolides do not respond to this test | Presence of cardiac glycosides |

sodium hydroxide (Figure 2.8). Alkaline fraction is treated with chloroform and evaporated till it is converted into dry mass. Afterwards column chromatography is used by using silica gel G as adsorbent, gradient solvent system (carbon tetrachloride followed by ethyl acetate and then methanol). Digitoxin is derived in methanol fraction with Rf: 0.486.

### 2.8.3.3    Identification of Phytochemicals

Cardiac glycosides are natural organic compounds derived from various medicinal plants. These bioactive compounds contain a steroid (aglycone) molecule which is either a cardenolide or a bufadienolide chemically linked to the hydrophilic carbohydrate unit [107, 108]. Thus, based on aglycone part, mainly different inferences are obtained during qualitative examination of these cardiac glycosides as mentioned in Table 2.1 and Figure 2.9.

### 2.8.3.4    Estimation of Cardiac Glycosides

Various methods such as spectroscopic, bioassay, and HPLC methods have been used to estimate the cardiac glycosides. In 1976, radioimmunological assay have been used to determine cardiac glycoside concentration in the plasma [109]. In 1973, digitoxin, gitaloxin, gitoxin, verodoxin, and strospesid were quantitatively determined in the leaves of *D. purpurea* by means of fluorescence [110]. In 1989, a systematic procedure for the estimation of cardiac glycosides in *D. purpurea* leaves by HPLC was developed, and it was found that this method is satisfactorily sensitive and reproducible to estimate secondary glycosides in *D. purpurea* leaves [111]. In 1996, a systematic procedure for the estimation of cardiac glycosides in *D. lanata* leaves by reversed-phase thin-layer chromatography (RP-TLC) was developed, and it was found that this is reliable and relatively

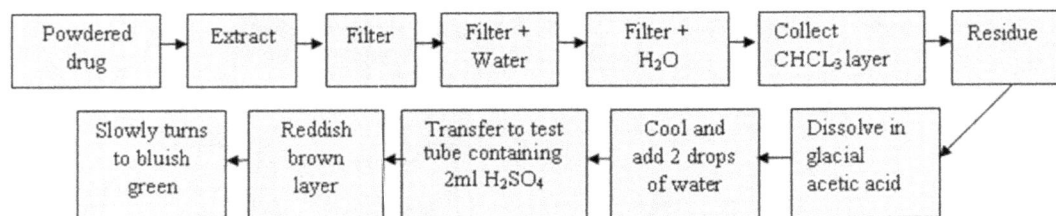

**FIGURE 2.9**    Keller Killani test (for deoxy sugar digitoxose).

simple for the determination of cardiac glycosides in *D. lanata* leaves [112]. In 2003, highly systematic, sensitive, and reproducible method for the estimation of digoxin as well as digitoxin in serum or plasma using combined process, HPLC, isotope-dilution MS as well as caesium-adduct formation is reported [113]. To assess the pharmacokinetics of digoxin in humans mainly drug-drug interaction (digoxin and imidafenacin) and the characterization of steady-state pharmacokinetics of digoxin in humans after oral administration a sensitive and specific LC/MS/MS method (with electro spray ionization tandem mass spectrometry in the positive ion-multiple reaction-monitoring [MRM] mode) was developed and validated. This method was found to be more sensitive, specific, accurate, and reproducible than common techniques such as radioimmunoassay (RIA) [114]. Generally spectroscopic procedures can also be used to estimate cardinolides by measuring absorbance based on digitoxose moiety and lactone ring. Estimation of cardiac glycosides depends on therapeutic effects, glycosidal sugar moiety as well as properties of the butanolide glycosidal moiety. Conventionally, the blood pressure assessment has been considered assay for Digitalis to offer the best sign of the glycosides in the complex form [115]. Estimation of cardinolides can be done by measuring the absorbance of digitoxose moiety as well as lactone ring. Absorbance of digitoxose moiety can be measured by using Keller Kiliani test at λmax—590 nm, whereas lactone ring can be measured by using dinitrobenzene reactions at λmax—620 nm. As per IP 1996, digitoxin can be estimated by measuring absorbance against blank at 495 nm. This can be done by dissolving digitoxin (40 mg) in ethanol (100 ml) and again dilute 5 ml of ethanolic solution with ethanol (up to 100 ml). Later on, 5 ml of the diluted solution is treated with alkaline picric acid solution (3 ml), eventually to measure the absorbance against blank at 495 nm [115]. HPLC method is considered one of the most common methods for the estimation of *D. purpurea* glycosides by using column (Octyl-, silicyl-bonded silica), solvent system [acetonitrile:methyl alcohol:water (10:15:18) at the flow rate of 0.5 ml/min as well as detected at wavelength of 220 nm. Sample preparation can be done by dissolving 50 mg of leaf powder sample in 50% of methanol (25 ml) having known quantity of lanatoside A as internal standard. During this procedure, 10 µl of sample is injected via sample injector into HPLC. After the completion of process, HPLC chromatogram representing peak areas of pupurea glycosides A, B, gitaloxin, digitoxin, gitoxin, and glucogitaloxin are documented and their respective concentrations in the samples can be estimated by their peak area ratios with the internal standard.

### 2.8.3.5   Bioassay of Digitalis and Its Preparations

To determine potency of digitalis and its related preparations, standardized biologically assays can be used. Digitalis and its preparations can be biologically assayed for their potency. As per IP, both pigeons and guinea pigs as an experimental animal's models can be used. In this bioassay usually biological activity of the test sample is compared with standard sample at same dilution. In this procedure by using ether, adult pigeons are anesthetized, to expose the alar vein and subsequently cannulated by means of venous cannula. Prescribed dose of diluted samples are administered at 5-min intervals until it pass away from cardiac arrest. The right internal jugular vein was cannulated for drug administration in case of the guinea pig. The effect of test and standard sample at different doses on the heart rate of the unanesthetized guinea pig was investigated by administration via chronic jugular cannula [116, 117].

### 2.8.4   CALCIUM SENNOSIDES

Anthraquinones are the organic compounds commonly present in medicinal plants as glycosides; e.g., are *O*-glycosides (the sennosides from *Cassia* sp.) and C-glycosides (the aloins from *Aloe* sp.). Apart from senna leaves, rhubarb also contains sennosides and anthrone glycosides. Senna, fast-acting laxative natural crude drug contain dimeric anthraquinone glycosides called sennosides. Sennosides (mainly sennoside A, sennoside B, sennoside C, and sennoside D) are dimeric anthraquinone glycosides derived from Senna. Sennosides A and B or anthracoids (homodianthrone glycosides) are the crystalline glycosides, responsible for the purgative action are considered the major glycosides of senna while sennosides C and D (heterodianthrone glycosides involving rhein and aloe-emodin) are considered minor constituents. Sennosides A and B hydrolyze by gut bacteria to form two glucose (glycine) molecules and the aglycones (sennidin A and B). Sennidin B is mesoform of Sennidin A (dextrorotatory) formed by intramolecular compensation. Emodin-8-$O$-sophoroside (a diglucoside), a new anthraquinone glycoside, has been isolated from dried Indian senna. C-C bond protects the anthrone from oxidation [118]. A new anthraquinone glycoside, emodin-8-$O$-sophoroside (a diglucoside), has been isolated in 0.0027% yield from dried Indian senna leaves [119].

Sennosides are obtained from the dried leaflets of *Cassia angustifolia* (Indian) and *C. angustifolia* (*Alexandrian senna*) (Leguminosae). Sennosides acids are converted into calcium sennosides to enhance their water solubility in order to enhance their absorption by the gastrointestinal tract (GIT). These glycosides show purgative effect, i.e., these organic compounds directly act on large intestine to increase the peristalsis movement. Human intestinal bacteria (hydrolytic enzyme, β-D-glucosidases, secreted by intestinal Bifidobacteria) convert sennosides in to rhein anthrone to liberate the active rhein anthrones in order to show purgative effects. Previous study suggested that glycone part of sennosides A and B remains unaltered in stomach and gut. However, it's hydrolyzed in the caecum by enzymatic action of bacteria to form dianthrones. Dianthrones are further hydrolyzed in the intestine to yield an anthrone and anthraquinone which are accountable for biological effects [120, 121].

### 2.8.4.1   Methods of Isolation

In 1913, aloe-emodin and rhein were first isolated by Tutin, and later on in 1941, Stoll et al. led down procedure for the isolation of crystalline glycosides (sennoside A and B) [122]. These glycosides after hydrolysis yield glycon part (2 mol. of glucose) and aglycone part (sennidin A and B). Time duration of extraction impacts ultimate yield; however, longer extraction can result into degradation. As per reports, short solvent treatment (0.5–5 min) has been suggested for the extraction of sennosides. Several procedures are available for the isolation of sennosides; however, sennosides are always isolated calcium sennosides due to their better water solubility and stability. Following procedures are usually employed for the isolation of sennosides:

- Stoll and Becker method
- Notherman and Lic method
- Modified Stoll and Becker method
- Method extraction of sennosides as calcium salts
- Mauzaram et al. method
- Ahmed and Samia method

Type of extraction procedure as well as extraction conditions always impact extraction efficiency in terms of yield and chemical profile of the product. Recently old (cold percolation and refluxing) as well as advance (ultrasound, microwave, and supercritical fluid-assisted extraction) have been employed to derive senna extract. It was found that microwave as well as ultrasound-assisted extraction techniques showed more promising results than cold percolation as well as refluxing methods [123]. One of the most common procedures used for the extraction of sennosides is by using benzene, ammonia, hydrochloric acid, methanol, and anhydrous calcium chloride (Figure 2.10). Dried powdered plant material obtained from senna leaves is subjected to extraction with benzene for almost 2 h. Marc is collected, dried, and extracted with methanol (70%) for 4 h. Marc is collected again and extracted with methanol (70%). Both extracts are mixed and their volume is reduced up to and to 1/8th part (Figure 2.10). Later on, extract was acidified with diluted hydrochloric acid (pH 3.2), and heated moderately for 2 h followed by the filtration. Filtrate is treated with add anhydrous calcium chloride and pH of the solution is adjusted (pH 8) by adding ammonia. Then at last precipitates are collected and dried as mentioned in Figure 2.10.

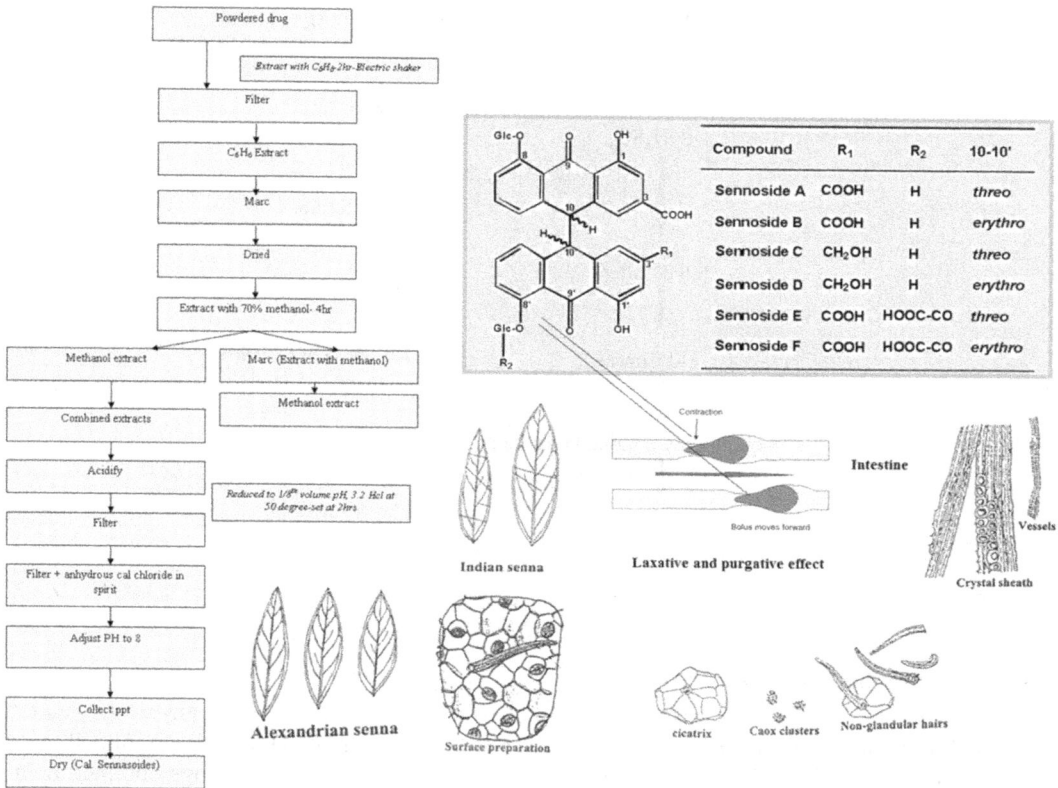

**FIGURE 2.10**  Method for the isolation of sennosides.

### 2.8.4.2  Identification Tests for Sennosides

Digoxin was initially isolated by Dr. Sydney Smith from *D. lanata* in 1930. The most common and reliable chemical test for the identification of anthraquinone glycosides is Bontrager's test. During this chemical test, the sample is boiled with sulphuric acid (1 ml) in a test tube for 5 min. Later on, solution is filtered (Figure 2.11). Filtrate is cooled and mixed well with equal volume of dichloromethane or chloroform. Subsequently lower layer of chloroform is separated and mixed with half volume of dilute ammonia. Appearance of rose red colour in an ammonical layer indicates the presence of free anthraquinone glycosides specifically *O*-glycosides. Borntrager's test distinguishes anthraquinones from anthrones and anthranols (Figure 2.11). Some C-glycosides such as aloin do not hydrolyze using mere acids/bases; thus, modified Borntrager's test is used by treating the sample with ferric chloride as well as hydrochloric acid via oxidative hydrolysis. However, modified Borntrager's test is used to identify the presence of C glycosides as mentioned in Figure 2.10, which require more extreme conditions for hydrolysis of anthraquinone glycosides. Therefore, the earlier test is modified by adding ferric chloride as well as hydrochloric acid to affect oxidative hydrolysis.

Anthraquinone glycosides are classified into four different types: O, N, S, C. As suggested above, Borntrager's test can only be used for O glycosides and modify Borntrager's test used for C glycosides. However, for A glycones or hydroxy anthraquinones, plant material is extracted with organic solvent followed by the treatment with $NH_4OH$ or KOH. Positive result indicated by rose red colour in the aqueous alkaline layer.

**Borntragers test      Modified Borntragers test**

5ml extract
+
5ml of $FeCl_3$
+
5ml of Dil. HCl.

$\Delta$ for 5 min.

3ml extract
+
Dil. $H_2SO_4$

Boil & filter                           Boil & filter

Add benzene   Cold filtrate          Cold filtrate Add benzene
/ $CHCl_3$                                     / $CHCl_3$
& shake                                        & shake

Separate the organic layer  Separate the organic layer

Add $NH_3$                              Add $NH_3$

Ammonia layer turns        Ammonia layer turns
pink / red                      pink / red

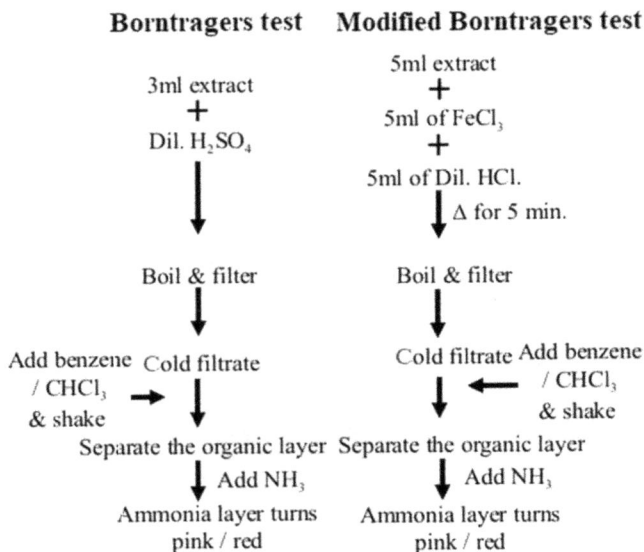

**FIGURE 2.11**   Identification test for anthraquinone glycosides.

### 2.8.4.3   Estimation of Sennosides

HPTLC procedure was used for assessment of sennosides (A, B, C, D) with no any derivatization in marketed preparations as well as plant crude material (pods and leaf). Chromatogram showed respective bands at 0.35, 0.25, 0.61, 0.46 for sennosides A, B, C, and D. Findings obtained demonstrated no significant difference in between spectrophotometric (BP) and spectrofluorimetric (USP) and HPTLC procedures for assessment of sennosides. Developed HPTLC procedure has merits such as simple, rapid precise, and accurate [124]. Another study presented spectrophotometric method for the estimation of Senna and its preparations, for both total sennosides and total rhein glycosides content. This method confirms complete removal of other minor non-carboxylic anthracene derivatives and flavonoidal contaminants and thus helpful in estimating the actual total sennosides content [125]. Another study demonstrated a liquid chromatographic-based procedure for the assessment of sennosides (B, A) in *C. angustifolia* leaves. It was found that detection limits were 10 µg/ml for sennoside B and 35 µg/ml for sennoside A in the sample. This procedure can be utilized for rapid screening of *C. angustifolia* samples in large numbers, which is needed in breeding/genetic engineering and genetic mapping experiments [126]. Similarly, a simple TLC method can also be utilized to identify sennosides from powdered drug senna using silica gel 60 F254 as absorbent and propanaol:ethyl acetate:water (4:4:3) as solvent, nitric acid KOH reagent, or visible/UV—366 nm for detection. Rf value of sennosides were found to be A—0.4, B—0.2, C—0.7, D—0.5. Determination of sennosides by calorimetry based on Borntrager's test can also be done by initially removal of free anthraquinone genins from extract followed by the oxidative hydrolysis of dianthrones. Further obtained fraction is subjected to acid hydrolysis to form monometric anthraquinone genins. Finally, resultant solution is estimated by calorimetry based using Borntrager's test at 513 nm (using methanol as blank) [127].

### 2.8.5   Glycyrrhizin

Glycyrrhizin is a glycoside derived from of a small leguminous shrub, *Glycyrrhiza glabra* root (Leguminosae). Glycyrrhizin is a triterpenoid saponin and it's a potassium and calcium salts of Glycyrrhizic acid. Glycyrrhizic acid on hydrolysis gives 18-β glycyrrhetinic acid. Licorice root has been used in traditional medicine to treat jaundice, gastritis, and bronchitis. The main chemical components of this crude drug are glycyrrhetinic acid (GRA), flavonoids, hydroxyl coumarins, and β-sitosterol [128–130]. Glycyrrhizin is responsible for the sweet taste of the drug. After collection, roots of *G. glabra* are dried to 10% moisture, processed, treated with aqueous ammonia, concentrated by evaporation followed by precipitation with $H_2SO_4$, and crystallization with alcohol (95%) [128–130]. This process forms crude ammonium glycyrrhizin (AG) which is further treated to form white, crystalline mono-AG (MAG). Both resultant compounds are differing in solubility and sensitivity to pH, however, have the same sweetness. AG is comparatively stable and highly soluble in hot or cold water and in alcohol. AG also remain stable at temperatures greater than 105°C for lesser duration and precipitates at pH values lesser than 4.5 [128–130]. It's widely known that glycyrrhizin is popular for its sweetness (50–100 times sweeter than sucrose), a thirst quencher and has a slow arrival of sweetness followed by licorice-like aftertaste. Glycyrrhizin is also known for corticoid activity, affecting steroid uptake in order to maintain blood pressure, volume as well as to control glucose/glycogen balance. In GIT, glycyrrhizin can be hydrolyzed by human intestinal microflora such as *Eubacterium* sp. strain GLH to 18-β-GRA as well as two molecules of glucuronic acid. 18-β-GRA directly binds to the plasma protein and go into the enterohepatic circulation where it's almost fully metabolized. Glycyrrhizinates produce mineralocorticoid-like effects as it inhibits enzyme (11-β-hydroxysteroid dehydrogenase enzyme type 2) accountable for deactivating cortisol to elevate sodium and reduce potassium levels (aldosterone-like action) [128–130]. Glycyrrhizin acid also shows anti-inflammatory, minerocorticoid-like, and antipeptic ulcer like effects. Apart from its medicinal properties, it's widely used as flavouring agent, expectorant, and demulcent. The US FEMA GRAS (generally regarded as safe) allowed utilization of liquorice extracts and ammoniated glycyrrhizin in various marketed products. It's important to understand the relationship between the pharmacokinetics of glycyrrhizic acid toxicity profile and its biologically active metabolite glycyrrhetinic acid. Absorption of glycyrrhizic acid is mediated via presystemic hydrolysis which leads to the formation of glycyrrhetinic acid. Toxicological profile of glycyrrhetinic acid is more important as its 200–1000 times more potent inhibitor of 11-β-hydroxysteroid dehydrogenase then glycyrrhizic acid; thus, it's important to study pharmacokinetic behaviour of glycyrrhetinic acid (Figure 2.12) [128–130].

In this report, the author has suggested that the rate of transportation of GIT contents via the small and large intestines mainly decides to what proportion glycyrrhetinic acid conjugates will be reabsorbed [131].

### 2.8.5.1  Isolation of Glycyrrhizin

Initially an acidic polysaccharide called glycyrrhizin GA was isolated from *G. glabra* (dried stolon) by Shimizu et al. in 1991, and glycyrrhizin GA was isolated from the stolon of *G. glabra* var. glandulifera by Takada et al. in 1992 [132, 133]. As reported by Fenwick et al. in 1990, licorice roots contain 3–5% glycyrrhizin [134]. To date, more than 400 compounds have been isolated from licorice [135]. Zeng et al. from 1990 to 1991 for the first time isolated various flavonoids, coumarins, and saponins, namely, liquiritigenin, licochalcone A, isoliquiritin, isoliquiritigenin, isoglycyrol, isoglycycoumarin, glycyrol, and glycycoumarin [136]; glycoumarin, glycyrrhizic acid, isoglycoumarin, isoliquiritin, liquiritigenin, liquiritin, and uralsaponin B [137]; and three saponins glycyrrhizic acid (S-I), uralsaponin B, and uralsaponin A [138]. Thus, Zeng et al. for the first time isolated glycyrrhizic acid and glycyrrhizic acid from *G. glabra* (Xinjiang) roots [136]. The isolation of glycyrrhizin from the roots of *G. glabra* is based on the solubility (Figure 2.13). Glycyrrhizin has varied solubility as it's completely insoluble in ether, soluble in hot water and alcohol and to some extent soluble in cold water (Figure 2.12). Based on the varied solubility, glycyrrhizin is isolated by three different methods as suggested below [139] (Figure 2.13):

- Acid precipitating method: This procedure is based on the treatment of plant material with acid ($H_2SO_4$/Conc HCl) (pH 2.8) to form glycyrrhizin precipitates from aqueous solution.
- Alcohol extraction method: This procedure relies on the varied solubility of glycyrrhizin in different solvent systems. As mentioned above, glycyrrhizin is experimentally insoluble in

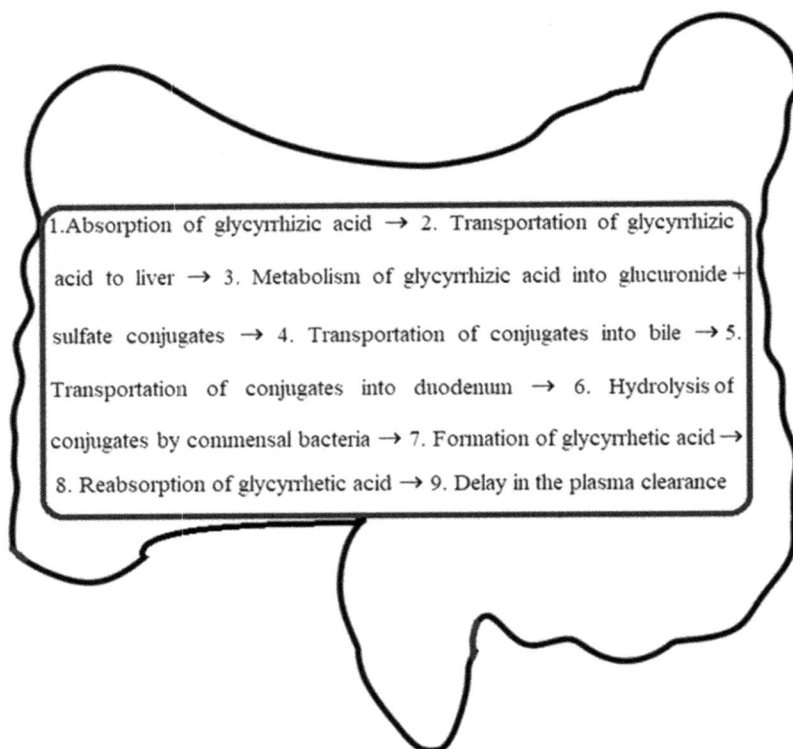

1.Absorption of glycyrrhizic acid → 2. Transportation of glycyrrhizic acid to liver → 3. Metabolism of glycyrrhizic acid into glucuronide + sulfate conjugates → 4. Transportation of conjugates into bile → 5. Transportation of conjugates into duodenum → 6. Hydrolysis of conjugates by commensal bacteria → 7. Formation of glycyrrhetic acid → 8. Reabsorption of glycyrrhetic acid → 9. Delay in the plasma clearance

**FIGURE 2.12**   Pharmacokinetics of glycyrrhizic acid.

non-polar solvents such as ether however soluble in alcohol. This parameter can be utilized to eliminate the undesirable components. Alcoholic solvent system is used to extract glycyrrhizin and after the removal of undesirable components alcoholic extract is extracted again with numerous solvents continuously to enhance the polarity and to get rid of undesirable components. Ultimately the alcoholic fraction is reduced and further isolated to get glycyrrhizin.

- Ammonia precipitating method: Once precipitates from acid precipitating procedure are obtained they are added in dilute ammonium hydroxide. Subsequently acetone is added to again precipitate the solution. Finally, ammonium glycyrrhizinate precipitates are recovered from the solution and then dried.
- In 2008, pressurized hot water extraction (using subcritical compressed liquid water at 50–150°C) for extraction of natural sweetener (mono-ammonium glycyrrhizinate) from licorice (*G. glabra*) roots by using dissolved ammonia and pressurized with carbon dioxide has been reported. It was found that maximum recovery of mono-ammonium glycyrrhizinate from licorice roots has been achieved at 110°C and 5 atm [139].

### 2.8.5.2   Estimation of Glycyrrhetinic Acid

Anti-3-monoglucuronyl GRA (one of the metabolites of glycyrrhizin responsible for pseudoaldosteronism) monoclonal antibody and an ELISA system to easily detect 3MGA in the plasma and urine of the patients has been developed. 22α-hydroxy-18β-glycyrrhetyl-3-*O*-sulphate-30-glucuronide from urine with the help of positive immunostaining of eastern blot was identified as new metabolite of GRA [140]. In 1999, urinary 18-β-GRA (metabolite of glycyrrhizin) was determined by sensitive and quantitative gas chromatography method. Findings obtained showed the detection limit of GC was 10 μg/l with a urine volume of 10 ml. A detection limit of 3 μg/l was obtained

**Acid precipitating method**

- 50g drug+300ml H2O
- Boiling with stirring
- Decant the supernatant liquid
- Filter the remaining residue and collect the filtrate
- Filter the decanted supernatant liquid
- Collect the filtrate and combine them
- Adjust pH 2.8 by the addition of acid
- Glycyrrhizin precipitates out
- Filter and collect the precipitate
- Wash the precipitate with cold water to make it free from acid
- Transfer the precipitates to china dish and heat gently to remove the water content

**Alcohol extraction method**

- 50g drug
- + 100ml methanol
- Allowed to stand for 24 Hr
- Filter
- Collect filtrate
- Extract this methaolic extract with 3 portions of pet ether
- Extract with benzene, ethyl acetate chromium and solvent ether
- Transfer methanolic layer in to china dish
- Evaporate on water bath to get glycyrrhizin

**Ammonia precipitating method**

- Drug extracted with hot water
- Filtrate
- Filtrate is acidified with conc $H_2SO_4$ to pH 2.8
- Precipitate is dissolved in dil NH40H
- Ammonium glycyrrhizinate precipitate
- PPt is dissolved in hot water
- Evaporated to get Ammonium glycyrrhizinate

Glycyrrhizin

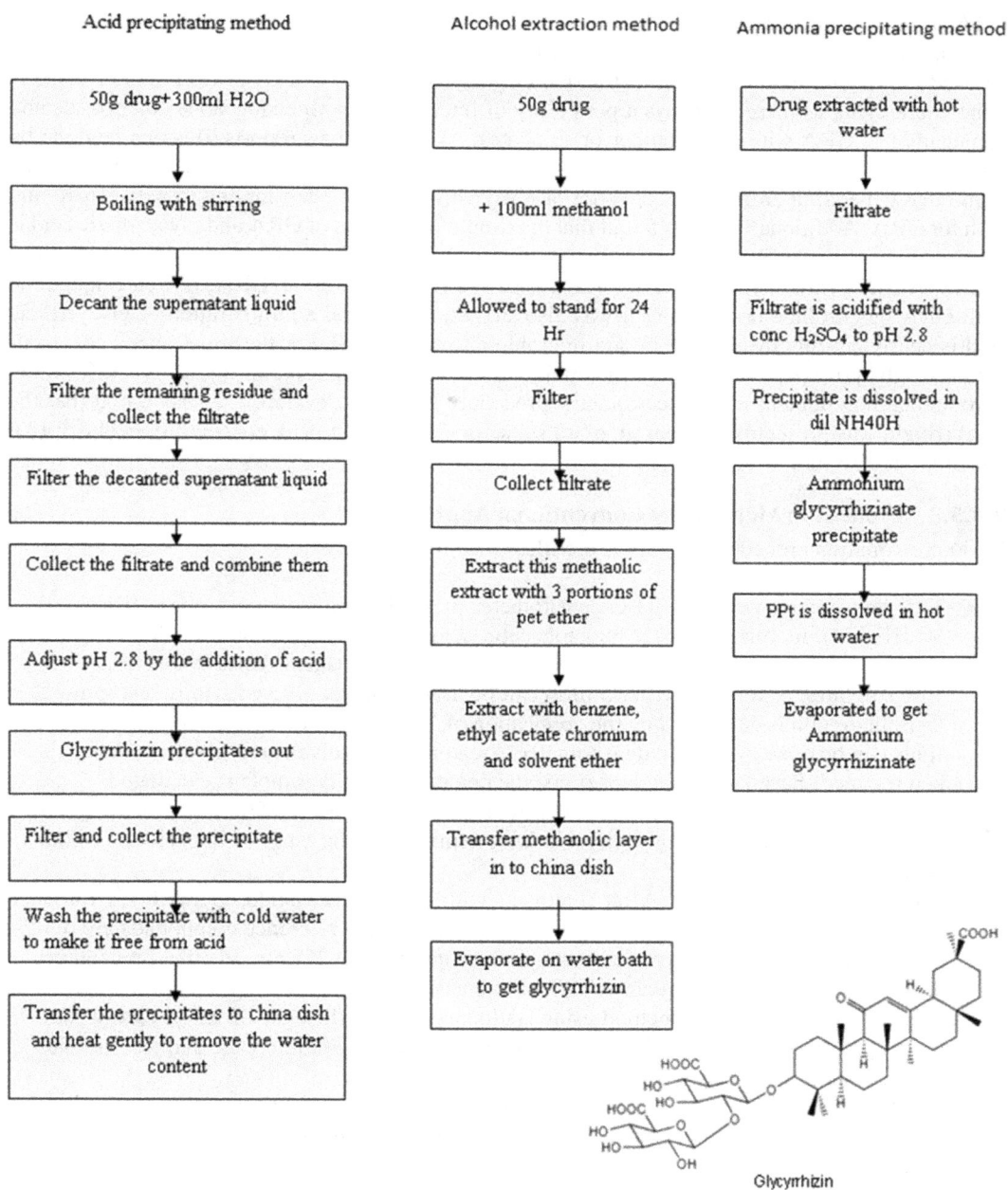

**FIGURE 2.13**  Isolation procedures for glycyrrhizin.

by performing GC-MS [141]. A simple, rapid, accurate, and reproducible capillary electrophoretic procedure has been developed for the estimation of glycyrrhizin and GRA in traditional Chinese medicinal preparations. Results showed that content of glycyrrhizin (0.04–2.00 mg/ml) and GRA (0.007–0.35 mg/ml) in Glycyrrhizae Radix and Glycyrrhizae Radix-containing Chinese medicinal preparations was successfully determined within 10 min [142]. High-performance liquid chromatography can be used to rapidly and precisely determine the content of glycyrrhizic acid and GRA. High-performance liquid chromatography has been used to assay the level of glycyrrhizic acid and

GRA in biological fluids and tissues from experimental animals and humans. From plasma and tissues, glycyrrhizic acid and GRA are extracted by organic solvents and the extracts can directly be used for HPLC. Estimation and extraction of glycyrrhizic acid and GRA from bile or urine is quite challenging as there is always a possibility of interference with endogenous compounds and conjugation of GRA with glucuronides or sulphates. As per previous reports, this can be done by ion-pairing followed by extraction with organic solvents or by solid-phase extraction. It was also found that depending on the dose, half-life of glycyrrhizic acid is 3.5 h for and in between 10 and 30 h for GRA. Additionally, it was found that maximum proportion of GRA and glycyrrhizic acid is eliminated by the bile. Report also suggested that glycyrrhizic acid can be excreted unmetabolized and experiences enterohepatic cycling, whereas before biliary excretion, GRA showed conjugation with GRA glucuronide or sulphate. It was also suggested that oral administration of glycyrrhizic acid is nearly entirely hydrolyzed by gut microbiota to GRA [143]. For the simultaneous determination of glycyrrhizin and its active metabolite GRA in human plasma, a highly sensitive liquid chromatography tandem mass spectrometry procedure has been developed. It was found that the glycyrrhizin was present in the plasma of all subjects with average peak concentration $24.8 \pm 12.0$ and $200.3 \pm 60.3$ ng/ml for GRA [144].

### 2.8.5.3  Estimation Methods by Conventional Approaches
Various estimation procedures for GRA are illustrated in Figure 2.14.

- TLC densitometer method: TLC densitometer method using stationary phase (Kieselgel 60 HF 254), mobile phase (N butanol:acetic acid:water; 7:1:2), standard solution (glycyrrhizin [0.5–2 mg/ml] in ethanol [50%]) and spraying reagent (ceric sulphate [1%] in 10% sulphuric acid at 105°C for 5 min) can be used to detect glycyrrhizin in test sample. Chromatogram is developed by the application of 10 μl std. and test. Visualizations of the spots can be done by scan in densitometer. Area of spots equivalent to glycyrrhetinic acid is integrated. Based on the area of spots, the percentage of the sample is calculated.
- Colorimetric method: After treatment with anisaldehyde (with $H_2SO_4$), glycyrrhizin gives purple colour. Quantity of glycyrrhizin is determined on the basis of intensity of colour developed at 556 nm.
- UV spectroscopic method: After treatment with sodium hydroxide, glycyrrhizin forms its sodium salt. In this method, weighed amount of test and reference compounds are dissolved in sodium hydroxide and absorptions are measured at 258 nm. In other procedures, glycyrrhizin is added into ethanol (80%) and measured at 250 nm.
- HPTLC method: HPTLC method using stationary phase (silica gel GF 254), solvent ratio [toluene (12.5):ethyl acetate (7.5):glacial acetic acid (0.5)] with reference and test samples can be used to detect glycyrrhizin at 254 nm, Rt: 0.41 min.
- HPLC method using reverse-phase C-18 column, solvent system [acetonitrile:water phosphoric acid (36:63:1)], 1 mg standard (in methanol), 1 g (test sample solution in 100 ml water) at flow rate: 1.4 ml/min detected at 250 nm, can be utilized to detect glycyrrhizin.

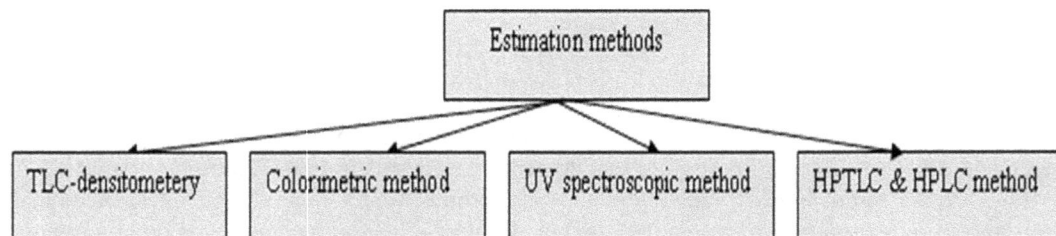

**FIGURE 2.14**   Estimation procedures for glycyrrhetinic acid.

### 2.8.5.4 Identification Tests

Various identification tests have been performed such as foam test, Salkowski test, and trichloro-acetic acid as mentioned below:

- Foam test: Test sample (containing glycyrrhizin) + water → vigorous shaking → froth formation (due to the presence of saponin)
- Salkowski test: Test sample (containing glycyrrhizin) + chloroform + Conc. $H_2SO_4$→Yellow colour
- Trichloroacetic acid: Test sample (containing glycyrrhizin) + trichloroacetic acid → Red coloration
- TLC method: TLC procedure by using stationary phase (Kiesel gel 60 HF 254), mobile phase (toluene:ethyl acetate:glacial acetic acid 12.5:7.5:0.5), reference standard (5 mg glycyrrhizin with 20 ml of 0.5 ml sulphuric acid), spraying reagent (anisaldehyde: sulphuric acid reagent), test sample (chloroform extract) has been used to detect glycyrrhetinic acid at 254 nm (UV). RF value of 0.41 min corresponds to glycyrrhetinic acid in both samples.

## 2.8.6 GYMNEMIC ACID

Gymnemic acids are the mixture of almost nine structurally related acidic glycosides of triterpene glycosides as well as 35-mer peptide named gurmarin derived from leaves of slow-growing, perennial, woody climber, i.e., *Gymnema sylvestre* (Asclepiadaceae). Gymnemic acids are considered the main chemical component of *G. sylvestre* that belongs to India and southern China. Gymnemic acids I–VI with a common gymnemagenin oleanane-type aglycone structure and a glucuronic acid moiety were isolated in 1989. Later on, gymnema saponins III–V (antisweet compounds) were discovered. In comparison with gymnemic acids I–VI, gymnema saponins III–V (nonacylated compounds) showed somewhat less potent sweetness-inhibitory activities. Afterwards gymnemic acids VIII–XVIII (antisweet compounds) have been identified. Gymnemic acids XIII and XIV were formerly called gymnemic acids VIII and IX. Gymnemic acids actively masks sweet detection, decreases caloric intake and inhibits the sensitivity of the sweet taste on the taste buds (if applied over tongue before meal) [145].

These acidic glycosides possess anti-saccharine (anti sweet) property via suppressing sweetness in humans. Traditionally this herb has been used in Ayurveda for almost 2000 years for the treatment of sugar diabetes called "honey urine". Vernacular (Indian) name of this plant is "Meshashringi" or "Gurmar" (sweet—destroyer) due to its wetness destroying property of its leaves. Once leaves are consumed orally, during mastication samples were sweetened by using sucrose flavour like water [146–148]. It's assumed that these organic natural compounds prevent the interaction of a sweet compounds to the sweet taste receptors. After the structure elucidation of isolated and purified gymnemic acid analogues along with various acyl groups from *G. sylvestre*, it was found that removal of the acyl group reduces the antisweetness effect. This chemical reaction used for the removal of acyl group destroys the sweetness of majority of sweeteners such as aspartame (artificial sweetener) and thaumatin (natural sweetener) [146–148]. Owing to its sweetness destroying capability this medicinal herb is usually utilized for the treatment and the management of diabetic condition in India and to control obesity in Japan by using Gymnema extracts. Adverse reactions associated with Gymnema in sensitive individuals can cause reflux and/or vomiting, to prevent these sensitive individuals can be prescribed for enteric-coated tablets or with meals [146–148].

Apart from *G. sylvestre*, other medicinal plants such as *Hovenia dulcis*, *Ziziphus jujuba*, *Gymnema alterniflorum*, *Stephanotis lutchuensis*, and *Styrax japonicus* have been reported for their anti-sweetness effects due to the presence of oleanane- and dammarane-type triterpenoid sweetness-inhibitory compounds. Furthermore, one more antisweet triterpenoidal compound (35-amino-acid peptide) namely gurmarin has been derived from the *G. sylvestre*. Anti-sweetness potential of plant triterpenoids has been assessed by placing test sample (1 mmol/l) in the mouth from 2 to 3 min. Later on rinse the mouth with distilled water and eventually varied dilutions of sucrose (0.1–1 mmol/l) are placed in the mouth. To determine sweetness effects, highest concentration of sucrose is selected (concentration at which complete inhibition of sweetness) is observed. Experimentally gymnemic acid I is considered a standard antisweet compound to assess the sweetness-inhibitory potency of any natural compound (in comparison with gymnemic acid I) [149].

G. sylvestre and its related products has been found to be effective in treating both T1-diabetes mellitus as well as T2-diabetes mellitus via either enhancing the synthesis or action of insulin or via stimulating regeneration of pancreas beta cells. G. sylvestre (Leaf extract) improves glycaemic control, decreases insulin necessities and fasting glucose in T1-diabetes mellitus. Gymenema extracts have been reported for their ability in suppressing glucose uptake from the intestine, enhance uptake of glucose, or enhance insulin release. Once applied to the tongue, these natural organic anti-sweet compounds inhibit sensation of sweetness. In comparison with the control, individuals who applied these organic compounds over the tongue consumed lesser calories at mealtimes. Surprising these compounds reduces frequent urge for sugar and allows patients with T2-diabetes mellitus to use a less-carbohydrate diet. As per the previous reports gymnemic acid also augmented enzymatic activity accountable for glucose metabolism and suppressed peripheral use of glucose by corticotrophin and somatotrophin [150].

### 2.8.6.1 Identification Tests for Gymnemic Acid

Qualitative chemical tests such as Salkowski test, Liebermann Burchardt test and Tschuggen test have been commonly used for the presence of Gymnemic acid. In Salkowski test, chloroform fraction of the test sample is mixed with conc. sulphuric acid. Appearance of yellow colour in lower layer at stationary phase indicates the presence of Gymnemic acid. In Lieberman Burchardt test chloroform fraction of the test sample treated with acetic acid and conc. sulphuric acid (1 ml). This leads to the formation of deep red colour at the junction of two layers. In Tschuggen test chloroform fraction of the test sample is allowed to treat with excess amount of acetyl chloride and zinc chloride. Solution obtained is subjected for heating on water bath which leads to the formation of Eosin red colour.

### 2.8.6.2 Safety Protocols and Possible Interactions of Gymnema

Gymnema is possibly safe when used as per the recommendations of physician up to 20 months; however, enough data about the safety of consuming Gymnema during Pregnancy and breastfeeding is not available. Using Gymnema with diabetes medications can cause considerable drop in blood sugar level; thus, periodic blood sugar test is required. Even interaction between Gymnema and antidiabetic medications are also found it's advisable to take physician opinion before its use especially for diabetic patients. Taking gymnema along with insulin can also cause considerable drop in blood sugar level.

### 2.8.6.3 Isolation of Gymnemic Acid

In the late-18th century, name "gymnemic acid" was coined for brittle, black, resinous substance having capability of destroying the sensitivity of the taste buds against sweet substances. During this period (1888) Hooper isolated Gymnemic acid (complex mixture of saponins with molecular formula $C_{32}H_{54}O_{12}$) for the first time from the leaves of G. sylvestre [151, 152]. Gymnemic acid exists as a potassium salt. Thus, it was isolated by treatment of the aqueous layer of the alcoholic extract with a mineral acid followed by the drying under hot air to ultimately precipitate potassium salt as mentioned in Figure 2.15. Later on in 1989, gymnemic acids (I–VI) with a common gymnemagenin oleanane-type aglycone structure and a glucuronic acid moiety were isolated. In 1992, structures of gymnemic acids-VIII and -IX were elucidated as 3'-O-β-D-arabino-2-hexulopyranosyl gymnemic acid-III and -IV, respectively [151, 152].

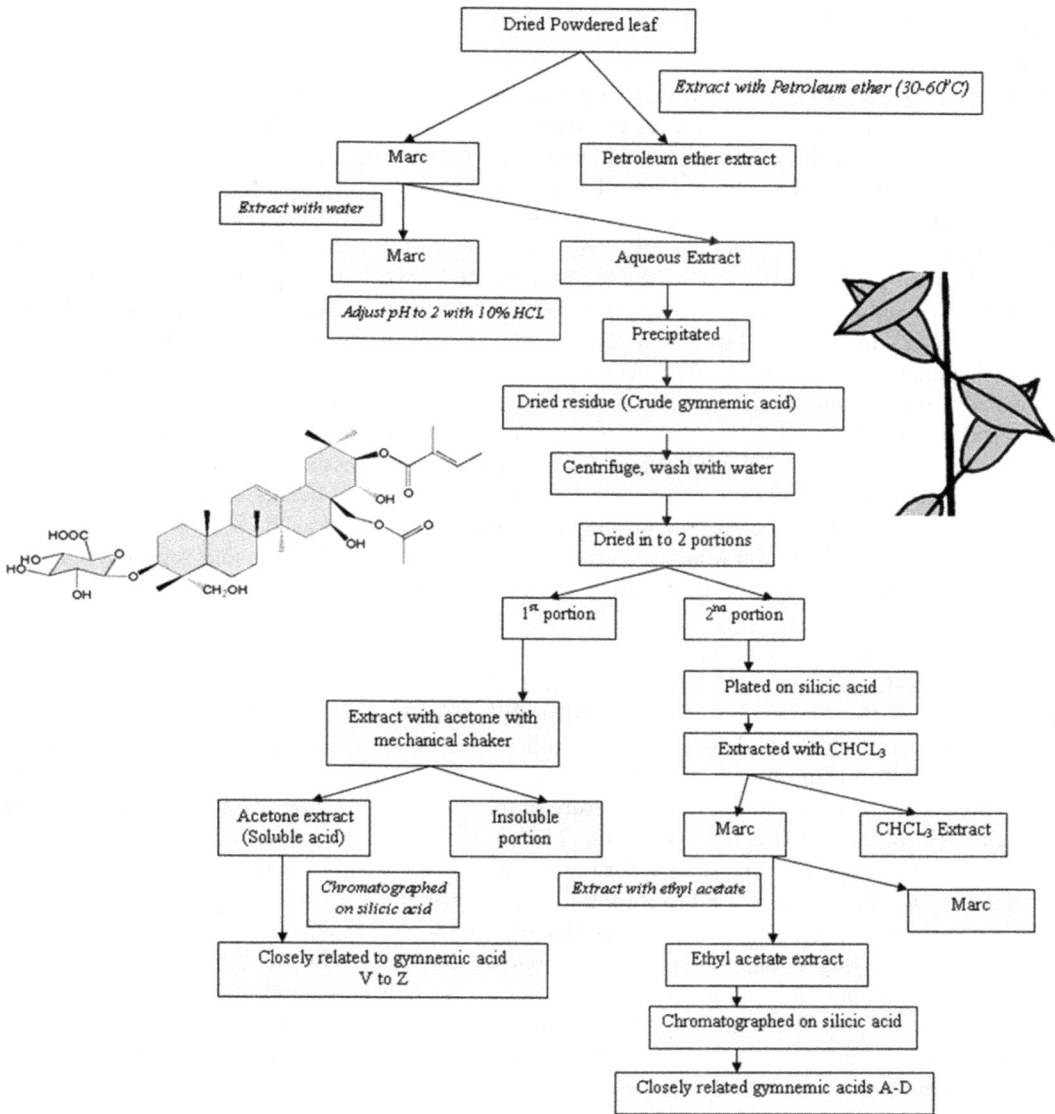

**FIGURE 2.15**  Scheme for the isolation of isolation of gymnemic acid.

### 2.8.6.4  Estimation of Gymnemic Acid

In 2006, gymnemic acids were extracted and quantified from *G. sylvestre* callus cultures using pre-coated silica gel 60 (GF254) plates by using chloroform as well as methanol [153]. In 2013, simple, rapid, with high sample throughput, a HPLC-ESI-MS/MS procedure has been used with MRM mode to quantify gymnemagenin in *G. sylvestre*, and its formulations has been developed. It was found that 0.056–4.77 w/w Gymnemagenin content was present in *G. sylvestre* and its marketed formulations [154]. A simple, precise, rapid, and reproducible HPLC method for the quantification of gymnemic acid (deacyl gymnemic acid) in *G. sylvestre* extracts and formulations has been developed using reverse-phase chromatography on a C18 column with isocratic elution of acetonitrile: buffer (23:77 v/v) at a flow rate of 2.0 ml/min. The linear range of method was found to be 50–800 µg/ml with correlation coefficient of 0.9998 [155].

### 2.8.7 ANDROGRAPHOLIDE

Andrographolide is the bitter diterpenoidal organic compound derived from dried leaves and tender shoots of *Andrographis paniculata* (Kalmegh) belonging to family Acanthaceae. *A. paniculata* is extensively used as bitter tonic, in fever and liver inflammation and regeneration. Chemical nature of Andrographolide is bicyclic diterpenoid lactone. It's extensively used in Indian traditional medicine as bitter tonic, for treatment of liver inflammation and for snake bite. Due to its extensive bitter taste and characteristic, *A. paniculata* is named as "king of bitters". Main chemical components of *A. paniculata* are andrographolide, andrographine, deoxyandrographiside, 14-deoxy-11-dehydroandrographolide, 14-deoxy-11-oxoandrographolide, neoandrographolide, panicoline, paniculide-A, paniculide-B, and paniculide-C. Andrographolide causes a considerable dose-dependent choleretic effect (increases in bile flow, bile salt, and bile acids) among experimental animal models (rats and pigs). Treatment of rats (type 1 DM-infected) with Andrographolide improved the glucose uptake in isolated soleus muscle via stimulating the expression of GLUT4. Reports also demonstrated that Andrographolide and *A. paniculata* (ethanolic extract) showed strong hypoglycaemic effects in insulin-resistant rats [156–158]. Nevertheless, inappropriate tissue distribution, short plasma half-life, and poor bioavailability have restricted its therapeutic applications; therefore, in several research papers, nano-carrier/liposomal-mediated delivery is encouraged. *A. paniculata* is a key component of several herbal traditional formulations available in India. Previous reports suggested role of andrographolide in liver regeneration and stimulation of insulin secretion [159].

#### 2.8.7.1 Isolation of Andrographolides

Andrographolide, active component of *A. paniculata*, was first isolated by Chakravarti and Chakravarti in 1951 [159]. Isolation of andrographolide ($C_{20}H_{30}O_5$) is based on its solubility (Figure 2.16). Andrographolide is soluble in methanol and ethanol. For isolation processed material obtained from Kalmegh was successively extracted with petroleum ether followed by the extraction with chloroform and methanol at a temperature of 60–80°C. Methanolic extract was collected and concentrated followed by the treatment with charcoal for 24 h. Solution was filtered, concentrated, and solution was kept it overnight in order to crystallize. At last crystals are collected by using filtration method followed by the refluxation for the recovery of charcoal by using methanol. Filtrate was kept overnight to promote the crystallization of compounds presents in solution. At last crystals were purified by using methanol to get andrographolide (melting point 228–229°C) (Figure 2.16).

In 2000, a simple and fast procedure for isolation of andrographolide from the leaves of *A. paniculata* was reported. In this procedure plant material was extracted with dichloromethane and methanol (1:1) by cold maceration. Isolation of andrographolide from final extract was done by recrystallization process [160]. In another report, three-phase partitioning (TPP) process was used for isolation of andrographolide from the leaves of *A. paniculata*. To improve the yield of andrographolide, optimization of process parameters has been done. In comparison with Soxhlet extract, it was found that three-phase partitioning showed 62.5% yield in 120 min while batch extraction produced 59.89% yield in 140 min [161].

#### 2.8.7.2 Identification Method for Andrographolide

- For the identification test sample (0.5 mg of amorphous residue) is added in methanol (5 ml), followed by the successive treatment with 2,4-dinitrophenylhydrazine (1 ml). Eventually 2M HCl (100 ml) added to the resultant solution. Appearance of orange colour indicates the presence of andrographolide.
- In another method test sample (0.5 mg of amorphous residue) is treated with alcoholic potassium hydroxide (10%) (Volume: 8 drops). Initial appearance of red colour after 15 min followed by the appearance of yellow colour indicate the presence of andrographolide.

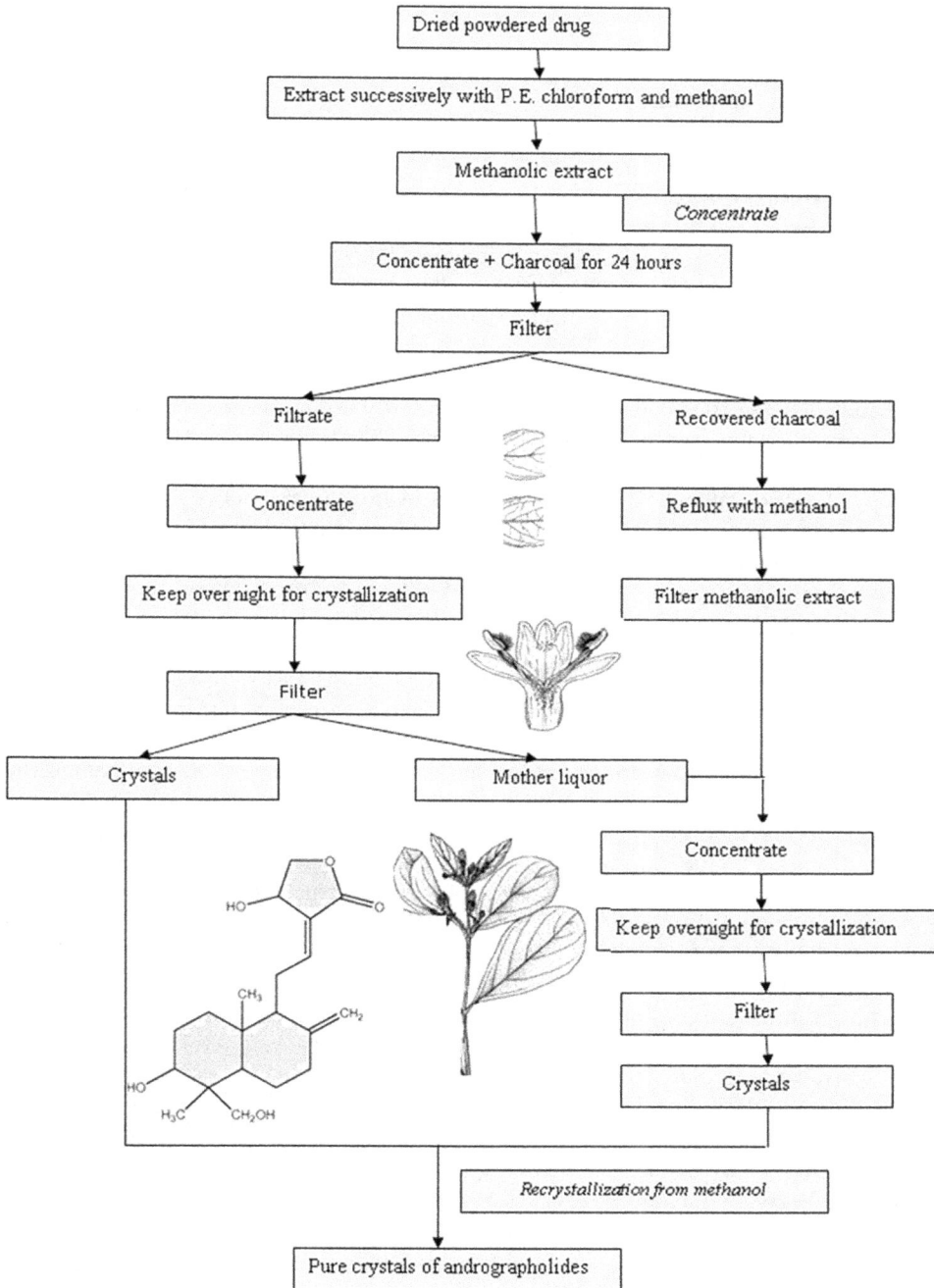

**FIGURE 2.16** Scheme for the isolation of andrographolide.

### 2.8.7.3 Estimation of Andrographolide

TLC method: A test sample (amorphous residue) and reference standard by using mobile phase, chloroform, methanol and ethyl acetate (7:2:1), and visualizing agent (3,5-dinitrobenzoic acid in ethanol [2% w/v solution] and KOH in ethanol [6% w/v solution]) was subjected to TLC analyses.

Rf value of residue (0.38) and reference andrographolide (0.37) indicates the presence of andrographolide in amorphous residue.

### 2.8.7.4  Estimation Method by HPLC

For estimating andrographolide content in *A. paniculata*, 20-µl sample was injected in to HPLC (Shimadzu Model-LC2010 CHT, Serial No. C-21254505638 using D2 lamp at 223 nm; methanol (65 V):35 water) using C18 column (250 nm × 4.6 nm). HPLC was performed till 15 min, as andrographolide Rt is 2.871. In another report, simple HPLC method has been developed to assay the andrographolide and 14-deoxy-11,12-dideoxyandrographolide content in a pooled urine of rat after an oral administration of *A. paniculata* leaf extract (1 g/kg). It was found that overall urinary excretion of andrographolide and 14-deoxy-11,12-dideoxyandrographolide in 24 h was 0.88 and 1.61% [162]. In 2007, quick procedure based on high-performance liquid chromatography/electrospray-mass spectrometry method for the quantitative determination of andrographolide in human plasma has been developed and validated by using isosorbide-5-mononitrate (IS-5-MN) was selected as the internal standard. It was found that correlation coefficient of the calibration curve was 0.998, in the range of 9.9–320.0 ng/ml [163]. HPLC coupled with on-line solid-phase extraction and ultraviolet detection was developed for determining andrographolide and dehydroandrographolide in rabbit plasma using methanol:acetonitrile:water (50:10:40; v/v) at 225 nm. It was found that calibration curves showed excellent linear relationship (R> or =0.9993) over the concentration range of 0.05–5.0 µg/ml [164].

### 2.8.8  DIOSGENIN

Chemically diosgenin is a steroidal sapogenin or also recognized as spirostanol saponin with triterpenoidal structure having two pentacyclic rings, have been used commercially as a precursor for more than 60% of the synthesis of cortisone, pregnenolone, progesterone, and other steroids. Diosgenin is obtained after the hydrolysis of the steroidal saponins namely dioscin (rhamno-rhamno glucoside), the most extensively distributed natural bioactive compound among various genus of plants such as *Dioscorea*, *Trigonella*, *Costus*, *Solanum*, and *Smilax* species. *Dioscorea*, *Costus*, and *Trigonella* are considered the key source for the production of diosgenin. Occurrence of dioscin is very common among rhizomes of various species of *Dioscorea* including *D. deltoidea*, *D. composita*, *D. floribunda* (Yam plant) (Figure 2.17, Table 2.2). In comparison with terpenoids, distribution of steroidal saponins is less common; however, these natural organic compounds are useful for the development of pharmaceuticals with steroidal activity. Currently hecogenin (from *Agave sisalana*) and diosgenin (from *Dioscorea* species) have been used for partial synthesis of anti-inflammatory corticosteroid drugs [166]. As reported earlier, diosgenin prevents cholesterol absorption to further prevent its conveyance to the liver. In liver, transported cholesterol interacts with hepatic cholesterol, followed by its release into the systemic lipoproteins. This initial step of the prevention of cholesterol absorption in the digestive tract via development of excessively big micelles by diosgenin further raises biliary cholesterol release as well as rise in excretion of bound lipids inhibits their reabsorption [167]. Diosgenin has been evidenced for its potential in reducing inflammation, proliferation as well as anti-tumour properties. Additionally, diosgenin has been also reported in decreasing plasma glucose levels in diabetic rats. In a cell-based study, diosgenin suppressed the expression of adhesion molecules in the cultured mouse vascular smooth muscle cells (MOVAS-1), responsible for the development of atherosclerosis. In other studies, antirheumatic and anti-diabetic (via reducing the actions of diabetes-associated enzymes) potential of diosgenin have been reported [168].

### 2.8.8.1  Detection of Diosgenin

First method: Test and reference samples were dissolved in methanol and applied over silica gel plate and mobile phase [toluene:ethyl acetate (7:3)] is allowed to run over the TLC plate (silica gel 6.1). Dried plates were treated with anisaldehyde and sulphuric acid reagent. This leads to the formation of dark green spot with Rf = 0.34–0.37. At last scan in a densitometer at 600 nm by

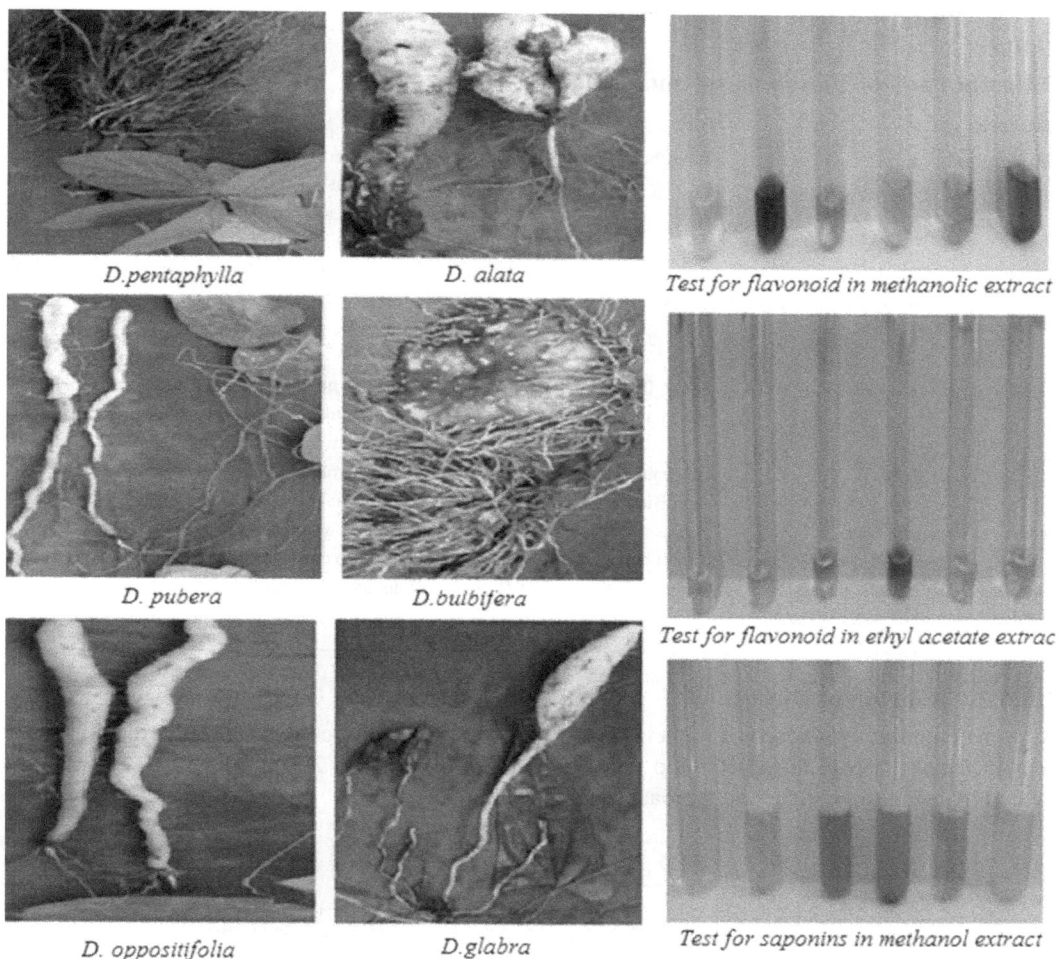

**FIGURE 2.17** Different root tubers of *Dioscorea* species and phytochemical test [165].

equating the areas related to diosgenin sample and standard preparation, to estimate the level of diosgenin in the sample. Second method: Standard was prepared in chloroform and sample preparation is illustrated in Figure 2.18.

Both samples were applied (known volume) on precoated HPTLC plates. Plates were dried and sprayed with Lieberman-Burchard reagent. After an application of spraying reagent slide was heated until the spot related to diosgenin turns black. Other methods: Spectrometry, GLC, IR, HPLC, HPTLC methods have been also used for estimation of diosgenin in different samples. The competitive enzyme-linked immunosorbent (indirect) procedure for the quantitative determination of diosgenin in herbal medicines by conjugation of diosgenin with bovine serum albumin for immunization has been developed. Results showed that developed method can be used as a rapid, simple, sensitive, and accurate procedure for quantitative estimation of samples containing diosgenin [169]. In another study, HPTLC-densitometric procedure for diosgenin estimation in fenugreek seeds has been reported by optimizing conditions for effective saponin extraction using three-step Soxhlet extraction and acid hydrolysis using $H_2SO_4$. Results showed that this validated method determined level of diosgenin in fenugreek which was ranged from 0.12 to 0.18% [170].

**TABLE 2.2**
**Different Parts of *Dioscorea* sp. and Their Uses [165]**

| *Dioscorea* sp. | Part | Uses | Ref. |
|---|---|---|---|
| *Dioscorea alata* Linn | Tuber/bulb | Tuber is edible, boiled, and cooked. Boiled bulb is taken orally twice a day for 15 days to cure piles and gonorrhoea; Tuber is also taken orally for treatment of piles and leprosy | [165] |
| *Dioscorea bulbifera* Linn. | Tuber | Boiled and cooked tuber is edible; one teaspoonful of tuber powder is given orally with water as single dose once only to cure abdominal pain | [165] |
| *Dioscorea pentaphylla* Linn. | Tuber, flowers, and young shoot | Taken as vegetables tubers are boiled and eaten; leaf paste mixed with mustard oil is rubbed on the effected part to treat rheumatism | [165] |
| *Dioscorea pubera* Bl | Tuberous rhizome and Bulbil | Tuber is eaten; tuberous rhizome and Bulbil are cooked and given to cure colic pain | [165] |
| *Dioscorea oppositifolia* L. | Leaves, flowers and tubers | Leaves paste is used as antiseptic for ulcers; The roots are chewed to cure toothache and aphthae | [165] |
| *Dioscorea glabra* Roxb Tuber Boiled and cooked tuber is edible | Tuber | Boiled and cooked tuber is edible | [165] |

### 2.8.8.2 Isolation of Diosgenin

At room temperature, Dioscorea plant is kept in water for three consecutive days. To avoid microbial growth, plant material must be stored in a closed container (Figure 2.19). Afterwards sample was treated by using three different approaches (Figure 2.19):

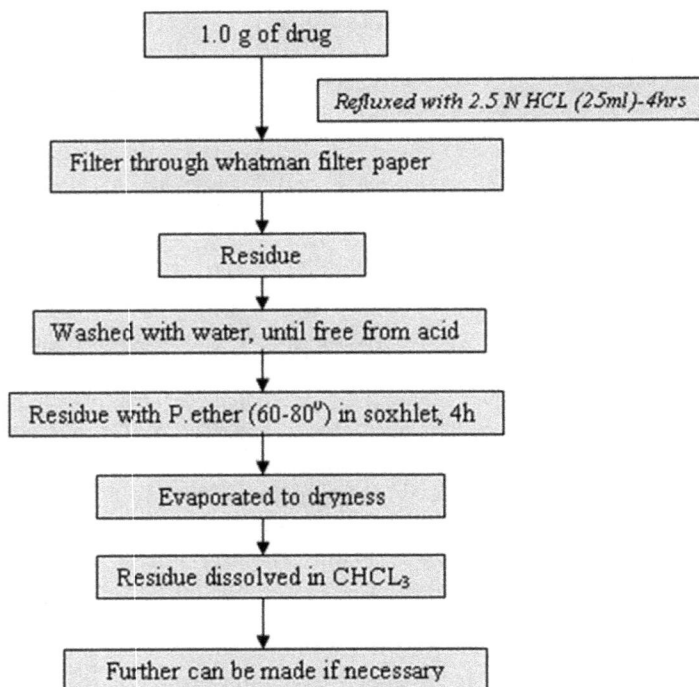

**FIGURE 2.18** Method for the detection of diosgenin.

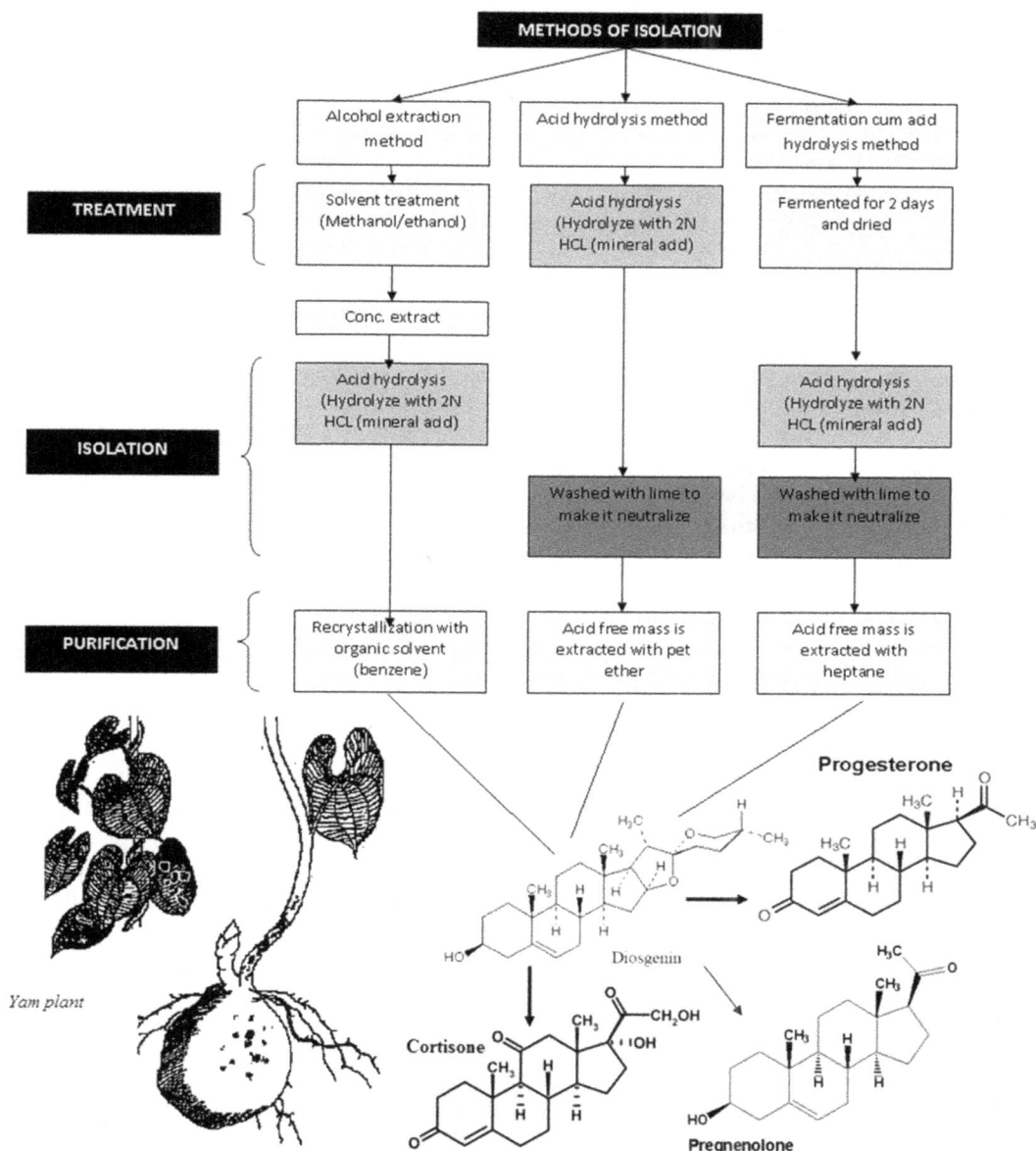

**FIGURE 2.19** Method for the isolation of diosgenin.

- Acid hydrolysis: Acid hydrolysis allows the treatment with 2N HCl followed by the treatment alkali to neutralize. Subsequently resultant solution was extracted with ether.
- Fermentation-based hydrolysis: Sample was fermented and then dried for 2 days followed by the treatment with 2N HCl for acid hydrolysis. Furthermore, acidic layer was neutralized by alkali and then extracted with heptane.
- Alcohol-mediated extraction: In this procedure, resulting plant material was treated with methanol/ethanol; extract was concentrated followed by the treatment with 2N HCl for acid hydrolysis. Furthermore, acidic layer was neutralized by alkali and recrystallize by using benzene.

### 2.8.9   PODOPHYLLOTOXIN

Podophyllotoxins are natural occurring cytotoxic lignin obtained from the rhizomes of *Podophyllum peltatum* and *Podophyllum hexandrum* (Berberidaceae) with important antimitotic and antiviral properties. It's also considered a precursor for the chemical synthesis of the semisynthetic drugs such as cytostatics etoposide and teniposide. Podophyllotoxins are considered important source for the synthesis of anticancer drugs, etoposide and teniposide. Podophyllotoxins derivatives (epipodophyllotoxins) such as teniposide (protein-bound drug), etoposide phosphate, etoposide is considered topoisomerase II inhibitors (modulate topoisomerase II) via interacting with DNA to induce DNA strand breaks in tumour cells and thus used to treat various types of tumours. Natural occurrence of podophyllotoxins in *P. peltatum* (0.25%) and *P. hexandrum* (about 4% however on the verge of extinction). Thus, alternative approach for the depodophyllotoxin production is reported from endophytic fungus, *Aspergillus fumigatus*. Culture conditions can be optimized to improve the yield of depodophyllotoxin from *A. fumigatus* to make it more suitable for scale up process [171, 172]. As far as the toxicity is concerned, podophyllotoxin is considered the main toxic ingredient of *Dysosma pleiantha* (Bajiaolian) rhizome which is the leading cause of toxicity among herbal medicine in Taiwan [173–175]. Previous investigations reported that genus Podophyllum contain diverse bioactive compounds such as 4-demethyldeoxypodophyllotoxin, 4-demethylpodophyllotoxin, 4-dimethyl podophyllotoxin glucoside, deoxypodophyllotoxin, isopicropodophyllone, kaempferol, peltatin and alpha peltatin, picropodophyllotoxin, picropodophyllotoxin, podophyllin, podophyllotoxin glucoside, podophyllotoxin, and quercetin.

#### 2.8.9.1   Isolation of Podophyllotoxin

PTOX was first isolated in 1880 by Podwyssotzki from May apple which is scientifically named as *Podophyllum peltatum* [176].

- First method: Plant material (root) of *P. peltatum* (4 g) was extracted (by heating and stirring) with ethanol (40 ml) for 30 min, cool and subjected to filtration. Filtrate was again extracted with high volume of ethanol (100 ml) and concentrate it up to 20 ml. Resultant concentrated extract was mixed with water (100 ml) and subsequently extracted with ethyl acetate (150 ml) and evaporated up to dryness. Isolation of podophyllotoxin can be done by preparative TLC or chromatographed over the stationary phase (neutral alumina-II) and run with chloroform to yield podophyllotoxin (40.8%). Other components such as deoxy-podophyllotoxin, β-peltatin, and α-peltatin have been also identified using this method.
- Second method: Commercially podophyllin is isolated by initially extracting *Podophyllum emodii* rhizome/roots by using methanol. This is followed by the reducing the whole solution under vacuum. Concentrated extract is subjected to acid hydrolysis (10-ml HCl in 100-ml water) which leads to the formation of precipitates. The precipitates are filtrated, decanted, and then washed. Gummy resultant material is subjected to drying to yield dark brown amorphous powder known as podophyllin. Resultant powdered material is treated with chloroform and subsequently purified via repeated recrystallization non-polar solvent such as benzene. After recrystallization, resultant solution is again treated with petroleum ether to form podophyllotoxin.
- Third method: In this procedure, chloroform fraction is dissolved in alcohol followed by the refluxation with neutral aluminium oxide to form yellow colour solution. Furthermore, resultant solution is mixed with benzene is added to yield podophyllotoxin (95–98%).
- Fourth method: In this procedure, podophyllotoxin is isolated by extracting plant material using alumina neutral using non-polar solvents such as benzene followed by recrystallization using solvents such as xylene, benzene, and toluene to yield genuine podophyllotoxin (95–97%).

### 2.8.9.2  Estimation of Podophyllotoxin

A sensitive, selective, and precise RP-HPTLC method (RP-18 F254 TLC plates) has been developed for the estimation of podophyllotoxin in *P. hexandrum* using mobile phase, acetonitrile-water (4:6, v/v). It was found that this procedure has given better resolution, with well separated, compact spots for podophyllotoxin (Rf value 0.41 ± 0.02). Densitometric analysis was carried out in the absorption remission mode at 217 nm [177]. In another report, HPTLC method has been developed for the quantification of podophyllotoxin (in roots of *P. hexandrum*) and etoposide (in a marketed formulation) using dichloromethane-methanol-formic acid (9.5:0.5:0.5 v/v/v) as the mobile phase followed by the densitometric evaluation. Developed method for the estimation of podophyllotoxin and etoposide was found to be simple, precise, specific, sensitive, and accurate [178].

### 2.8.10  SOLASODINE

Solasodine is derived from solamargine or solasonine a steroidal glycoalkaloid, on hydrolysis forms an aglycone part (solasodine) and glycine part (mannose, glucose, and galactose). Solasodine ($C_{27}H_{43}NO_2$; mol. wt. 413.6 g/mol) is an aglycone of steroidal glycoalkaloid sapogenin present in various nightshade plants belonging to family Solanaceae. This spironolone-structured *Solanum* alkaloid usually found in the *S. laciniatum*, *S. aculeatissimum*, *S. sisymbriifolium*, *S. aviculare*, *S. dulcamara*, *S. elaeagnifolium*, *S. jasminoides*, *S. khasianum*, *S. nigrum*, *S. verbascifolium*, *S. xanthocarpum*, *S. aviculare*, *S. xanthocarpum*, and others. Solasodine offers broad range of biological properties such as anti-oxidant, anti-infection, and neurogenesis promotion. Recent report suggested anticancer properties of solasodine via causing a programmed cell death as well as cell cycle arrest in different tumour cells such as basal cell carcinoma, breast cancer, chronic myelogenous leukaemia, oral epidermoid carcinoma, and prostate cancer [179–186].

### 2.8.10.1  Isolation of Solasodine

Processed plant material (dried berries in powder form) is treated with petroleum ether for defatting to defatted material and greenish yellow oil. Defatted material is treated with ethanol and concentrated to reduce volume (1/10th) followed by the treatment with hydrochloric acid till the resultant concentration reaches up to 5–6%. For the complete hydrolysis of glycoalkaloid, the resultant mass is refluxed (6 h). Furthermore, acidified fraction is neutralized by ammonia and refluxed (1 h). Solution obtained is cooled and filtered followed by washing of the residual material. Dried residual mass is treated with chloroform and then subjected to filtration followed by solvent evaporation to form the mass containing solasodine. Purification of resultant residual mass containing solasodine is done by methanol as mentioned in Figure 2.20.

### 2.8.10.2  TLC of Solasodine

Test and standard (1 mg each) dissolved into methanol separately and chromatographed over the silica gel-G plate using mobile phase [toluene (7):ethyl acetate (2):diethyl amine (1)] and spots were visualized by spraying reagent (Dragendorff's reagent). TLC plates were dried followed by the treatment with sulphuric acid (10%) in methanol. Spot (orange to red colour) matching to solasodine at Rf value 0.60 appeared and compared with standard.

### 2.8.11  VASICINE

Vasicine [molecular formula ($C_{11}H_{12}N_2O$), molecular weight (188.23), melting point (210°C)] is a bitter pyrrolazoquinazoline alkaloid from the leaves, roots, and flowers of Ayurvedic and Unani medicinal plant, *Adhatoda vasica* (Common Indian name: Vasaka); family Acanthaceae. In India, *A. vasica* has been used in the folk medicine from last 2000 years [187]. Traditionally it's utilized used for the treatment of different respiratory disorders (both upper and lower RTI infections),

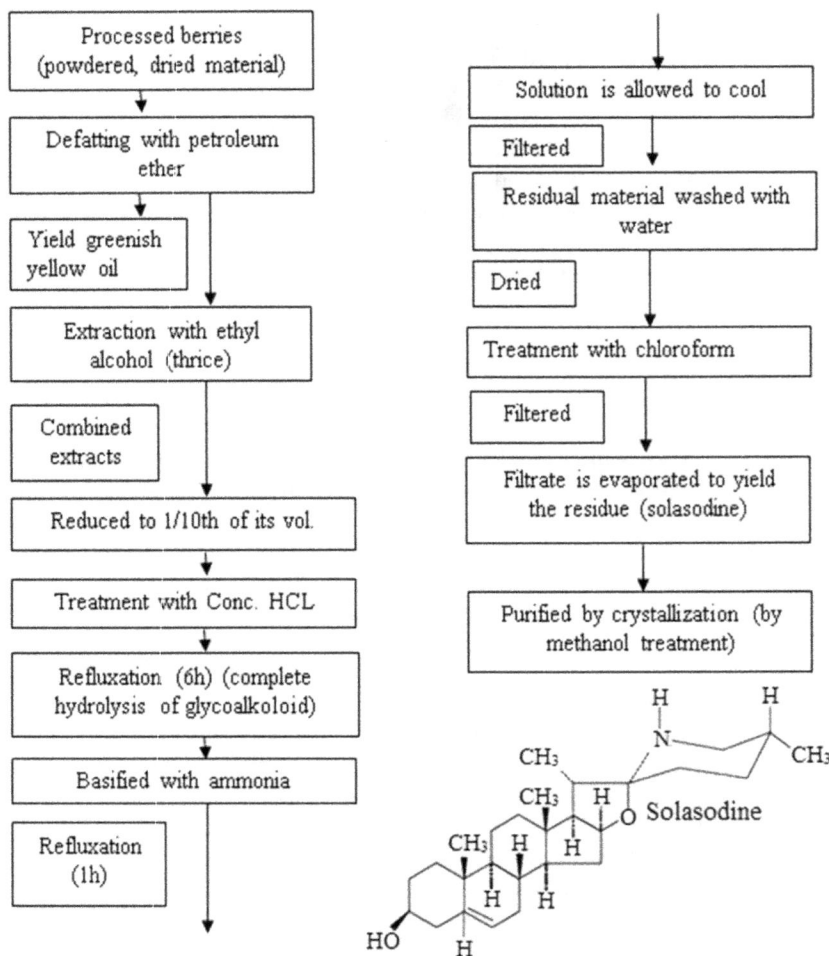

**FIGURE 2.20**   Scheme for the isolation of solasodine.

including cough, bronchitis, asthma, and tuberculosis. During 1977 it was found that vasicine retains uterine stimulating effects as like oxytocin [187]. Content of vasicine varies significantly and this variation directly correlated with seasonal variation as it was reported that the flowering phase represents highest level of vasicine throughout the year (especially during July to September). Apart from vasicine, the leaves also contain several other chemical components such as adhatodine, adhatodine, adhavasinone, anisotine and peganine, vasicinol, vasicinone which are alkaloidal in nature. As suggested earlier, pharmacological activity of vasicine is mainly bronchodilatory effects (in vitro and in vivo); however, the key metabolite of vasicine (vasicinone) which is also found in *A. vasica* showed bronchoconstricting activity, in vivo. Overall these two bitter alkaloids stimulate bronchodilation (in vitro and in vivo). In addition to bronchiodilatory effects, vasicine showed uterotonic stimulatory effects in an experimental animal models and human beings [187]. Recently, without using transition metal catalyst, this bitter organic compound called vasicine has been as an organocatalyst for direct C-H arylation of unactivated arenes with aryl iodides/bromides [188].

### 2.8.11.1   Isolation of Vasicine

Processed (dried and powdered) plant material is treated with alkali solution (ammonia) to basify its pH (pH 9) followed by the extraction with chloroform. Chloroform fraction is subjected to washing with water followed by drying over $Na_2SO_4$ (anhy.). Furthermore, solvent is evaporated to derive the residual part containing vasicine which is further subjected to crystallization in order to get purified vasicine.

### 2.8.11.2 Identification of Vasicine

Identification of vasicine can be done by TLC using test and standard samples (vasicine (1 mg)/ methanol 91 ml)). These samples are chromatographed over silica gel-G plate by using mobile phase [toluene (1):methanol (1):dioxane (2.5):ammonia (0.5)] and visualized by using spraying reagent (Dragendorff's reagent) which gives orange-coloured spot after reaction with vasicine.

### 2.8.12 VINCA ALKALOIDS

Vinca bis-indole dimeric alkaloids such as vinblastine, vincristine, vindesine, and vinorelbine are considered cytotoxic agents obtained from whole plant of *C. roseus* (Apocynaceae) commonly known as periwinkle plant. Vinca alkaloids also include some other important alkaloids such as catharanthine (indole nucleus) and vindoline (dihydroindole nucleus). Their cytotoxicity is based on their specificity and selectivity towards cancerous cells in order to inhibit their ability to multiply. However, these alkaloids also cause neurotoxicity (neutropenia) and can cause mortality when administered through intrathecal route [189]. From the early 1960s, vinca or Catharanthus alkaloids have been used against lymphomas and leukaemias. Vincristine is used in combination with other medication against non-Hodgkin's lymphoma, whereas vinblastine is mainly used in the treatment of Hodgkin's disease. Their main mechanism of action is based on supression of tubulin dynamics by interacting with vinca site. These alkaloids disturb the mitotic apparatus after binding to tubulin, resulting in mitotic arrest in metaphase. These cell-specific mitotic blockers bind to tubulin to further inhibit polymerization from soluble dimers into microtubules. This step leads to the disruption of typical mitotic spindle function resulting in malorientation of microtubules and neurofilaments. These microtubules also involved in a cellular process called axonal transport or axoplasmic flow, accountable for movement of biomolecules or substances or organelles between the axon tip and cell body in nerve cells. Thus, this initial physiological injury to microtubules via binding of alkaloids causes neuropathy via interference with axoplasmic flow. It was also suggested that that capability of these compounds in suppressing angiogenesis could play an important role in the antiproliferative activity of these alkaloidal cytotoxic compounds. Vinorelbine (approved by FDA in United States) is an important antimicrotubule antineoplastic agent derived from vinca plant. This alkaloid has been investigated in both first- and second-line treatments of NSCLC and now approved by FDA in United States for NSCLC treatment. Vincristine from *C. roseus* is another cytotoxic agent and considered an important chemotherapeutics which binds to tubulin (critical component of microtubules) to inhibit polymerization of the subunits to form microtubules. This results in the inhibition of the metaphase of mitosis, thus often called mitotic spindle inhibitor. These dimeric indole alkaloids are employed usually to treat different types of cancers by inhibiting microtubule polymerization; however, they cause serious side effects such as myelosuppression, neurotoxicity, and neuropathy. Several vinca alkaloids have been widely studied; however, out of all, only three alkaloids (vincristine, vinblastine, and vinorelbine) have been officially permitted for use in the United States [190, 191].

Vinblastine, vincristine, vindesine, and navelbine have been extensively used as antimitotic agents. As far as pharmacokinetics is concerned, these antimitotic alkaloids (vinblastine, vincristine, vindesine and navelbine) are extensively used either as alone or with other antitumour medication to treat different types of cancers. Clinical studies suggested that oral administration and intravenous administration of vinca alkaloids showed rapid total plasma clearance, extended terminal half-life and large apparent total distribution volume. It was found that in humans, elimination of vinca alkaloids through faecal route is considered the main route of elimination while excretion of these alkaloids via urine is usually low [192–203]. Additionally, pharmacokinetics of vinca alkaloids are time- and dose-dependent. In vitro studies related with metabolites formation of vinca alkaloids in human liver cells showed that these alkaloids widely metabolized into a number of metabolites [192–203]. These cytotoxic agents are extensively used for several forms of cancer (leukaemias, lymphomas, sarcomas, and other solid tumours) and as a result of poor penetration across BBB, these agents cause less frequently CNS toxicity [192–203].

### 2.8.12.1  Isolation of Vincristine and Vinblastine

Method 1: Processed (dried) plant material obtained from *Vinca rosea* is procured, weighed, and subjected to extraction with hot ethanol-water-acetic acid (9:1:1). Resultant solution obtained is condensed by evaporation followed by the treatment with hot HCl solution (2%). For the removal (precipitation) of non-alkaloidal content, solution is acidified (pH 4), followed by the separation by filtration. Subsequently acidic solution is neutralized at pH 7, followed by the extraction with benzene. Neutralized solution is condensed by evaporation of benzene to further to derive vinblastine and other vinca alkaloids. Considering the isolation of vinblastine and vincristine, initially phenolic content present in the solution is separated via rinsing the extract with alkali (dil.). Fraction obtained was submitted for column chromatography using adsorbent alumina as well as eluted by using mobile phase [benzene-methylene chloride (65:35)] followed by methylene chloride. Monitor the column fractions with TLC (silica gel G) by using mobile phase [ethyl acetate:methanol (9:1)]. Gradient elution technique is used to isolate vinblastine first followed by the isolation of vincristine.

Method 2: In this procedure, processed plant material (dried powdered vinca leaves) is subjected to extraction with aqueous alcoholic acid solution. Fraction is concentrated by evaporation followed by the treatment of concentrated extract with HCl (2%) (pH 4). Acidified fraction is extracted with benzene. Resultant benzene fraction is neutralized by using NaOH (pH 7) and again extracted with benzene. Neutralized benzene solution is concentrated by evaporating benzene and further dissolved in benzene (65%): methylene chloride (35%). Fraction obtained is concentrated and subjected to column chromatography (neutral alumina) and eluted with methylene chloride as well as benzene and as mobile phase using gradient elution method. Initial fraction obtained from column chromatography results in the separation of vinblastine which is recrystallized by using ethanol followed by the separation of vincristine from later fraction (further recrystallized by using ethanol).

### 2.8.12.2  TLC of Vincristine

Test and standard samples are solublized in a solution of methanol:water solution (75:25). After dissolution, samples are chromatographed over Silica gel-G plate using mobile phase [solvent, acetonitrile (30):benzene (70)], visualizing reagent [ceric ammonium sulphate (1%) in phosphoric acid (85%)]. Standard sample usually show Rf value of 0.39, which can be compared with test to check the presence of vincristine in test sample.

### 2.8.13  STRYCHNINE AND BRUCINE

Indole alkaloids are the nitrogen-containing organic compounds that are obtained from more than 30 families such as Passifloraceae, Rubiaceae, Apocynaceae, Loganiaceae, as well as numerous fungi [204]. These compounds are different from each other as per their distribution among various plants (Figure 2.21).

*Strychnos nux-vomica*, also named as Nux-vomica (Maqianzi), a prime source of various poisonous alkaloids, contains 190 species, and more than 300 alkaloids are mainly distributed among Asian countries. Majority of species are found in Asia, whereas some are also found in South America, Africa, and Australia such as *Strychnos panamensis*, *Strychnos tienningsi*, and *Strychnos icaja*. These alkaloids have important pharmacological properties activities and strong medicinal impact however CNS toxicity restricts its clinical applications. Strychnine as well as brucine are closely related alkaloids derived from the *S. nux-vomica* seeds (Loganiaceae). *S. nux-vomica*

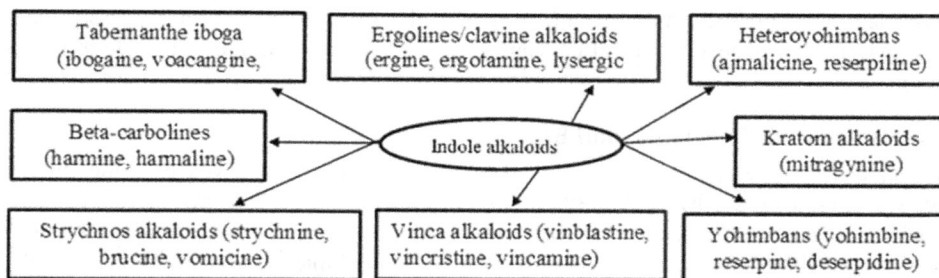

| Tabernanthe iboga (ibogaine, voacangine, | Ergolines/clavine alkaloids (ergine, ergotamine, lysergic | Heteroyohimbans (ajmalicine, reserpiline) |

**FIGURE 2.21** Distribution of indole alkaloids among various medicinal plants [204].

generally contains strychnine and brucine (indole alkaloids), in a range of almost 1.8–5.3%. Usually *S. nux-vomica* contains brucine 1.55%, whereas level of strychnine is relatively less (1.23%). In addition, some other alkaloids in less amount are also present such as 3-methoxyicajine, icajine, isostrychnine, novacine, *N*-oxystrychnine, protostrychnine, pseudo-strychnine, vomicine, α-colubrine, and β-colubrine [205].

*S. nux-vomica* L., *Strychnos ignatii*, and *Strychnos wallichiana*, all contain strychnine as an active chemical component. These all plants are classified into Logan plant genus with high level of alkaloids such as strychnine, brucine, loganin, and curare. Strychnine and brucine are almost similar (isostructural to each other), as the only difference in structural feature is strychnine has methoxy groups, instead of hydrogen (at 9- and 10-positions). Overall *S. nux-vomica* contains bitter alkaloid 3% due to which this plant is extremely bitter in taste and high toxicity. In human's strychnine showed $LD_{50}$ (oral) 30 mg/kg with systemic toxicity at 0.1 mg/dl and also showed high toxicity in rats. Likely fatal dose of Brucine in adult is 1 g [206], whereas in animals, $LD_{50}$ 50.10 mg/kg. Brucine is found to be less toxic in comparison with strychnine; however, its consumption by a human greater than 2 mg in pure form can almost cause the same symptoms similar to strychnine poisoning. Biologically strychnine is more active in comparison with brucine thus seeds of nux vomica always evaluated for strychnine and not other alkaloids [207]. Strychnine and brucine both have narrow therapeutic window; thus, difference in dose or systemic concentration can lead to serious adverse effects or morbidity. From earlier times, both alkaloids have been used as a rodenticide [207]. In the beginning of 18th century (1818), strychnine (marketed as strychnine sulphate, melting point: 286–288°C) was isolated by Pierre Joseph Pelletier and Joseph-Bienamé Caventou followed by its structure evaluation was done by Woodward in 1948. Strychnine shows poor solubility in water, whereas its salts form (nitric acid as well as hydrochloride) are soluble in water. Strychnine provokes an extreme bitter taste by activating bitter taste receptors TAS2R10 (threshold is 30 times higher than TAS2R46) and TAS2R46 with a threshold of 0.000 001 6.115 while brucine bitterness threshold 1:220,000 [208, 209].

Rapid absorption of strychnine through gastrointestinal system, the respiratory tract [210], and intact skin [211] results in the early detection (within minutes) of strychnine in biological fluids such as blood and urine [212]. A dilute solution of strychnine in 80% sulphuric acid gives a reddish-violet to bluish-purple colour on the addition of a trace amount of potassium dichromate solution (Otto reaction).

Strychnine (m.pt at 275–285°C) as high melting point in comparison to brucine (m.pt at 178°C). Strychnine is a colourless and odourless base having potent convulsant effects because of the intervention with the postsynaptic inhibition which is mediated by the amino acid glycine via blocking chloride channel related with inhibitory receptors (glycine receptor) [213–218] found in the spinal cord and brainstem [213–218]. Glycine receptors are responsible for inhibitory neurotransmission (via glycine-mediated inhibitory transmission) in the spinal cord (especially to motor neurons and interneurons). Strychnine blocks the inhibitory effects of glycine as the glycine receptors via acting as selective competitive antagonist [213–218]. Although brucine has an impressive profile in pharmacology research, severe central nervous system toxicity is the main obstacle to its clinical application.

### 2.8.13.1    Isolation of Strychnine and Brucine

Strychnine, an odourless, colourless, and crystalline powder that gives extreme bitter taste in water was first isolated by French chemists (Joseph Bienaimé Caventou and Pierre-Joseph Pelletier) in the early-18th century (1818) from Ignatius bean, a plant closely related to the strychnine tree [219–223]. Commercially strychnine is available in nitrate, sulphate, and phosphate salt forms. In the mid of the 19th century (1946), for the first time, Sir Robert Robinson determined structure of strychnine and further structural characteristics were first evaluated by Robert B [219–223] and by Woodward in 1954. Just after one year (1818) of isolation of strychnine, a weak alkaline indole alkaloid, brucine, was first isolated from the nux vomica by Pelletier and Caventou. Structure of brucine was first determined by chemist Hanssen (1884) when he established the fact that brucine is structurally similar to strychnine [219–223]. It is also found in the Ignatius bean, a plant closely related to the strychnine tree, and it was from this source that the compound was first isolated in 1818 by two French chemists. In 1819, it was first isolated brucine from the nux vomica in 1819 [219–223].

Weighed amount (100 g) of processed, dried plan material (nux vomica seeds) are mixed carefully with calcium hydroxide solution (10%) and kept overnight at 37°C (Figure 2.22). Dried semi-solid mass is subjected to extraction with chloroform by using soxhlation (3 h). Resultant chloroform solution is then subjected to extraction again with sulphuric acid (5%) followed by the basification with aq. NaOH (10%). Solution obtained is allowed to cool and filtered to get the crystals. Crystals is allowed to dissolved in ethanol (50%) followed by the refluxation till majority of the solid mass has dissolved (Figure 2.22). Afterwards solution is treated with charcoal followed by the filtration. Crystals obtained are washed with ethanol (50%) to obtain mother liquor which can be further used for the brucine isolation. Impure strychnine mass obtained is added into boiling water (9 vol.) followed by the acidification by sulphuric acid (15%). Sulphuric acid is added gradually till the solution becomes slightly acid to Congo red (Figure 2.22). Furthermore, resultant solution is treated with activated charcoal and further subjected to refluxation (1 h) to get the hot filtrate. Again crystals collected from the previous step is treated with water (15 vol.) followed by heating (80°C) and treatment with charcoal (Figure 2.22). This step is followed by the basification with $Na_2CO_3$ (10%) and then solution is filtered hot. Basification followed by cooling resulted into precipitation of strychnine. The precipitates of strychnine are subjected to filtration and washing with cold water. Ultimately precipitates recrystallized by ethanol (Figure 2.22).

Mother liquor obtained during the isolation of strychnine is used for the isolation of brucine. Initially mother liquor is concentrated by vacuum by using water bath to remove most of the alcohol. Resultant mass obtained is acidified (pH 6) by using $H_2SO_4$ and then volume is reduced (3–4 ml) by evaporation. Residual mass is kept in a refrigerator followed by the filtration of the product and washing with cold water. Solution obtained is treated with hot distilled water (4.5 vol.) and subjected to boiling (1 h) after the addition of charcoal. Resultant solution containing brucine sulphate is purified as well transferred for refrigeration for numerous days and further recovered by using the same method used for strychnine. At last it's subjected to recrystallization using aq. acetone.

### 2.8.13.2    Identification Tests

* Test sample containing strychnine gives purple red colour after treatment with sulphovanadic acid.
* Test sample containing strychnine gives purple colour, which gradually turns into red after treatment with sulphuric acid and potassium dichromate. However, test sample containing brucine gives immediate red colour.

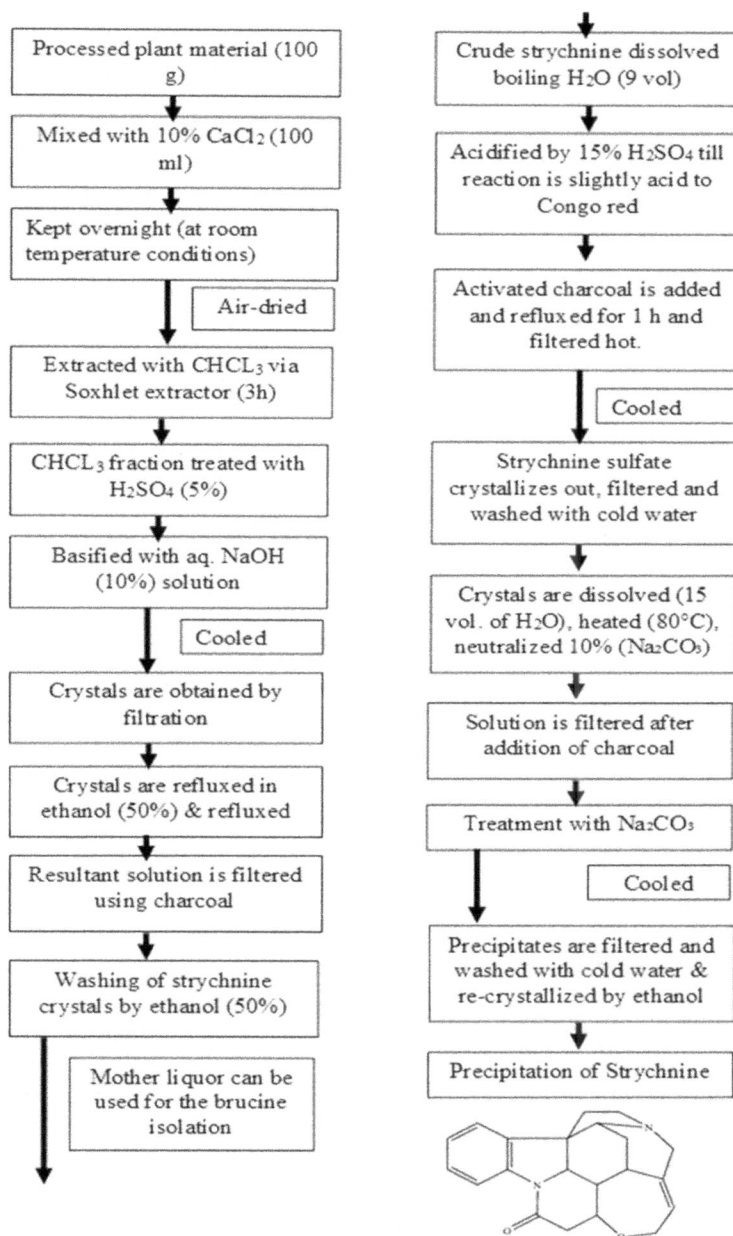

**FIGURE 2.22**    Isolation for strychnine.

- Test sample containing brucine gives blood red colour after treatment with nitric acid. This colour if cleared by the treatment with stannous chloride indicates the presence of brucine.

### 2.8.13.3    TLC of Brucine as well as Strychnine

Strychnine and brucine were solubilized in methanol and chromatographed over TLC plate (silica gel-G) using mobile phase [benzene:chloroform:diethylamine (9:4:1)] and spraying reagent (Dragendroff's reagent).

### 2.8.14  Piperine

Piperine, an amide alkaloid, forms strong pungent taste of Indian medicinal plant (black pepper) scientifically known as *Piper nigrum* (Piperaceae), and *Piper longum* L., commonly known as long pepper. Apart from piperine, these plants chavicine (cis-cis isomer), isochavicine (trans-cis isomer), and isopiperine (cis-trans isomer) and piperanine (trans-trans isomer) are also found in plants. Earlier chavicine was assumed for the characteristic pepper taste, nevertheless, later on, piperine was found to be the actual constituent responsible for strong pungent taste. Piperine alkaloid can be isolated from oleoresin of long pepper (*P. longum*), the kernel of the ripe fruit of *P. nigrum* (white pepper), from unripe fruit (black pepper), seeds of *Cubeba censii* (Piper fainechotti and *Piper chaba*) and from the fruit of aschanti (*Piper clusii*). The level of piperine in black pepper fluctuates from 6 to 9%. *P. nigrum* (white pepper) contains 2–7.4%; however, others species of Piper (Piperaceae) contain high level of piperine (9% in black pepper), 4% of long pepper (*P. longum* L.) fruits and *Piper retrofractum* (4.5% in Balinese long pepper fruits) [224, 225]. The level of piperine is impacted by various aspects such as weather, cultivation, geographical, and habitat conditions [224].

It was found that piperine offers more antimicrobial effects against bacteria (Gram-positive) then bacteria (Gram-negative). Also piperine improves antimicrobial effects of certain antimicrobials when used in combination with such as ciprofloxacin against *Staphylococcus aureus* [226, 227]. Piperine has been also reported to induce modification in intestinal brush border membrane fluidity and also affects permeation characteristics resulting in increase in microvilli length, followed by the increase in intestinal absorptive surface. Due to the presence of these or alike organic compounds spices like dietary black pepper/piperine, red pepper/capsaicin, and ginger change the intestinal membrane dynamics and permeability characteristics to ultimately impact the absorption of nutrients [228]. Piperine ($C_{17}H_{19}NO_3$), slightly water soluble solid organic yellow crystalline powder with a melting point of 128–130°C, belongs to the vanilloid family of compounds, tasteless initially, however induce a burning pungent taste afterwards. This vanilloid family also includes capsaicin, fount in chilli peppers. This compound has been alleged recognized for anti-inflammatory potential and potential to improve digestion. Also piperine has been reported for improving bioavailability of some medicines and for dietary components [229].

### 2.8.14.1  Isolation of Piperine

Piperine is a major alkaloid present in black pepper. In the beginning of 18th century (1819), piperine was isolated from *P. nigrum* by Danish philosopher Hans Christian Ørsted. This weak base alkaloid upon hydrolysis (acid or alkali) lead to the formation of basic piperine, called piperidine (volatile), and piperic acid. Piperine is a weak base, which after treatment with acid or alkali hydrolysed to piperidine (volatile compound), as well as piperic acid.

Processed plant material [dried, weighed (20 g)] is subjected to soxhlation (2 h) by using ethanol (95%) and the resultant solution is allowed to filter. Filtrate obtained is concentrated to reduce the volume by using vacuum (water bath) at 60°C. Furthermore, resultant concentrated solution is basified (by 20-ml alcoholic KOH). Afterwards it's decanted from the insoluble residue and kept overnight, as a result of which yellow-coloured needle-shaped crystals of piperine (approximately 0.3 g) are formed.

### 2.8.14.2  TLC of Piperine

Test sample such as extract containing piperine is chromatographed over TLC plate Silica gel-G by using mobile phase [benzene (2): ethyl acetate (1)]. Spots were noticed at UV-365, where piperine displays blue fluorescence. This can be further confirmed by using spraying reagent (anisaldehyde sulphuric acid) followed by heating at 110°C for 10 min. Yellow colour spot Rf 0.25 indicates the presence of piperine on the test sample.

## 2.8.15 COLCHICINE

Colchicine ($C_{22}H_{25}NO_6$) is a tropolone alkaloid which considered mitotic poison (antimitotic agent) and first known tubulin binding agent derived from ornamental plants such as autumn crocus or meadow saffron (*Colchicum autumnale* L.), other species of colchicum and flame lily (*Gloriosa superba* L.) belonging to family Liliaceae. From many years, colchicine has been used to treat gout nevertheless; till 2009, it was not accepted by the FDA for this ailment. In human, colchicine has been used to treat gout while in animals, it has been used as to prevent fibrosis (antifibrotic agent) which can further lead to hepatic failure [230–240]. However, the mechanism of action and pathways involved, as well as safety profile for its capability in reducing liver fibrosis has been criticized still. Colchicine, a toxic alkaloid, is also used to induce polyploidy in plants, and also in humans, it is used in determination of the chromosome number [230–240]. Therapeutically it's mainly used in the treatment of gout, familial Mediterranean fever, and many other diseases such as atrial fibrillation. Colchicine is highly toxic and thus used in the treatment of various types of cancer. Antimitotic action of colchicine is based on its capability in interacting with tubulin. Primary target of colchicine is microtubule which is an important part of cell cytoskeleton and also required for various cell process such as mitosis as well as trafficking of intracellular organelle/vesicle and signal transduction, gene expression, migration, and secretion. Microtubules are the hollow cylinders, made up of protofilaments, comprising heterodimers which are polymerized end to end [230–240].

Heterodimers (α- and β-tubulin) are responsible for the formation of dynamic polymers called microtubules. These microtubules are responsible for the elongation and contraction as filaments to further regulate structure and function of the cytoskeleton [230–240]. Colchicine binds to soluble nonpolymerized tubulin heterodimers to form tubulin dimer-colchicine complex. This crucial step results in a drastic change in structural conformation which prevent its further growth and elongation, thus completely destroying the dynamics of that microtubulin. Colchicine an antifibrotic compound, at high dose prevents polymerization of microtubules. Due to this capability, colchicine is potentially utilized in suppressing collagen synthesis and fibrosis in experimental animal models [230–240]. On the other side, colchicine also interferes with neutrophil function by interfering with various inflammatory pathways and mediators of neutrophil activation. Neutrophil function is compromised by colchicine via impacting inflammatory pathways and mediators of neutrophil activation to further inhibit neutrophil migration and suppress neutrophil superoxide synthesis. Considering the anti-inflammatory effects, colchicine reduced the level of proinflammatory cytokines IL-1β, IFNγ, IL-18, and IL-6 [230–240].

As far as toxicokinetics is concerned colchicine, a lipid-soluble drug is quickly absorbed via GIT, resulting in attainment of peak serum levels in 30–120 min after therapeutic dosing. In liver, colchicine hydrolyzes by metabolic reactions such as deacetylation and hydrolysis [230–240]. Colchicine shows quick early distribution stage in tissue, with a plasma half-life of 19 or 20 min, signifying rapid acceptance by the tissues. It was found that approximately 40% of colchicine is excreted in the urine, with almost 20–30% in unchanged form. Colchicine shows an average elimination half-life of 20 h as maximum part enters into enterohepatic recirculation and is excreted via bile and faeces [240].

### 2.8.15.1 Isolation of Colchicine

In 1820, colchicine, a gout suppressant tricyclic alkaloid, was first isolated from roots of *C. autumnale* by French chemists and pharmacists Pierre-Joseph Pelletier and Joseph Bienaimé Caventou [241–245]. Nevertheless, initially they initially identified this compound as veratrine, a steroid-derived alkaloid derived from plants in the death camas (Melanthiaceae). During 1833, this organic compound was further examined by the Philipp Lorenz Geiger who named this compound as colchicine [241–245]. Furthermore, in the later 18th century (1884), French pharmacist, Houde produced pure crystallized colchicine for the first time in granules form. Then in the mid-19th century (1952), King and co-workers [241–245], based on the hypothesis of Dewar, elucidated the structure of compound by using the X-ray crystallography, suggested presence of two 7-membered rings of which one is a tropolone [241–245].

Plant processed material (dried powdered colchicum seeds) are subjected to extraction by using ethanol. Ethanolic fraction is reduced to dried syrupy mass by vacuum. In order to precipitate the insoluble fats and resins, reduced mass is dissolved in water [241–245]. Aqueous fraction obtained after the filtration is then subjected to repetitive extraction with chloroform or digestion with $PbCO_3$ solution. Resultant filtered solution is again evaporated to reduce volume and subjected to extraction with chloroform. Afterwards colchicine in crystalline form is separated by using chloroform. After the evaporation of the residual solvent, chloroform is removed to get the amorphous colchicine and subsequently colchicine (amorphous) subjected for crystallization by using ethyl acetate to derive pale yellow crystals [241–245].

### 2.8.15.2   TLC of Colchicine

Test sample solution is spotted over TLC plate silica gel-G and chromatographed by using mobile phase [chloroform-diethyl amine (9:1)] and spraying reagent. Rf value of 0.41 can indicate the presence of colchicine.

### 2.8.16   CURCUMIN

Turmeric name originated from the Latin word terra merita, representing the colour of ground turmeric. Turmeric, also called Haldi, has at least 53 different names in Sanskrit. Turmeric has been used in Vedic culture from last 4000 years where it was used as culinary spice and had certain spiritual importance. Use of in Ayurvedic medicine as a treatment for inflammatory conditions has been evidenced from earlier Vedic times [246]. Turmeric scientifically named as *Curcuma longa*, family Zingiberaceae, is as rhizomatous herbaceous perennial plant. India exports high quality of turmeric with superior inherent qualities and high content of curcumin. Due to these features, Indian turmeric is considered to be the best in the global market. South Indian city, Erode, is the world's largest producer of turmeric, thus named as "Yellow City". There are more than hundred components present in turmeric which are broadly classified in to two categories volatile oil (constituting turmerone) and colouring agents named as curcuminoids in turmeric. Standard turmeric contain curcumin (5–6.6%) and volatile oils (<3.5%). Percentage of content in standard turmeric rhizome varies between 60 and 70% curcumin, 20 and 27% demethoxycurcumin, and 10 and 15% bisdemethoxycurcumin [247]. The FDA has approved turmeric and its active component curcumin as GRAS [248]. Curcumin, main active chemical component of turmeric which is accountable for its vibrant yellow colour was first identified in 1910 by Lampe and Milobedzka. The low molecular weight and hydrophobic nature of this compound facilitates its penetration across various biological membranes such as blood-brain barrier efficiently.

Water-insoluble organic compound, curcumin, or diferuloylmethane are the diaryl heptanoid natural polyphenolic components derived from the dried rhizomes of turmeric, *C. longa* as well as others *Curcuma* spp. (Zingiberaceae). Due to its various colouring and medicinal properties such as antioxidant, anti-inflammatory [249], antimutagenic, antimicrobial [250], and anticancer properties [251, 252], *C. longa* is utilized since earlier times in Asian nations. The rhizome of *C. longa* contains approximately 5% of colouring compound which contain almost 50–60% curcuminoids specifically curcumin, desmethoxycurcumin, and bisdesmethoxycurcumin [253].

There are several pharmacological properties attributed to curcumin. However, the antioxidant and anti-inflammatory effects are considered the most common effects reported so far. Despite of its proven medicinal benefits, poor bioavailability [254], due to poor absorption, rapid metabolism, and rapid elimination, appears to be the major problems faced by curcumin. Also dietary intake of turmeric in form of cooked food with an object to have the healing effects on body could be the wrong approach as after cooking the effect of curcumin mainly antioxidant and anti-inflammatory effects are almost destroyed. Several phytochemicals with a proven track record as bioavailability enhancer have been assessed so far to increase its bioavailability, such as piperine derived from black pepper [255] drastically increase bioavailability of curcumin (2000% rise in the bioavailability) [256–259]. Owing to their excellent safety as well as better tolerability at 4000–8000 mg/day and at 12,000 mg/day, they showed safe profile [259–261]. Also these compounds have been approved by US Food and Drug Administration (FDA) as "Generally Recognized As Safe".

### 2.8.16.1 Isolation of Curcuminoids

Turmeric contains three curcuminoids [curcumin (m.pt:183°C), desmethoxycurcumin (m.pt:168°C), and bisdesmethoxycurcumin (m.pt:224°C)]; however, curcumin is considered the most pharmacologically active compound which was first isolated by Vogel and Pelletier in 1815 [262]. Later on, in the late-18th century (1870), curcumin was purified as crystalline compound by Daube [263]. During 1910, structure of curcumin was first elucidated by Polish scientists, Miłobędzka [264]. There are different procedures involved in the isolation of curcumin. Processed plant material (turmeric rhizome dried powdered material) is subjected to soxhlation by using alcohol. Resultant alcoholic fraction is reduced under pressure and dried. Dried mass is subjected to extraction with hexane and then with acetone. Resultant fraction volume is reduced and further dried in order to get curcumin. Another approach to isolate curcumin is by extracting the plant material with hot ethanol and reduce the volume further to obtain concentrated mass. Concentrated solution is then filtered and mixed the kerosene (upper grade) to obtain solid mass. The volume is separated from kerosene by using petroleum ether and further recrystallization is done by using hot ethanol to obtain orange-red needles of curcumin.

### 2.8.16.2 Thin Layer Chromatography of Curcumin

Test and standard samples solution were prepared (1-mg sample in 1-ml methanol). Samples were marked over TLC plates (silica gel-G) and chromatographed once the mobile phase [chloroform-ethanol-glacial acetic acid (94:5:1)] eluted. TLC plates are dried and visualized under light. Curcumin displays an appearance of spot (bright yellow fluorescent) at Rf 0.79 indicated the presence of curcumin, whereas spots at Rf 0.60 as well as 0.43 representing the presence of desmethoxycurcumin and bisdesmethoxycurcumin.

### 2.8.17 EMETINE

Emetine is an isoquinoline alkaloid containing monoterpenoid-tetrahydroisoquinoline skeleton and considered one of the main alkaloids found in roots as well as rhizomes of *Carapichea ipecacuanha* and *Cyanea acuminata* (Rubiaceae), and major component of ipecac syrup with emetic property, mainly used to induce vomiting during the management of poisoning. Several plants belonging to families such as Alangiaceae, Icacinaceae, and Rubiaceae contain emetine and its analogues [265]. Roots of *Psychotria ipecacuanha* (also known as *Cephaelis ipecacuanha*) belonging to family Rubiaceae are considered a major source of emetine. The root contains two major alkaloids, emetine, and cephaeline, which forms almost 90% of the alkaloids found in ipecac [266]. Apart from emetine and cephaeline, *C. ipecacuanha* also contains other alkaloids such as psychotrine, psychotrine methyl ether, and emetamine.

From several years, emetine has been utilized for the treatment of severe intestinal extraintestinal amebiasis. A synthetic analogue of emetine dihydrochloride known as dehydroemetine has been used in replacement of metronidazole, when treatment with metronidazole becomes ineffective [267–270]. Emetine, a potential amoebicidal agent, has the potential to cause destruction to trophozoite forms of the amoebas directly; however, this alkaloid is not effective against infective stage (cyst form) of *Entamoeba histolytica*. Mechanism of action is most probably based on inhibition of protein synthesis in the pathogenic Entamoeba [267–270]. As its oral administration lead to various toxicities, emetine is usually administered through s.c. or i.m. route. Emetine has been also reported as myotoxic as well as cardiotoxic, thus causing serious side effects such as inflammation in muscles at the site of administration, low blood pressure, tachycardia, chest pain, shortness of breath, and complications on electrocardiogram. Emetine most likely inhibits protein synthesis in the parasite [267–270]. Emetine is usually administered s.c. or i.m. In mammalian cells, emetine supresses protein synthesis at the translocation stage; thus, tumour cells die due to lack of ribosomes. Cardiotoxicity is one of the major complications associated with emetine which can lead to abnormal ECG, dysrhythmias, hypotension, and tachycardia. Cardiotoxicity could be due to the effects of emetine on the cellular permeability of sodium and calcium ions or effect over intracellular magnesium concentrations [267–270].

### 2.8.17.1 Isolation of Emetine

In the beginning of 18th century (1817), French chemists Pelletier and Magendie isolated emetine for the first time in history which was identified as an alkaloid in 1823. There are two procedures commonly used for the isolation of emetine and cephaline. Processed plant material is subjected to extraction with ethanol (70%), and further volume of the resultant solution is reduced, concentrated and dissolved in water (Figure 2.23). The obtained solution is basified intensely with ammonia and extracted with di-isopropyl ether. Subsequent fraction is weakly basified by using aq. KOH (10–15%) to derive cephaeline. The extract is subjected to evaporation to derive emetine. Further evaporated fraction has been separated by dihydrobromide salt. Salts obtained from the previous step are treated with hydrochloride to neutralized the basified fraction. In next procedure, processed plant material is treated with ammonia followed by the treatment with ether. Ether fraction is further treated with dilute sulphuric acid to obtain alkaloids. Acidified extract is basified and washed with ether (Figure 2.23). This is followed by treatment with strong alkali and ether. Since emetine is more soluble in ether, it can be obtained from ether fraction while cephaeline can be obtained from aqueous phase. Further volume of ether fraction is reduced and solubilized in methanol. At last solubilized fraction is treated with hydrobromic acid to obtain emetine hydrobromic acid (Figure 2.23).

### 2.8.17.2 TLC of Emetine

Test and standard samples containing emetine HCl was solubilized in methanol or water and marked over TLC plates. Samples are chromatographed over TLC plates using solvent system [chloroform–methanol (85:15)]. Spots were identified under UV or by using spraying reagent (Dragendorff's reagent). Rf value (0.3–0.5) indicates the presence emetine.

### 2.8.17.3 Chemical Test of Emetine

Powdered drug (0.5) is treated with hydrochloric acid (20 ml) and 5-ml water. Resultant solution is filtered and 2 ml of filtrate is treated with potassium chloride (0.01g). Appearance of yellow colour, which if kept undisturbed for 1 h gradually turns to red, indicates the presence of emetine.

### 2.8.18 Guggulsterone ((4,17(20)-Pregnadiene-3,16-Dione)

Plant-derived sterols which are abundantly present in nature namely "Phytosterols" are natural organic compounds, mainly derived from nuts, vegetable oils as well as grains, structurally almost analogous to mammalian cell-derived cholesterol. These organic compounds, chemically present in plant as esterified as well as free alcohol forms. Phytosterols such as brassicasterol, campesterol, diosgenin, guggulsterone, stigmasterol, and β- and γ-sitosterol has been reported for their lipid lowering effects. Regular consumption of phytosterols can prevent atherosclerosis and also reduce low-density lipoprotein-cholesterol levels in serum [271]. Also excessive consumption of phytosterols in diet decreases cholesterol absorption across the intestine via regulating cholesterol homeostasis genes to regulate transcription factors called sterol regulatory elements binding proteins in the liver.

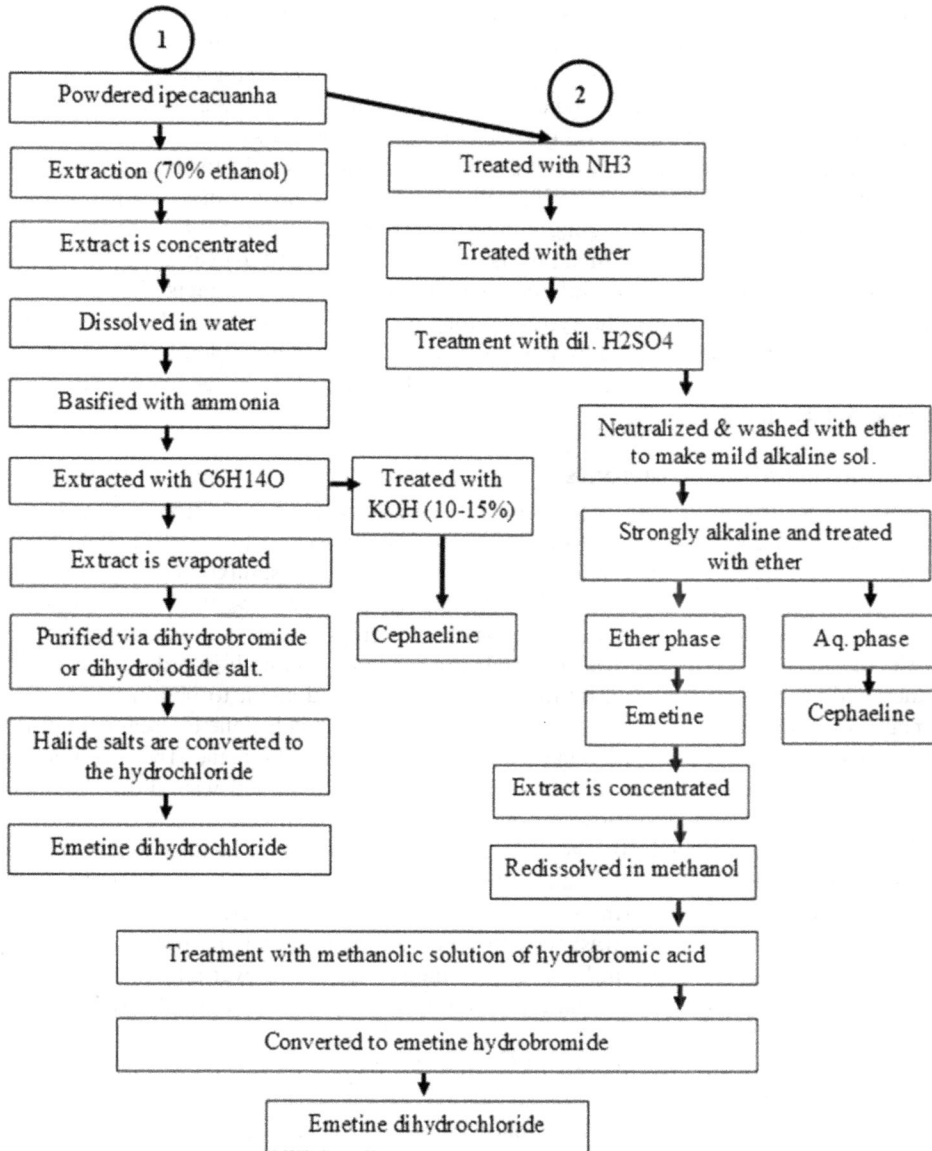

**FIGURE 2.23** Schemes for the isolation of emetine.

Also excessive consumption of phytosterols in diet escalate faecal excretion of cholesterol. It was also found that in the intestinal lumen, phytosterols can prevent cholesterol absorption via competitive solubilization of mixed micelle formation of cholesterol [272–277].

Guggulsterone or gugulipid is the plant-sterol-based organic compounds actively present in the neutral fraction of oleogum resin (guggul or gum guggul) derived from tree *Commiphora wightii* syn. *Commiphora mukul* (guggul tree or Guggul or Indian bdellium; family: Burseraceae), diversely found in Asian countries mainly India, Pakistan, and Bangladesh [272–277]. Guggulsterones are present in lipid soluble fraction of the gugulipid in the form of stereoisomers (E- and Z-), whereas neutral extract contains guggulsterol and other steroidal compounds. Guggulsterone stereoisomeric forms (E-guggulsterone and Z-guggulsterone) show similar hypolipidemic activities, and its hypolipidemic properties is equivalent to clofibrate (hypolipidemic drug) [272–277]. It is a plant natural steroid thus named as phytosteroid, possessing anti-inflammatory, lipid-lowering, and antioxidant effects. Guggelsterone is a potential farnesoid X receptor antagonist and thus can effectively regulate bile acid synthesis and carbohydrate metabolism [272–277]. Gum resin derived from *C. mukul* has been used for several decades in various holistic systems such as Ayurveda to treat hyperlipidaemia, obesity, and inflammatory conditions such as arthritis. In 1966, this oleogum resin was initially introduced by Indian researcher, G. V. Satyavati by an investigation based on studying the hypolipidemic effects of guggul in an experimental animal model (rabbits) [272–277]. Afterwards several preclinical and clinical trials have been conducted to evaluate its hypolipidemic effects. These all scientific studies mainly corroborated the biological effects claimed for guggul and described in the ancient Ayurvedic system. Along with proven efficacy and safety information, guggul in 1986 was officially accepted for marketing in India as a hypolipidemic drug. In 1987, after the approval from the competent regulatory body in India, gum resin has been extensively utilized to treat hyperlipidaemia [272–277].

### 2.8.18.1   Isolation of Guggulsterones

In the late-19th century (1971), Dr. Sukh Dev isolated Z- and E-guggulsterones from guggulipid [278]. For isolation of guggulsterone, processed gum guggul (coarsely powdered), which is derived from *C. mukul*, is extracted with organic solvent (ethyl acetate). Resultant ethyl acetate fraction is concentrated by evaporation in vacuum (at 50°C) to form guggulsterone containing gummy residue (dark brownish). Residual material is again treated with ethyl acetate followed by the extraction with hydrochloric acid (3 N). Resultant materials are treated with alkali in order to basify or neutralize the residual fraction followed by the treatment with acid again to obtain non-carbonyl as well as neutral ketonic portion (via below mentioned method). This is done by treatment of neutralized resultant solution with semi-carbazide (10%) over silica gel and toluene. Treatment is followed by the stirring, heating (60–62°C) till 14 h. Further solution obtained was allowed to cool and subjected to filtration. Eventually silica gel was treated with toluene to obtain toluene soluble non-carbonyl portion. Toluene treated silica gel was again treated with oxalic acid as well as toluene, followed by the stirring, refluxation (2.5 h) and at last solution obtained is allowed to cool and filtered. Subsequently silica gel residue is rinsed with ethyl acetate three times. Overall extract obtained after ethyl acetate washing is again washed with water and subsequently reduced as well as further dried to give neutral ketonic portion. Fraction obtained is separated over silica gel using mobile-phase benzene and ethyl acetate to ultimately form portions comprising guggulsterone (E-and Z-).

### 2.8.18.2   Thin Layer Chromatography of Guggulsterone

TLC is considered one of the easy and reliable methods for the separation of guggulsterone. To separate or identify guggulsterone, test and reference samples (1 mg each) are dissolved in ethyl acetate (1 ml). Both test and reference samples are chromatographed over TLC silica gel-G plate using solvent system [toluene-ethyl acetate (80:20)] and spraying reagent (1% vanillin $H_2SO_4$), followed by the heating for 10–15 min at 110°C. After this step, bluish violet spots appear on guggulsterone with Rf 0.45 with standard sample, which can be compared with test sample to check the presence of guggulsterone.

### 2.8.19   HESPERIDIN (3,5,7-TRIHYDROXYFLAVANONE-7-RHAMNOGLUCOSIDE)

Hesperidin ($C_{28}H_{34}O_{15}$) (Vitamin P) (melting point: 252–254°C) is a flavanone glycoside abundantly present in citrus species such as sweet orange, lemon, as well as in some other fruits, vegetables, and various traditional formulations [279–281]. Hesperidin and other flavonoids such as hesperetin, naringenin, nobiletin, and naringin called citrus flavonoids possess wide range of pharmacological properties. Among all, methoxylated and polymethoxylated citrus flavonoids are present in level in the peel of *Citrus* sp. Hesperidin chemically contains an aglycone part [hesperetin (methyl eriodictyol)] linked to glycone part (rutinose). Glycone part (rutinose, a disaccharide) which contains rhamnose and glucose [279–281]. This flavanone glycoside is considered one of the harmless and pharmacologically active compounds which retains various biological activities such as hypolipidemic, antidiabetic, antihypertensive, and cardioprotective effects. Medicinal properties of this bioflavonoid are due to its antioxidant effects and attenuation of pro-inflammatory cytokine production. Hesperidin metabolite, hesperetin ($C_{16}H_{14}O_6$) has better bioavailability [279–281]. Hesperetin is inactively absorbed via intestine to reach in systemic circulation. Majority of the flavonoids exist in the glycoside form as hesperidin. Due to its chemical nature, hesperidin cannot be freely available to be absorbed through intestine. Hesperidin is usually metabolized to hesperidin by microbiota enzyme named as β-glucosidase before its absorption [279–281]. Hesperidin show high bioavailability as after the intake it's excreted in the urine in 24 h. However, the bioavailability and solubility of majority of citrus flavonoids is poor due to which high doses are usually prescribed to attain therapeutic plasma levels in the brain. These properties restrict clinical application of the citrus flavonoids. Therefore, to improve the solubility as well as bioavailability of citrus flavonoids usually different delivery systems in form of formulations have been assessed to mainly achieve maximum therapeutic effects [279–281].

### 2.8.19.1 Isolation of Hesperidin

In the beginning of 18th centuries, French chemist Lebreton, isolated Hesperidin from the albedo tissue of oranges. Hesperidin is diversely present in different citrus species such as *Clonorchis sinensis*, *Caryota mitis*, and *Citrus aurantium* (Rutaceae) (Figure 2.24). Hesperidin could be isolated by two methods:

Procedure 1: Peeled, powdered, and dried plant material (200 g) is transferred in round bottom flask (RBF) connected with refluxation condenser, followed by the addition of petroleum ether (1 l) to start refluxation at 40–60°C for 1 h. Refluxation is followed by the filtration of solution by using Buchner funnel. Residual mass obtained after filtration is dried and again subjected to refluxation by using methanol (1 l) for 3 h (Figure 2.24). Methanolic solution is filtered, followed by the washing of the marc with hot methanol (200 ml). Volume of resultant filtrate is reduced to form concentrated syrupy mass containing hesperidin. Hesperidin is further crystalized by using dil. acetic acid, to form needles of hesperidin (Figure 2.24).

Procedure 2: In the second procedure, chopped orange peel material (200 g) in flask (2 l) was treated with calcium hydroxide solution (10%) and left undisturbed for 12 h. Result at solution obtained is filtered using Buchner funnel. Yellow orange colour filtrate obtained from the last step is treated with conc. HCl to reduce between pH 4 and 5 (Figure 2.24). After acidification, hesperidin separates and collected as amorphous powder. This step is followed by washing of the powder with water using Buchner funnel and further subjected to recrystallization after treating with aqueous formamide (Figure 2.24).

### 2.8.19.2 Identification Test for Hesperidin

- Ferric chloride test: Treatment of sample containing hesperidin with ferric chloride solution present a wine red colour.
- Magnesium-HCl reduction test: Treatment of ethanolic solution containing hesperidin as well as magnesium with conc. HCl presents bright violet colour.

## 2.8.20 NICOTINE

Nicotine is a pyridine alkaloid and a tertiary amine (contain pyridine and pyrrolidine ring), bitter in taste diversely present among various nightshade plants (Solanaceae), mainly from the leaves of *Nicotiana tabacum* (tobacco plant) (2–8% in tobacco leaves) where it exists as malic or citric

**Procedure 1**

Powdered peels (200g)

↓

Refluxation with petroleum ether (1L, 1h)

↓

Filtration followed by drying

↓

Dried powder treatment with methanol (1L)

↓

Refluxation (3h)

Filtered

↓

Marc treatment with hot methanol (200 ml)

↓

Filtrate volume is reduced

↓

Concentrated residue crystallized by dil. acetic

**Procedure 2**

Chopped orange peel (200 g)

↓

Treated with calcium hydroxide solution

Left overnight at room temperature

↓

Filtered through a large buchner funnel

↓

Yellow orange color filtrate

↓

Acidified by conc. HCL (pH 4-5)

↓

Hesperidin (amorphous powder)

↓

Washing of hesperidin by using buchner funnel

↓

Re-crystallization by using aq. formamide

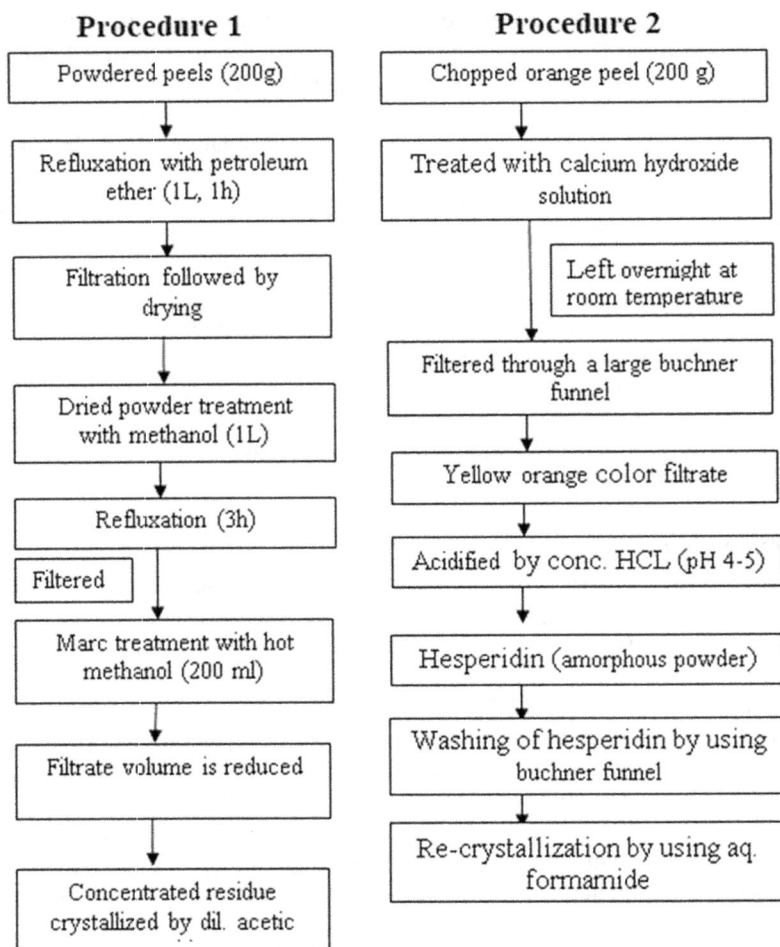

**FIGURE 2.24**   Schemes for the isolation of hesperidin.

acid salts. Nicotine has more than 30 derivatives, which are available in different forms. In humans (adults), LD50 of nicotine was found to be 0.5–1.0 mg/kg. In mice, LD50 of nicotine is 3 mg/kg, which is more toxic than cocaine (LD50 of 95.1 mg/kg). Nicotine shows agonistic effects at most nicotinic acetylcholine receptors (nAChRs). However, it shows antagonistic effects over two nicotinic receptor subunits (nAChR$\alpha$9 and nAChR$\alpha$10). At small dose, nicotine shows stimulatory effects, whereas at high dose (50–100 mg), it can cause toxicity or harmful effects [282]. Nicotine is important for the synthesis of nicotinic acid (niacin). Nicotine-based addiction also activates certain neurochemical pathways that cause pleasure and euphoria. Its capability of causing addition and other ill effects lead to various forms of disability and premature death. Nicotine allows neurotransmitter release via binding to nicotinic cholinergic receptors (muscle receptors and neuronal receptors) [283]. Nicotine dependence stimulates excitatory glutamate neurotransmission, and further chronic nicotine reduces inhibitory gamma-aminobutyric acid (GABA) neurotransmission. Thus, glutamate, dopamine, and GABA pathways mediated by nicotine are responsible for the nicotine dependence, whereas corticotropin-releasing factor seems to play an important role in withdrawal of nicotine [283].

Nicotine is a weak base which is freely absorbed via the lung, skin, gastrointestinal tract, nasal mucosa, oral, and route; however, lung and kidneys metabolize nicotine but to a much lesser extent. Nicotine experiences a first-pass metabolism (80–90%) where it metabolizes by the liver via C-oxidation of nicotine (regulated by CYP2A6) as well as other metabolic pathways such as N-oxidation, and glucuronidation of nicotine, cotinine, and trans-3-hydroxycotinine. Main metabolite of nicotine is cotinine (70–80%) [284]. Even after showing its solubility in both water and lipids, its ionization determines its absorption across the membranes. Nicotine (non-ionized form) is absorbed rapidly across the mucous membranes of the mouth and the bronchial tree, whereas ionized form does not rapidly cross membranes [284].

### 2.8.20.1 Isolation of Nicotine

In the 17th century, tobacco was first introduced as an insecticide. This natural alkaloid name "Nicotine" was coined after the discovery of plant *N. tabacum*. In the beginning of 18th century, German physician Wilhelm Heinrich Posselt and chemist Karl Ludwig Reimann isolated nicotine from the tobacco plant and identified it as poison [285, 286]. Occurrence of nicotine is very common among nightshade plants (*Nicotiana rustica*, *N. tabacum*, *Duboisia hopwoodii*, *Asclepias syriaca*) belonging to family Solanaceae. Nicotine (boiling point: 247°C) is volatile liquid, colourless to pale yellow, sharp taste, unpleasant and pungent odour, and hygroscopic in nature [285, 286]. To isolate nicotine, plant material is treated with lime followed by the extraction with water. The aqueous solution later obtained is further treated with chloroform. Chloroform fraction containing nicotine is allowed to treat with dilute $H_2SO_4$ to form nicotine sulphate in the solution. Furthermore, nicotine sulphate layer can be separated due to density differences. Further using ion exchange chromatography using cation exchanger, nicotine can be recovered from aqueous solution by eluting with a right solvent. Cation exchanger can be reused after washing with dilute acid. This procedure can yield a better quality of nicotine (insecticide grade) from tobacco waste.

### 2.8.20.2 Identification Test for Nicotine

TLC of nicotine can be used for the identification test for nicotine by 1 gram of test sample in methanol (1 ml). Similarly standard sample is prepared and applied over TLC silica gel-G plate. Further elution is done by using solvent system [chloroform-methanol-ammonium hydroxide: 60:10:1] and spraying reagent [*p*-aminobenzoic acid (2%); phosphate buffer (0.1 M)]. This step is followed by the treatment with bromine cyanide vapour (visualizing reagent), using Konig reaction. Nicotine over TLC showed Rf value 0.77.

### 2.8.21 OPIUM ALKALOIDS

Utilization of opium (*Papaver somniferum*) has been evidenced from prehistoric times. In the 19th century, opium was actually introduced in India by the Arabs. Opium is a central nervous system depressant and narcotic analgesic; it is not a hallucinogen. Opium acts via interacting specifically with opiate receptors (including mu, delta, kappa, nociceptin receptor) [287]. Opium alkaloids include morphine, codeine, thebaine (the phenanthrenes), papaverine, and noscapine (isoquinolines). Melting point of morphine, codeine, thebaine are 254, 193, and 154–156°C, respectively. In the beginning, the use of opioids in the form of raw opium (afeem) and poppy husk (doda) became historically as well as culturally acceptable. In Europe and the United States, opium consumption was common in the 18th century. Earlier, synthetic opioids including pentazocine as well as meperidine were primarily presented as safe analgesics. Based on the availability of different forms of opium, Banys (2014) classified them as illustrated in Figure 2.25 [288]. Among these, codeine as well as morphine are considered main isoquinoline compounds found in the latex derived by incision using equipment named as "Nushtar" or "Nishtar" to unripe capsules of *P. somniferum* (Papaveraceae). Opium is considered source for the production of several drugs, both legal (morphine, codeine, oxycodone, hydrocodone, hydromorphone, etc.) and illegal (heroin). Opium was utilized in different forms (raw opium: afeem; poppy husk: doda). Also opioids are conventionally classified in to different types as mentioned in Figure 2.25 [289, 290].

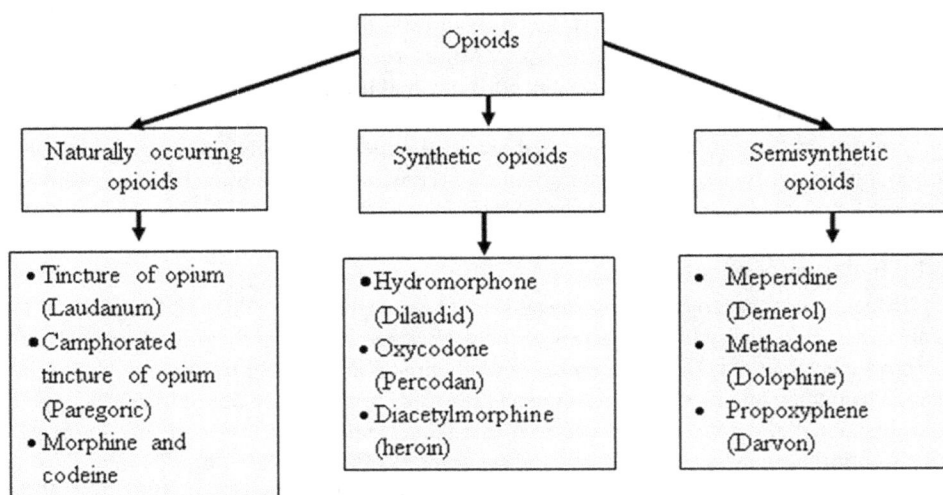

**FIGURE 2.25** Types of opioids.

### 2.8.21.1 Isolation for Opium Alkaloids

In 1805, a German pharmacist (Friedrich W. Sertürner) has isolated the main alkaloid and he named it "morphine". Later on, various researchers elucidated the relationship between its structure and analgesic activity of morphine. These pharmacology-based studies revealed analgesic property of morphine which can be sequentially blocked by an opioid antagonist medication named as naloxone. This discovery has revealed the perception of an endogenous opioid system that was later on corroborated by the finding of specific opioid-binding receptors in the CNS [291, 292].

Processed plant material is extracted with hot water followed by the treatment with calcium chloride. Resultant solution volume is reduced by heating to derive codeine as well as morphine salts as well as crystal form. Concentrated solution was treated with chloroform to get insoluble fraction (morphine) as well as soluble fraction (codeine). Again, processed plant material is allowed to be treated with calcium chloride and followed by treatment of filtrate with sodium hydroxide (10%). Filter the solution to get the marc (narcotine, papaverine, thebaine) and filtrate (morphine, codeine). Filtrate solution is treated with chloroform to form chloroform fraction (codeine) and aqueous fraction (morphine and narceine). Aqueous fraction is treated with acid followed by the treatment with alkali (ammonia), which leads to the precipitation of morphine. Further narceine can be obtained from aqueous fraction. To derive narcotine, papaverine and thebaine marc is treated with in alcohol followed by the treatment with acid (acetic acid) and further to the resultant solution add hot water (3 vol). This step leads to the formation of narcotine and papaverine. Both narcotine and papaverine are separated by using oxalic acid (0.3%) and allowed to cool down which leads papaverine crystal formation. At last to separate noscapine, aqueous layer is treated with ammonia to precipitate noscapine and recrystallization by using water is done as mentioned in Figure 2.26.

### 2.8.21.2 Identification Tests

- Presence of opium alkaloids can be tested by treatment of test sample with ferric chloride solution, which leads to the appearance of deep reddish purple colour indicates the presence of opium alkaloids. This colour, deep reddish purple, still stays after its treatment with hydrochloric acid.
- The test sample when mixed with conc. $H_2SO_4$ as well as formaldehyde results in dark violet colour, which indicates the presence of morphine.

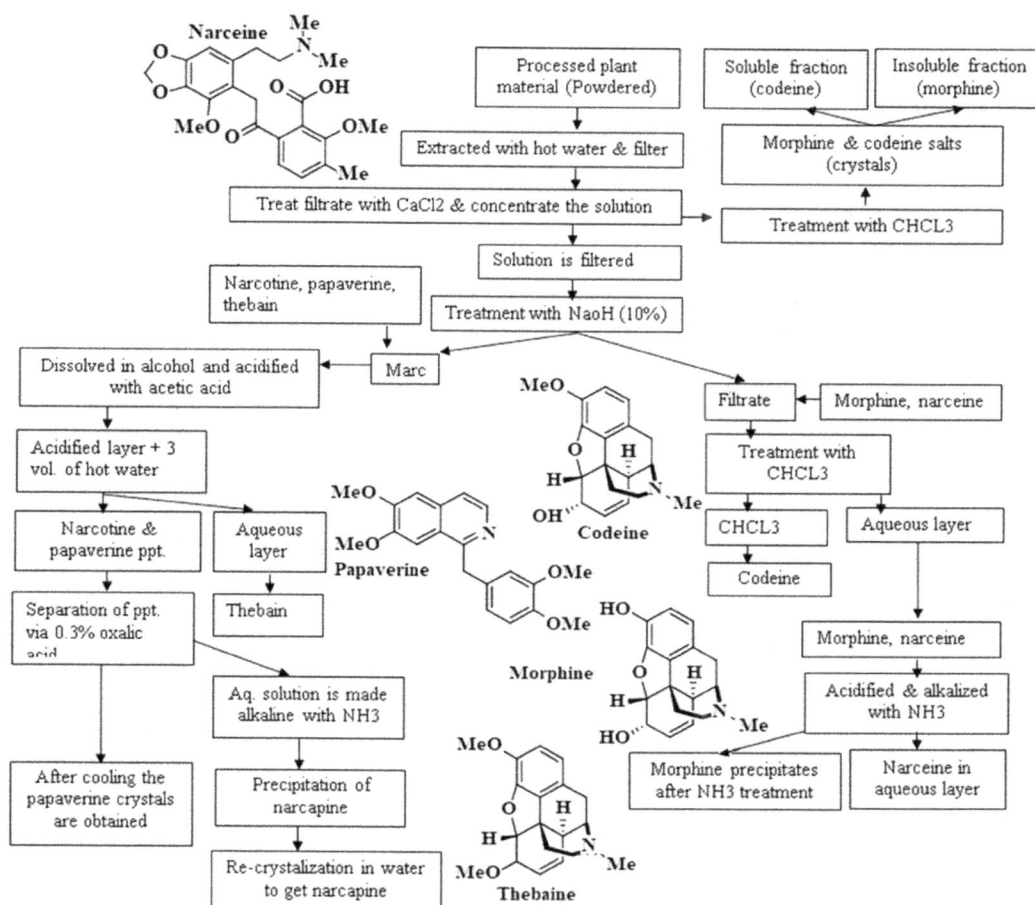

**FIGURE 2.26** Isolation methods for opium alkaloids.

# REFERENCES

1. Genovese S, Curini M, Epifano F. Chemistry and biological activity of azoprenylated secondary metabolites. *Phytochemistry*. 2009;70:1082–1091.
2. Rea G, Antonacci A, Lambreva M, Margonelli A, Ambrosi C, Giardi MT. The basic research and biotechnological programs on nutraceutical. In: Bio-Farms for Nutraceuticals: Functional Food and Safety Control by Biosensors (eds. MT Giardi, G Rea, B Berra). Springer, 2010;698:1–16.
3. Krzyzanowska J, Czubacka A, Oleszek W. Dietary phytochemicals and human health. In: Bio-Farms for Nutraceuticals: Functional Food and Safety Control by Biosensors (eds. MT Giardi, G Rea, B Berra). Springer, 2010;698:74–99.
4. Parsaeimehr A, Sargsyan E, Vardanyan A. Expression of secondary metabolites in plants and their useful perspective in animal health. *ABAH Bioflux*. 2011;3:115–124.
5. Ferrari S. Biological elicitors of plant secondary metabolites: Mode of action and use in the production of nutraceutics. In: Bio-Farms for Nutraceuticals: Functional Food and Safety Control by Biosensors (eds. MT Giardi, G Rea, B Berra). Springer, 2010;698:152–166.
6. Lattanzio V, Lattanzio VMT, Cardinali A. Role of phenolic in the resistance mechanisms of plants against fungal pathogens and insects. In: Phytochemistry: Advances in Research (ed. F Imperato). Research Signpost, Kerala, India; 2006;23–67.
7. Morales LO, Tegelberg R, Brosché M, Keinänen M, Lindfors A, Aphalo PJ. Effects of solar UV-A and UV-B radiation on gene expression and phenolic accumulation in *Betula pendula* leaves. *Tree Physiol*. 2010;30:923–934.

8. Saviranta HK, Julkunen-Tiitto R, Oksanen E, Karjalainen RO. Leaf phenolic compounds in red clover (*Trifolium pratense* L.) induced by exposure to moderately elevated ozone. *Environ Pollut.* 2010;158:440–446.

9. Magalhães STV, Guedes RNC, Demuner AJ, Lima ER. Effect of coffee alkaloids and phenolics on egg-laying by the coffee leaf miner *Leucoptera coffeella*. *Bull Entomol Res.* 2008;98:483–489.

10. Lattanzio V, Lattanzio VMT, Cardinali A. Role of phenolic in the resistance mechanisms of plants against fungal pathogens and insects. In: Phytochemistry: Advances in Research (ed. F Imperato). Research Signpost, Kerala, India; 2006:23–67.

11. Bassoli BK, Cassolla P, Borba-Murad GR, Constantin J, Salgueiro-Pagadigorria CL, Bazotte RB, Da Silva RS, De Souza HM. Chlorogenic acid reduces the plasma glucose peak in the oral glucose tolerance test: Effects on hepatic glucose release and glycemia. *Cell Biochem Funct.* 2008;26:320.

12. Yao LH, Jiang YM, Shi J, Tomas-Barberan FA, Datta N, Singanusong R, Chen SS. Flavonoids in food and their health benefits. *Plant Foods Human Nutr.* 2004;59:113–122.

13. Hounsome N, Hounsome B, Tomos D, Edwards-Jones G. Plant metabolites and nutritional quality of vegetables. *J Food Sci.* 2008;73:P48–P65.

14. Winks M, Schimmer O. Modes of action of defensive secondary metabolites. Function of plant SMs and their exploitation in biotechnology. *Annu Plant Rev.* 1999;39:17–133.

15. Paiva PMG, Gomes FS, Napoleão TH, Sá RA, Correia MTS, Coelho CBB. Antimicrobial activity of secondary metabolites and lectins from plants. In: Current Research, Technology and Education Topics in Applied Microbiology and Microbial Biotechnology. (ed. A Méndez-Villas). FORMATEX; 2010:396–406.

16. Khadem S, Marles RJ. Chromone and flavonoid alkaloids: Occurrence and bioactivity. *Molecules* 2012;17:191–206.

17. Angelova S, Buchheim M, Frowitter D, Schierhorn A, Roos W. Overproduction of alkaloid phytoalexins in California poppy cells is associated with the co-expression of biosynthetic and stress-protective enzymes. *Mol Plant.* 2010 Sep;3(5):927–939.

18. Lee JS, Latimer LJ, Hampel KJ. Coralyne binds tightly to both T.A.T- and C.G.C(+)-containing DNA triplexes. *Biochemistry.* 1993 Jun 1;32(21):5591–5597.

19. Baumann TW, Gabriel H. Metabolism and excretion of caffeine during germination of *Coffea arabica* L. *Plant Cell Physiol.* 1984:25;1431–1436

20. Ashihara H, Sano H, Crozier A. Caffeine and related purine alkaloids: Biosynthesis, catabolism, function and genetic engineering. *Phytochemistry.* 2008 Feb;69(4):841–856.

21. Aires A, Mota VR, Saavedra MJ, Rosa EA, Bennett RN. The antimicrobial effects of glucosinolates and their respective enzymatic hydrolysis products on bacteria isolated from the human intestinal tract. *J Appl Microbiol.* 2009 Jun;106(6):2086–2095.

22. Iriti M, Faoro F. Chemical diversity and defence metabolism: How plants cope with pathogens and ozone pollution. *Int J Mol Sci.* 2009 Jul 30;10(8):3371–3399.

23. Rohmer M. The discovery of a mevalonate-independent pathway for isoprenoid biosynthesis in bacteria, algae and higher plants. *Nat Prod Rep.* 1999 Oct;16(5):565–574.

24. Liu WK, Xu SX, Che CT. Anti-proliferative effect of ginseng saponins on human prostate cancer cell line. *Life Sci.* 2000 Aug 4;67(11):1297–1306.

25. Bhatia S. Natural Polymer Drug Delivery Systems: Nanoparticles, Plants, and Algae. Springer Nature, Basingstoke, UK; 2016:117–127.

26. Bhatia S. Introduction to Pharmaceutical Biotechnology, Volume 2: Enzymes, Proteins and Bioinformatics. IOP Publishing Ltd, Bristol, UK; 2018:1.

27. Bhatia S. Introduction to Pharmaceutical Biotechnology, Volume 1: Basic Techniques and Concepts. IOP Publishing Ltd, Bristol, UK; 2018:2.

28. Bhatia S. Introduction to Pharmaceutical Biotechnology, Volume 3: Animal Tissue Culture Technology. IOP Publishing Ltd, Bristol, UK; 2019:3.

29. Bhatia S. Natural Polymer Drug Delivery Systems: Nanotechnology and Its Drug Delivery Applications. Springer International Publishing, Switzerland; 2016:1–32.

30. Bhatia S. Natural Polymer Drug Delivery Systems: Nanoparticles Types, Classification, Characterization, Fabrication Methods and Drug Delivery Applications. Springer International Publishing, Switzerland; 2016:33–93.

31. Bhatia S. Natural Polymer Drug Delivery Systems: Natural Polymers vs Synthetic Polymer. Springer International Publishing, Switzerland; 2016:95–118.

32. Bhatia S. Natural Polymer Drug Delivery Systems: Plant Derived Polymers, Properties, and Modification & Applications. Springer International Publishing, Switzerland; 2016:119–184.

33. Bhatia S. Natural Polymer Drug Delivery Systems: Marine Polysaccharides Based Nano-Materials and Its Applications. Springer International Publishing, Switzerland; 2016:185–225.

34. Altemimi A, Lakhssassi N, Baharlouei A, Watson DG, Lightfoot DA. Phytochemicals: Extraction, isolation, and identification of bioactive compounds from plant extracts. *Plants (Basel)*. 2017 Sep 22;6(4):42.

35. Zhang QW, Lin LG, Ye WC. Techniques for extraction and isolation of natural products: A comprehensive review. *Chin Med*. 2018 Apr 17;13:20.

36. Altemimi A, Lakhssassi N, Baharlouei A, Watson DG, Lightfoot DA. Phytochemicals: Extraction, isolation, and identification of bioactive compounds from plant extracts. *Plants (Basel)*. 2017 Sep 22;6(4):42.

37. Bhatia S. Systems for Drug Delivery: Mammalian Polysaccharides and Its Nanomaterials. Springer International Publishing, Switzerland; 2016:1–27.

38. Bhatia S. Systems for Drug Delivery: Microbial Polysaccharides as Advance Nanomaterials. Springer International Publishing, Switzerland; 2016:29–54.

39. Bhatia S. Systems for Drug Delivery: Chitosan Based Nanomaterials and Its Applications. Springer International Publishing, Switzerland; 2016:55–117.

40. Bhatia S. Systems for Drug Delivery: Advance Polymers and Its Applications. Springer International Publishing, Switzerland; 2016:119–146.

41. Bhatia S. Systems for Drug Delivery: Advanced Application of Natural Polysaccharides. Springer International Publishing, Switzerland; 2016:147–170.

42. Bhatia S. Systems for Drug Delivery: Modern Polysaccharides and Its Current Advancements. Springer International Publishing, Switzerland; 2016:171–188.

43. Bhatia S. Systems for Drug Delivery: Toxicity of Nanodrug Delivery Systems. Springer International Publishing, Switzerland; 2016:189–197.

44. Bhatia S, Sharma K, Bera T. Structural characterization and pharmaceutical properties of porphyran. *Asian J Pharm*. 2015;9:93–101.

45. Bhatia S, Sharma A, Sharma K, Kavale M, Chaugule BB, Dhalwal K, Namdeo AG, Mahadik KR. Novel algal polysaccharides from marine source: Porphyran. *Pharmacogn Rev*. 2008;4:271–276.

46. Bhatia S. Nanotechnology in Drug Delivery Fundamentals, Design, and Applications: Part 1: Protein and Peptide-Based Drug Delivery Systems. Apple Academic Press, Palm Bay, FL; 2016:50–204.

47. Bhatia S. Nanotechnology in Drug Delivery Fundamentals, Design, and Applications: Part 2: Peptide-Mediated Nanoparticle Drug Delivery System. Apple Academic Press, Palm Bay, FL; 2016:205–280.

48. Bhatia S. Nanotechnology in Drug Delivery Fundamentals, Design, and Applications: Part 3: CPP and CTP in Drug Delivery and Cell Targeting. Apple Academic Press, Palm Bay, FL; 2016:309–312.

49. Bhatia S. Systems for Drug Delivery: Safety, Animal, and Microbial Polysaccharides. Springer Nature, Basingstoke, UK; 2016:122–127.

50. Bhatia S. Stem cell culture. In: Introduction to Pharmaceutical Biotechnology, Volume 3: Animal Tissue Culture and Biopharmaceuticals. IOP Publishing Ltd, Bristol, UK; 2019;3:1–24.

51. Bhatia S. Organ culture. In: Introduction to Pharmaceutical Biotechnology, Volume 3: Introduction to Animal Tissue Culture Science. IOP Publishing Ltd, Bristol, UK; 2019;3:1–28.

52. Bhatia S. Animal tissue culture facilities. In: Introduction to Pharmaceutical Biotechnology, Volume 3: Animal Tissue Culture and Biopharmaceuticals. IOP Publishing Ltd, Bristol, UK; 2019;3:1–32.

53. Bhatia S. Characterization of cultured cells. In: Introduction to Pharmaceutical Biotechnology, Volume 3: Animal Tissue Culture and Biopharmaceuticals. IOP Publishing Ltd, Bristol, UK; 2019;3;1–47.

54. Bhatia S. Introduction to genomics. In: Introduction to Pharmaceutical Biotechnology, Enzymes, Proteins and Bioinformatics. IOP Publishing Ltd, Bristol, UK; 2018;2:1–39.

55. Bhatia S. Bioinformatics. In: Introduction to Pharmaceutical Biotechnology, Enzymes, Proteins and Bioinformatics. IOP Publishing Ltd, Bristol, UK; 2018;3:1–16.

56. Bhatia S. Protein and enzyme engineering. In: Introduction to Pharmaceutical Biotechnology, Enzymes, Proteins and Bioinformatics. IOP Publishing Ltd, Bristol, UK; 2018;2:1–15.

57. Bhatia S. Introduction to genomics. In: Introduction to Pharmaceutical Biotechnology, Enzymes, Proteins and Bioinformatics. IOP Publishing Ltd, Bristol, UK; 2018;2:1–39.

58. Bhatia S. Industrial enzymes and their applications. In: Introduction to Pharmaceutical Biotechnology, Enzymes, Proteins and Bioinformatics. IOP Publishing Ltd, Bristol, UK; 2018;2:21.

59. Bhatia S. Introduction to enzymes and their applications. In: Introduction to Pharmaceutical Biotechnology, Enzymes, Proteins and Bioinformatics. IOP Publishing Ltd, Bristol, UK; 2018;2:1–29.

60. Bhatia S. Biotransformation and enzymes. In: Introduction to Pharmaceutical Biotechnology, Enzymes, Proteins and Bioinformatics. IOP Publishing Ltd, Bristol, UK; 2018;3:1–13.

61. Bhatia S. Modern DNA science and its applications. In: Introduction to Pharmaceutical Biotechnology, Volume 1: Basic Techniques and Concepts. IOP Publishing Ltd, Bristol, UK; 2018;1(3):1–70.

62. Bruce-Chwatt LJ. Cinchona and quinine: A remarkable anniversary. *Interdisciplinary Sci Rev.* 2013;15:87–93.

63. Murauer A, Ganzera M. Quantitative determination of major alkaloids in Cinchona bark by Supercritical Fluid Chromatography. *J Chromatogr A.* 2018 Jun 15;1554:117–122.

64. Prabhakaran NKP. Cinchona (*Cinchona* sp.). In: The Agronomy and Economy of Important Tree Crops of the Developing World (ed. NKP Prabhakaran), Elsevier, 2010:111–129.

65. Galloway JH, Marsh ID, Forrest AR. A simple and rapid method for the estimation of quinine using reversed-phase high-performance liquid chromatography with UV detection. *J Anal Toxicol.* 1990 Nov–Dec;14(6):345–347.

66. Meineke I, Rohde S, Gundert-Remy U. An inexpensive and sensitive method for the determination of quinidine in plasma by high-performance liquid chromatography with ultraviolet detection. *Ther Drug Monit.* 1995 Feb;17(1):75–78.

67. Karbwang J, Na Bangchang K, Molunto P, Bunnag D. Determination of quinine and quinidine in biological fluids by high performance liquid chromatography. *Southeast Asian J Trop Med Public Health.* 1989 Mar;20(1):65–69.

68. Babalola CP, Bolaji OO, Dixon PA, Ogunbona FA. Column liquid chromatographic analysis of quinine in human plasma, saliva and urine. *J Chromatogr.* 1993 Jun 23;616(1):151–154.

69. Edstein MD, Prasitthipayong A, Sabcharoen A, Chongsuphajaisiddhi T, Webster HK. Simultaneous measurement of quinine and quinidine in human plasma, whole blood, and erythrocytes by high-performance liquid chromatography with fluorescence detection. *Ther Drug Monit.* 1990 Sep;12(5):493–500.

70. Karawya MS, Diab AM. Colorimetric assay of quinine and quinidine in raw materials, formulations, and biological fluids. *J Pharm Sci.* 1977;66:1317–1319.

71. Zakhari N, Ibrahim F, Kovar KA. Non-aqueous titration of quinine and quinidine sulphates by use of barium perchlorate. *Talanta.* 1989 Jul;36(7):780–782.

72. Soliman SA, Abdine H, Zakhari NA. Nonaqueous titration of sulfates of quinine and quinidine using barium acetate. *J Pharm Sci.* 1976 Mar;65(3):424–427. doi: 10.1002/jps.2600650327. PMID: 1263094.

73. Ramsden W, Lipkin IJ. Detection and estimation of quinine in blood and urine. *Ann Trop Med Parasitol.* 1918;11(4):443–464.

74. Lee MR. The history of Ephedra (ma-huang). *J R Coll Physicians Edinb.* 2011;41(1):78–84.

75. Henry A. Spiller. Speed. In: Encyclopedia of Toxicology (Second Edition). Elsevier, 2005:69–71.

76. Vaidya VS, Mehendale HM. Ephedra. In: Encyclopedia of Toxicology (Second Edition) (ed. P Wexler). Academic Press, 2005:223–228.

77. Vaidya VS, Mehendale HM. Ephedra. Encyclopedia of Toxicology (Third Edition) (ed. P Wexler). Academic Press, 2014:426–430.

78. Hausner EA, Poppenga RH. Chapter 26: Hazards associated with the use of herbal and other natural products. Small Animal Toxicology (Third Edition) (eds. ME Peterson, PA Talcott), W.B. Saunders, 2013:335–356.

79. Blumenthal M. FDA announces ban on ephedra supplements: Federal move follows bans by California, Illinois and New York. *HerbalGram* 2004;61:64–55. https://www.herbalgram.org/resources/herbalgram/issues/61/table-of-contents/article2644/.

80. Levy JL, Cabrera J, Thomas S, Brennan FH. Chapter 49 – Ergogenic aids. In: The Sports Medicine Resource Manual (eds. PH Seidenberg, AI Beutler), W.B. Saunders, 2008:598–610.

81. Medvedovskiĭ AO. Tytrymetrychnyĭ metod vyznachennia efedrynu hidrokhlorydu na osnovi utvorennia midnykh kompleksiv [Titration method of determination of ephedrine hydrochloride on the basis of copper complex formation]. *Farm Zh.* 1971;26(5):70–73.

82. Midha KK, Cooper JK, McGilveray IJ. Simple and specific electron-capture GLC assay for plasma and urine ephedrine concentrations following single doses. *J Pharm Sci.* 1979 May;68(5):557–560.

83. Lin ET, Brater DC, Benet LZ. Gas-liquid chromatographic determination of pseudoephedrine and norpseudoephedrine in human plasma and urine. *J Chromatogr.* 1977 Oct 21;140(3):275–279.

84. Marchei E, Pellegrini M, Pacifici R, Zuccaro P, Pichini S. A rapid and simple procedure for the determination of ephedrine alkaloids in dietary supplements by gas chromatography-mass spectrometry. *J Pharm Biomed Anal.* 2006 Aug 28;41(5):1633–1641.

85. Trujillo WA, Sorenson WR. Determination of ephedrine alkaloids in human urine and plasma by liquid chromatography/tandem mass spectrometry: collaborative study. *J AOAC Int.* 2003 Jul–Aug;86(4):643–656.

86. Roman MC. Determination of ephedra alkaloids in urine and plasma by HPLC-UV: collaborative study. *J AOAC Int.* 2004 Jan–Feb;87(1):15–24.

87. Evans WC, Evans D. Chapter 5: A taxonomic approach to the study of medicinal plants and animal-derived drugs. In: Trease and Evans' Pharmacognosy (Sixteenth Edition) (eds. WC Evans, D Evans). W.B. Saunders, 2009:18–44.

88. Akinmoladun AC, Olaleye MT, Farombi EO, 13 – Cardiotoxicity and cardioprotective effects of African medicinal plants. In: Toxicological Survey of African Medicinal Plants (ed. V Kuete). Elsevier, 2014:395–421.

89. Bhatia S. Introduction to genetic engineering. In: Introduction to Pharmaceutical Biotechnology, Volume 1: Basic Techniques and Concepts. IOP Publishing Ltd, Bristol, UK; 2018;1(3):1–63.

90. Bhatia S. Applications of stem cells in disease and gene therapy. In: Introduction to Pharmaceutical Biotechnology, Volume 1: Basic Techniques and Concepts. IOP Publishing Ltd, Bristol, UK; 2018;1:1–40.

91. Bhatia S. Transgenic animals in biotechnology. In: Introduction to Pharmaceutical Biotechnology, Volume 1: Basic Techniques and Concepts. IOP Publishing Ltd, Bristol, UK; 2018;1:1–67.

92. Bhatia S. History and scope of plant biotechnology. In: Modern Applications of Plant Biotechnology in Pharmaceutical Sciences. Academic Press, 2015:1–30.

93. Bhatia S. Plant tissue culture. In: Modern Applications of Plant Biotechnology in Pharmaceutical Sciences. Academic Press, 2015:31–107.

94. Bhatia S. Laboratory organization. In: Modern Applications of Plant Biotechnology in Pharmaceutical Sciences. Academic Press, 2015:109–120.

95. Bhatia S. Concepts and techniques of plant tissue culture science. In: Modern Applications of Plant Biotechnology in Pharmaceutical Sciences. Academic Press, 2015:121–156.

96. Bhatia S. Application of plant biotechnology. In: Modern Applications of Plant Biotechnology in Pharmaceutical Sciences. Academic Press, 2015:157–207.

97. Bhatia S. Somatic embryogenesis and organogenesis. In: Modern Applications of Plant Biotechnology in Pharmaceutical Sciences. Academic Press, 2015:209–230.

98. Bhatia S. Classical and nonclassical techniques for secondary metabolite production in plant cell culture. In: Modern Applications of Plant Biotechnology in Pharmaceutical Sciences. Academic Press, 2015:231–291.

99. Bhatia S. Plant-based biotechnological products with their production host, modes of delivery systems, and stability testing. In: Modern Applications of Plant Biotechnology in Pharmaceutical Sciences. Academic Press, 2015:293–331.

100. Bhatia S. Edible vaccines. In: Modern Applications of Plant Biotechnology in Pharmaceutical Sciences. Academic Press, 2015:333–343.

101. Bhatia S. Microenvironmentation in micropropagation. In: Modern Applications of Plant Biotechnology in Pharmaceutical Sciences. Academic Press, 2015:345–360.

102. Bhatia S. Micropropagation. In: Modern Applications of Plant Biotechnology in Pharmaceutical Sciences. Academic Press, 2015:361–368.

103. Bhatia S. Laws in plant biotechnology. In: Modern Applications of Plant Biotechnology in Pharmaceutical Sciences. Academic Press, 2015:369–391.

104. Bhatia S. Technical glitches in micropropagation. In: Modern Applications of Plant Biotechnology in Pharmaceutical Sciences. Academic Press, 2015:393–404.

105. Bhatia S. Plant tissue culture-based industries. In: Modern Applications of Plant Biotechnology in Pharmaceutical Sciences. Academic Press, 2015:405–417.

106. Moore WN, Taylor LT. Extraction and quantitation of digoxin and acetyldigoxin from the *Digitalis lanata* leaf via near-supercritical methanol-modified carbon dioxide. *J Nat Prod.* 1996 Jul;59(7):690–693.

107. Mandal SC, Mandal V, Das AK, Chapter 9 – Qualitative phytochemical screening. In: Essentials of Botanical Extraction (eds. SC Mandal, V Mandal, AK Das). Academic Press, 2015:173–185.

108. Sasidharan S, Chen Y, Saravanan D, Sundram KM, Latha LY. Extraction, isolation and characterization of bioactive compounds from plants' extracts. *Afr J Tradit Complement Altern Med.* 2011;8(1):1–10.

109. Bodem G. Radioimmunologische Bestimmung von Herzglykosiden im Plasma [Radioimmunological estimation of cardiac glycoside concentration in the plasma]. *Prakt Anaesth.* 1976 Apr;11(2):97–100.

110. Lugt CB. Quantitative determination of digitoxin, gitaloxin, gitoxin, verodoxin and strospesid in the leaves of *Digitalis purpurea* by means of fluorescence. *Planta Med.* 1973 Mar;23(2):176–181.

111. Fujii Y, Ikeda Y, Yamazaki M. High-performance liquid chromatographic determination of secondary cardiac glycosides in *Digitalis purpurea* leaves. *J Chromatogr.* 1989 Oct 6;479(2):319–325.

112. Ikeda Y, Fujii Y, Umemura M, Hatakeyama T, Morita M, Yamazaki M. Quantitative determination of cardiac glycosides in *Digitalis lanata* leaves by reversed-phase thin-layer chromatography. *J Chromatogr A*. 1996 Oct 11;746(2):255–260.

113. Kaiser P, Kramer U, Meissner D, Kress M, Wood WG, Reinauer H. Determination of the cardiac glycosides digoxin and digitoxin by liquid chromatography combined with isotope-dilution mass spectrometry (LC-IDMS) – a candidate reference measurement procedure. *Clin Lab*. 2003;49(7–8):329–343.

114. Hashimoto Y, Shibakawa K, Nakade S, Miyata Y. Validation and application of a 96-well format solid-phase extraction and liquid chromatography-tandem mass spectrometry method for the quantitation of digoxin in human plasma. *J Chromatogr B: Analyt Technol Biomed Life Sci*. 2008 Jun 15;869(1–2):126–132.

115. Evans WC, Evans D. Chapter 23 – Saponins, cardioactive drugs and other steroids. In: Trease and Evans' Pharmacognosy (Sixteenth Edition) (eds. Evans WC, Evans D). W.B. Saunders, 2009:304–332.

116. Nesher M, Shpolansky U, Viola N, Dvela M, Buzaglo N, Cohen Ben-Ami H, Rosen H, Lichtstein D. Ouabain attenuates cardiotoxicity induced by other cardiac steroids. *Br J Pharmacol*. 2010 May;160(2):346–354.

117. Kalyoncu NI, Ozyavuz R. Ketanserin inhibits digoxin-induced arrhythmias in the anaesthetized guinea-pig. *Fundam Clin Pharmacol*. 1999;13(6):646–649.

118. Evans WC, Evans D. Chapter 21 – Phenols and phenolic glycosides. Trease and Evans' Pharmacognosy (Sixteenth Edition) (eds. WC Evans, D Evans). W.B. Saunders, 2009:219–262.

119. Kinjo J, Ikeda T, Watanabe K, et al. An anthraquinone glycoside from *Cassia angustifolia* leaves. *Phytochemistry*. 1994:37(6):1685.

120. Aldred EM, Buck C, Vall K. Chapter 24 – Glycosides. In: Pharmacology (eds. EM Aldred, C Buck, K Vall). Churchill Livingstone, 2009:181–185.

121. Hattori M, Namba T, Akao T, Kobashi K. Metabolism of sennosides by human intestinal bacteria. *Pharmacology*. 1988;36(Suppl 1):172–179.

122. Stoll A, Kussmaul W, Becker B. *Verh Schweiz Natf Ges*. 1941:235.

123. Dhanani T, Singh R, Reddy N, Trivedi A, Kumar S. Comparison on extraction yield of sennoside A and sennoside B from senna (*Cassia angustifolia*) using conventional and non-conventional extraction techniques and their quantification using a validated HPLC-PDA detection method. *Nat Prod Res*. 2017 May;31(9):1097–1101.

124. Shah SA, Ravishankara MN, Nirmal A, Shishoo CJ, Rathod IS, Suhagia BN. Estimation of individual sennosides in plant materials and marketed formulations by an HPTLC method. *J Pharm Pharmacol*. 2000 Apr;52(4):445–449.

125. Habib AA, El-Sebakhy NA. Spectrophotometric estimation of sennosides and rhein glycosides in senna and its preparations. *J Nat Prod*. 1980 Jul–Aug;43(4):452–458.

126. Srivastava A, Pandey R, Verma RK, Gupta MM. Liquid chromatographic determination of sennosides in *Cassia angustifolia* leaves. *J AOAC Int*. 2006 Jul–Aug;89(4):937–941.

127. Wahbi AM, Abu-Shady H, Soliman A. A colorimetric method for the determination of sennosides. *Planta Med*. 1976 Nov;30(3):269–272.

128. Glória MBA. Sweeteners. In: Encyclopedia of Food Sciences and Nutrition (Second Edition) (ed. B Caballero), Academic Press, 2003:5695–5702.

129. Chan K, Lin TX. Chapter 48: Treatments used in complementary and alternative medicine. In: Side Effects of Drugs Annual (ed. JK Aronson), Elsevier, 2009;31:745–756.

130. Ramos-Tovar E, Muriel P. Chapter 9 – Phytotherapy for the Liver. Dietary Interventions in Liver Disease (eds. RR Watson, VR Preedy). Academic Press, 2019:101–121.

131. Ploeger B, Mensinga T, Sips A, Seinen W, Meulenbelt J, DeJongh J. The pharmacokinetics of glycyrrhizic acid evaluated by physiologically based pharmacokinetic modeling. *Drug Metab Rev*. 2001 May;33(2):125–47.

132. Shimizu N, Tomoda M, Satoh M, Gonda R, Ohara N. Characterization of a polysaccharide having activity on the reticuloendothelial system from the stolon of *Glycyrrhiza glabra* var. glandulifera. *Chem Pharm Bull*. 1991;39(8):2082–2086.

133. Takada K, Tomoda M, Shimizu N. Core structure of glycyrrhizin GA, the main polysaccharide from the stolon of *Glycyrrhiza glabra* var. glandulifera; anti-complementary and alkaline phosphatase-inducing activities of the polysaccharide and its degradation product. *Chem Pharm Bull*. 1992;40(9):2487–2490.

134. Fenwick GR, Lutomski J, Nieman C. Liquorice, *Glycyrrhiza glabra* L. – composition, uses and analysis. *Food Chem*. 1990;38(2):119–143.

135. Bao F, Bai HY, Wu ZR, Yang ZG. Phenolic compounds from cultivated *Glycyrrhiza uralensis* and their PD-1/PD-L1 inhibitory activities. *Nat Prod Res*. 2021 Feb;35(4):562–569.

136. Zeng L, Zhang RY, Meng T, Lou ZC. Determination of nine flavonoids and coumarins in licorice root by high-performance liquid chromatography. *J Chromatogr.* 1990;513:247–254.

137. Zeng L, Lou ZC, Zhang RY. Quality evaluation of Chinese licorice. *Yaoxue Xuebao.* 1991;26(10):788–793.

138. Zeng L, Zhang RY, Lou ZC. Quantitative determination of three saponins in licorice root by high-performance liquid chromatography. *Yaoxue Xuebao.* 1991;26(1):53–58.

139. M Mukhopadhyay, P Panja. A novel process for extraction of natural sweetener from licorice (*Glycyrrhiza glabra*) roots. *Sep Purif Technol.* 2008;63(3):539–545.

140. Morinaga O, Ishiuchi K, Ohkita T. et al. Isolation of a novel glycyrrhizin metabolite as a causal candidate compound for pseudoaldosteronism. *Sci Rep.* 2018;8:15568.

141. Guillaume CP, van der Molen JC, Kerstens MN, Dullaart RP, Wolthers BG. Determination of urinary 18 beta-glycyrrhetinic acid by gas chromatography and its clinical application in man. *J Chromatogr B: Biomed Sci Appl.* 1999 Aug 20;731(2):323–334.

142. Chen HR, Sheu SJ. Determination of glycyrrhizin and glycyrrhetinic acid in traditional Chinese medicinal preparations by capillary electrophoresis. *J Chromatogr A.* 1993 Oct 29;653(1):184–188.

143. Krähenbühl S, Hasler F, Krapf R. Analysis and pharmacokinetics of glycyrrhizic acid and glycyrrhetinic acid in humans and experimental animals. *Steroids.* 1994 Feb;59(2):121–126.

144. Suzuki T, Tsukahara M, Akasaka Y, Inoue H. A highly sensitive LC-MS/MS method for simultaneous determination of glycyrrhizin and its active metabolite glycyrrhetinic acid: Application to a human pharmacokinetic study after oral administration. *Biomed Chromatogr.* 2017 Dec;31(12): doi:10.1002/bmc.4032

145. Kinghorn AD, Chin YW, Pan L, Jia Z. Natural products as sweeteners and sweetness modifiers. In: Comprehensive Natural Products II (eds. HW Liu, L Mander), Elsevier, 2010:269–315.

146. Bone K, Mills S. Optimising Safety: Principles and Practice of Phytotherapy (Second Edition), Churchill Livingstone, 2013:100–117.

147. Izawa K, Amino Y, Kohmura M, Ueda Y, Kuroda M. Human–environment interactions – Taste. In: Comprehensive Natural Products II (eds. HW Liu, L Mander). Elsevier, 2010:631–671.

148. Mitscher LA. Traditional medicines. In: Comprehensive Medicinal Chemistry II (eds. JB Taylor, DJ Triggle). Elsevier, 2007:405–430.

149. Kinghorn AD, Chin YW, Pan L, Jia Z. Natural products as sweeteners and sweetness modifiers. Comprehensive Natural Products II (eds. H-W (Ben) Liu, L Mander). Elsevier, 2010:269–315.

150. Pizzorno JE, Murray MT, Joiner-Bey H. Diabetes mellitus. The Clinician's Handbook of Natural Medicine (Third Edition) (eds. JE Pizzorno, MT Murray, H Joiner-Bey). Churchill Livingstone, 2016:249–286.

151. Liu HM, Kiuchi F, Tsuda Y. Isolation and structure elucidation of gymnemic acids, antisweet principles of *Gymnema sylvestre*. *Chem Pharm Bull (Tokyo).* 1992 Jun;40(6):1366–1375.

152. Hooper D. Gymnemic acid. *J Am Chem Soc.* 1888;10:179–198.

153. Kanetkar PV, Singhal RS, Laddha KS, Kamat MY. Extraction and quantification of gymnemic acids through gymnemagenin from callus cultures of *Gymnema sylvestre*. *Phytochem Anal.* 2006 Nov–Dec;17(6):409–413.

154. Kamble B, Gupta A, Patil D, Janrao S, Khatal L, Duraiswamy B. Quantitative estimation of gymnemagenin in *Gymnema sylvestre* extract and its marketed formulations using the HPLC-ESI-MS/MS method. *Phytochem Anal.* 2013 Feb;24(2):135–140.

155. Devi K, Jain N. A validated HPLC method for estimation of Gymnemic acids as deacyl gymnemic acid in various extracts and formulations of *Gymnema sylvestre*. *Int J Phytomed* 2014:6;165–169.

156. Wynn SG, Fougère BJ, Veterinary herbal medicine: A systems-based approach. Veterinary Herbal Medicine (eds. SG Wynn, BJ Fougère). 2007:291–409 doi: 10.1016/B978-0-323-02998-8.50024-X. Epub 2009 May 15. PMCID: PMC7151902.

157. Brahmachari, G. Andrographolide: A molecule of antidiabetic promise. In: Natural Product Drug Discovery, Discovery and Development of Antidiabetic Agents from Natural Products (ed. G Brahmachari). Elsevier, 2017:1–27.

158. Dev K, Ramakrishna E, Maurya R. Chapter 4 – Glucose transporter 4 translocation activators from nature. In: Natural Product Drug Discovery, Discovery and Development of Antidiabetic Agents from Natural Products (ed. G Brahmachari). Elsevier, 2017:113–145.

159. Chakravarti RN, Chakravarti D. Andrographolide, the active constituent of *Andrographis paniculata* Nees; a preliminary communication. *Ind Med Gaz.* 1951;86:96–97.

160. Rajani M, Shrivastava N, Ravishankara MN. A rapid method for isolation of andrographolide from *Andrographis paniculata* Nees (Kalmegh). *Pharm Biol.* 2000;38(3):204–209.

161. Rao PR, Rathod VK. Rapid extraction of andrographolide from *Andrographis paniculata* Nees by three phase partitioning and determination of its antioxidant activity. *Biocatal Agric Biotechnol.* 2015; 4(4):586–593.

162. Akowuah GA, Zhari I, Mariam A. Analysis of urinary andrographolides and antioxidant status after oral administration of *Andrographis paniculata* leaf extract in rats. *Food Chem Toxicol*. 2008 Dec;46(12):3616–3620.

163. Gu Y, Ma J, Liu Y, Chen B, Yao S. Determination of andrographolide in human plasma by high-performance liquid chromatography/mass spectrometry. *J Chromatogr B: Analyt Technol Biomed Life Sci*. 2007 Jul 1;854(1–2):328–331.

164. Chen L, Yu A, Zhuang X, Zhang K, Wang X, Ding L, Zhang H. Determination of andrographolide and dehydroandrographolide in rabbit plasma by on-line solid phase extraction of high-performance liquid chromatography. *Talanta*. 2007 Nov 15;74(1):146–152.

165. Sheikh N, Kumar Y, Misra AK. Pfoze L. Phytochemical screening to validate the ethnobotanical importance of root tubers of *Dioscorea* species of Meghalaya, North East India. *J Med Plants Stud* 2013;1(6):62–69.

166. Jesus M, Martins AP, Gallardo E, Silvestre S. Diosgenin: Recent highlights on pharmacology and analytical methodology. *J Anal Methods Chem*. 2016;2016:4156293.

167. Irene D. Chapter 14 – Spices and herbs as therapeutic foods. In: Handbook of Food Bioengineering, Food Quality: Balancing Health and Disease (eds. AM Holban, AM Grumezescu). Academic Press, 2018:433–469.

168. Sung B, Prasad S, Gupta SC, Patchva S, Aggarwal BB. Regulation of inflammation-mediated chronic diseases by botanicals. In: Advances in Botanical Research (eds. LF Shyur, ASY Lau), Academic Press, 2012;62:57–132.

169. Li J, Yang D, Yu K, He J, Zhang Y. Determination of diosgenin content in medicinal plants with enzyme-linked immunosorbent assay. *Planta Med*. 2010 Nov;76(16):1915–1920.

170. Król-Kogus B, Lamine KM, Migas P, Boudjeniba M, Krauze-Baranowska M. HPTLC determination of diosgenin in fenugreek seeds. *Acta Pharm*. 2018 Mar 1;68(1):97–107.

171. Theobald RJ. Podophyllotoxin. In: Reference Module in Biomedical Sciences. Elsevier, 2016.

172. Varsha K, Sharma A, Kaur A, Madan J, Pandey RS, Jain UK, Chandra R. Chapter 28 – Natural plant-derived anticancer drugs nanotherapeutics: A review on preclinical to clinical success, In: Micro and Nano Technologies, Nanostructures for Cancer Therapy (eds. A Ficai, AM Grumezescu). Elsevier, 2017:775–809.

173. Hao DC. Chapter 5 – Drug metabolism and disposition diversity of Ranunculales phytometabolites. In: Ranunculales Medicinal Plants (ed. D-C Hao). Academic Press, 2019:175–221.

174. Nadumane V, Venkatachalam P, Gajaraj B. Chapter 19 – Aspergillus applications in cancer research. In: New and Future Developments in Microbial Biotechnology and Bioengineering (ed. VK Gupta). Elsevier, 2016:243–255.

175. Nadumane VK, Venkatachalam P, Gajaraj B. Aspergillus applications in cancer research. In: New and Future Developments in Microbial Biotechnology and Bioengineering (ed. VK Gupta). Elsevier, 2016:243–255.

176. Kumari A, Singh D, Kumar S. Biotechnological interventions for harnessing podophyllotoxin from plant and fungal species: Current status, challenges, and opportunities for its commercialization. *Crit Rev Biotechnol*. 2017;37:739–753.

177. Mishra N, Gupta AP, Singh B, Kaul VK, Ahuja PS. A rapid determination of podophyllotoxin in *Podophyllum hexandrum* by reverse phase high performance thin layer chromatography. *J Liq Chromatogr Related Technol* 2005;28(5): 677–691.

178. Kamal A, Singh M, Ahmad FJ, Saleem K. Ahmad S. A validated HPTLC method for the quantification of podophyllotoxin in *Podophyllum hexandrum* and etoposide in marketed formulation. *Arab J Chem*. 2017;10(2):S2539–S2546.

179. Sharma T, Airao V, Panara N, et al. Solasodine protects rat brain against ischemia/reperfusion injury through its antioxidant activity. *Eur J Pharmacol*. 2014;725:40–46.

180. Lecanu L, Hashim A, McCourty A, et al. The naturally occurring steroid solasodine induces neurogenesis in vitro and in vivo. *Neuroscience*. 2011;183:251–264.

181. Li Y, Chang WQ, Zhang M, et al. Natural product solasodine-3-*O*-D-glucopyranoside inhibits the virulence factors of *Candida albicans*. *FEMS Yeast Res*. 2015;15:1–8.

182. Cui CZ, Wen XS, Cui M, et al. Synthesis of solasodine glycoside derivatives and evaluation of their cytotoxic effects on human cancer cells. *Drug Discov Ther*. 2012;6:9–17.

183. Punjabi S, Cook L, Kersey P, et al. Solasodine glycoalkaloids: A novel topical therapy for basal cell carcinoma. A double-blind, randomized, placebo-controlled, parallel group, multicenter study. *Int J Dermatol*. 2008;47:78–82.

184. Cham BE, Chase TR. Solasodine rhamnosyl glycosides cause apoptosis in cancer cells. Do they also prime the immune system resulting in long-term protection against cancer? *Planta Med.* 2012;8:349–353.

185. Xu XH, Zhang LL, Wu GS, et al. Solasodine induces apoptosis, affects autophagy, and attenuates metastasis in ovarian cancer cells. *Planta Med.* 2017;83:254–260.

186. Lecanu L, Hashim AI, McCourty A, et al. The naturally occurring steroid solasodine induces neurogenesis in vitro and in vivo. *Neuroscience.* 2011;183:251–264.

187. Claeson UP, Malmfors T, Wikman G, Bruhn JG. *Adhatoda vasica*: A critical review of ethnopharmacological and toxicological data. *J Ethnopharmacol.* 2000 Sep;72(1–2):1–20.

188. Sharma S, Kumar M, Kumar V, Kumar N. Vasicine catalyzed direct C–H arylation of unactivated arenes: Organocatalytic application of an abundant alkaloid. *Tetrahedr Lett.* 2013:54(36):4868–4871.

189. Kuete V. 21 – Health effects of alkaloids from African Medicinal Plants. In: Toxicological Survey of African Medicinal Plants (ed. V Kuete). Elsevier, 2014:611–633.

190. Rowinsky E. The vinca alkaloids. In: Holland-Frei Cancer Medicine (Sixth Edition) (eds. DW Kufe, RE Pollock, RR Weichselbaum, et al.). BC Decker, Hamilton, ON; 2003.

191. Zhou XJ, Rahmani R. Preclinical and clinical pharmacology of vinca alkaloids. *Drugs.* 1992;44 Suppl 4:1–16; discussion 66-9.

192. Rathore R, Weitberg AB. Vinca alkaloids and epipodophyllotoxins. In: Encyclopedia of cancer (Second Edition) (ed. JR Bertino). Academic Press, 2002:509–512.

193. Stevens GHJ, Lizbeth R. Neuromuscular complications. In: Supportive oncology (eds. MP Davis, PC Feyer, P Ortner, C Zimmermann). W.B. Saunders, 2011:283–291.

194. Postma TJ, Heimans JJ. Neurological complications of chemotherapy to the peripheral nervous system. In: Handbook of Clinical Neurology (eds. W Grisold, R Soffietti). Elsevier, 2012;105:917–936.

195. Montgomery B, Lin DW. Chapter 10: Toxicities of chemotherapy for genitourinary malignancies. In: Complications of Urologic Surgery (Fourth Edition) (ed. SS Taneja). W.B. Saunders, 2010:117–123.

196. Waller DG, Sampson AP. Chemotherapy of malignancy. In: Medical Pharmacology and Therapeutics (Fifth Edition) (eds. DG Waller, AP Sampson). Elsevier, 2018:631–656.

197. Stevens GHJ, Robles L. Neuromuscular complications. In: Supportive Oncology (eds. MP Davis, PC Feyer, P Ortner, C Zimmermann). W.B. Saunders, 2011:283–291.

198. Newton HB. Neurological complications of chemotherapy to the central nervous system. In: Riccardo Soffietti, Handbook of Clinical Neurology (ed. W Grisold). Elsevier, 2012:105:903–916.

199. Alexa-Stratulat T, Luca A, Bădescu M, Bohotin CR, Alexa ID. Nutritional modulators in chemotherapy-induced neuropathic pain. In: Nutritional Modulators of Pain in the Aging Population (eds. RR Watson, S Zibadi). Academic Press, 2017:9–33.

200. Kuete V. Health effects of alkaloids from African medicinal plants. In: Toxicological Survey of African Medicinal Plants (ed. V Kuete). Elsevier, 2014:611–633.

201. Rahmani R, Zhou XJ. Pharmacokinetics and metabolism of vinca alkaloids. *Cancer Surv.* 1993;17:269–281.

202. Moudi M, Go R, Yien CY, Nazre M. Vinca alkaloids. *Int J Prev Med.* 2013 Nov;4(11):1231–1235.

203. Arora RD, Menezes RG. Vinca alkaloid toxicity. 2020 Sep 20. In: StatPearls [Internet]. StatPearls Publishing, Treasure Island, FL; 2021 Jan–. PMID: 32491774.

204. Kukula-Koch WA, Widelski J. Chapter 9: Alkaloids In: Pharmacognosy (eds. S Badal, R Delgoda). Academic Press, 2017:163–198.

205. Evans WC, Evans D. Alkaloids. In: Trease and Evans' Pharmacognosy (Sixteenth Edition) (eds. WC Evans, D Evans). W.B. Saunders, 2009:353–415.

206. Gosselin RE, Smith RP, Hodge HC. Clinical Toxicology of Commercial Product (Fifth Edition). Williams & Wilkins, Baltimore/London; 1984;II-249, III-375-379.

207. Izawa K, Amino Y, Kohmura M, Ueda Y, Kuroda M. Human–environment interactions – Taste. In: Comprehensive Natural Products II (eds. H-W (Ben) Liu, L Mander). Elsevier, 2010:631–671.

208. Merck Index Rahway (Eleventh Edition). Merck & Co., Inc., New Jersey, USA; 1989:201.

209. O'Callaghan WG, Joyce N, Counihan HE, Ward M, Lavelle P, O'Brien E. Unusual strychnine poisoning and its treatment: report of eight cases. *Br Med J (Clin Res Ed)* 1982;285 (6340):478.

210. Makarovsky I, Markel G, Hoffman A, Schein O, Brosh-Nissimov T, Tashma Z, Dushnitsky T, Eisenkraft A. Strychnine—A killer from the past. *Isr Med Assoc J.* 2008;10(2):142–145.

211. Dittrich K, Bayer MJ, Wanke LA. A case of fatal strychnine poisoning. *J Emerg Med.* 1984;1(4):327–330.

212. Chen J, Hu W, Qu YQ, Dong J, Gu W, Gao Y, Fang Y, Fang F, Chen ZP, Cai BC. Evaluation of the pharmacodynamics and pharmacokinetics of brucine following transdermal administration. *Fitoterapia.* 2013;86:193–201.

213. Dorandeu F, Calas G, Dal Bo G, Fares R. Models of chemically-induced acute seizures and epilepsy: Toxic compounds and drugs of addiction. In: Models of Seizures and Epilepsy (Second Edition) (eds. A Pitkänen, PS Buckmaster, AS Galanopoulou, SL Moshé). Academic Press, 2017:529–551.

214. Velíšková J, Shakarjian MP, Velíšek L. Chapter 34 – Systemic chemoconvulsants producing acute seizures in adult rodents. In: Models of Seizures and Epilepsy (Second Edition) (eds. A Pitkänen, PS Buckmaster, AS Galanopoulou, SL Moshé). Academic Press, 2017:491–512.

215. Rivera HL, Barrueto F. Strychnine. In: Encyclopedia of Toxicology (Third Edition) (ed. P Wexler). Academic Press, 2014:407–408.

216. Aronson JK. Strychnine: Meyler's Side Effects of Drugs (Sixteenth Edition). Elsevier, 2016:500.

217. Chen J, Yan GJ, Hu RR, Gu QW, Chen ML, Gu W, Chen ZP, Cai BC. Improved pharmacokinetics and reduced toxicity of brucine after encapsulation into stealth liposomes: Role of phosphatidylcholine. *Int J Nanomedicine*. 2012;7:3567–3577.

218. Guo R, Wang T, Zhou G, Xu M, Yu X, Zhang X, Sui F, Li C, Tang L, Wang Z. Botany, phytochemistry, pharmacology and toxicity of *Strychnos nux-vomica* L.: A review. *Am J Chin Med*. 2018;46(1):1–23.

219. Wormley T. Micro-Chemistry of Poisons Including Their Physiological, Pathological, and Legal Relations: Adapted to the Use of the Medical Jurist, Physician, and General Chemist. W. Wood, New York, NY; 1869.

220. Buckingham J. Bitter Nemesis: The Intimate History of Strychnine. CRC Press, 2007:225.

221. Pelletier PP, Caventou JB. "Note sur un nouvel alkalai" [Note on a new alkali]. *Annales de Chimie et de Physique (in French)* 1818;8:323–324.

222. Pelletier PP, Caventou JB. "Mémoire sur un nouvel alcali vegetal (la strychnine) trouvé dans la feve de Saint-Ignace, la noix vomique, etc" [Memoir on a new vegetable alkali (strychnine) found in the St. Ignatius bean, the nux-vomica, etc)]. *Annales de Chimie et de Physique (in French)*. 1819;10:142–176.

223. Tang H, Zhang L, Li X, Sun C, Ma P. The experimental study of different solvent extracting total alkaloid from semen strychni. *Guiding J Tradit Chin Med Pharm*. 2009;15(11):52–53. doi: 10.13862/j.cnki.cn43-1446/r.2009.11.039

224. Peter KV Handbook of Herbs and Spices. Woodhead Publishing, Sawston, UK; 2006.

225. Agarwal OP. Chemistry of Organic Natural Products. Goel Publishing House, Meerut, India; 2010.

226. Izawa K, Amino Y, Kohmura M, Ueda Y, Kuroda M. Human–environment interactions – Taste. In: Comprehensive Natural Products II (eds. H-W (Ben) Liu, L Mander). Elsevier, 2010:631–671.

227. Kumar S, Bhandari C, Sharma P, Agnihotri N. Role of piperine in chemoresistance. In: Cancer Sensitizing Agents for Chemotherapy, Role of Nutraceuticals in Cancer Chemosensitization (eds. AC Bharti, BB Aggarwal). Academic Press, 2018;2:259–286.

228. Srinivasan K. Spices and flavoring crops: Uses and health effects, Encyclopedia of Food and Health, (eds. B Caballero, PM Finglas, F Toldrá). Academic Press, 2016:98–105.

229. Gorgani L, Mohammadi M, Najafpour GD, Nikzad M. Piperine: The bioactive compound of black pepper: From isolation to medicinal formulations. *Compr Rev Food Sci Food Saf*. 2017;16:124–140.

230. Tjio JH, Levan A. The chromosome number of man. *Hereditas*. 1956;42:1–6.

231. Terkeltaub R. Management of gout and hyperuricemia. In: Rheumatology (Sixth Edition) (eds. MC Hochberg, AJ Silman, JS Smolen, ME Weinblatt, MH Weisman). Mosby, 2015:1575–1582.

232. Roubille F, Kritikou E, Busseuil D, et al. Colchicine: An old wine in a new bottle? *Antiinflamm Antiallergy Agents Med Chem*. 2013;12:14–23.

233. Slobodnick A, Shah B, Pillinger MH, et al. Colchicine: Old and new. *Am J Med*. 2015;128:461–470.

234. Komlodi-Pasztor E, Sackett DL, Fojo AT. Inhibitors targeting mitosis: Tales of how great drugs against a promising target were brought down by a flawed rationale. *Clin Cancer Res*. 2012;18:51–63.

235. Dalbeth N, Lauterio TJ, Wolfe HR. Mechanism of action of colchicine in the treatment of gout. *Clin Ther*. 2014;36:1465–1479.

236. Cronstein BN, Molad Y, Reibman J, et al. Colchicine alters the quantitative and qualitative display of selectins on endothelial cells and neutrophils. *J Clin Invest*. 1995;96:994–1002.

237. Asako H, Kubes P, Baethge BA, et al. Colchicine and methotrexate reduce leukocyte adherence and emigration in rat mesenteric venules. *Inflammation*. 1992;16:45–56.

238. Chia EW, Grainger R, Harper JL. Colchicine suppresses neutrophil superoxide production in a murine model of gouty arthritis: A rationale for use of low-dose colchicine. *Br J Pharmacol*. 2008;153:1288–1295.

239. Marques-da-Silva C, Chaves MM, Castro NG, et al. Colchicine inhibits cationic dye uptake induced by ATP in P2X2 and P2X7 receptor-expressing cells: Implications for its therapeutic action. *Br J Pharmacol*. 2011;163:912–926.

240. Spiller HA, Colchicine. In: Encyclopedia of Toxicology (Third Edition) (ed. P Wexler). Academic Press, 2014:1007–1008.

241. Karamanou M, Tsoucalas G, Pantos K, Androutsos G. Isolating colchicine in 19th century: An old drug revisited. *Curr Pharm Des.* 2018;24(6):654–658.
242. Pelletier PJ, Caventou J. Examen chimique de plusieurs végétaux de la famille des colchicées et du principe actif qu'ils renferment. *Ann de Chim et de Phy.* 1820;14:69–83.
243. Geiger PL. Ueber einige neue giftige organische Alkalien. *Ann Pharm* 1833;7:269–280.
244. King MV, deVries JL, Pepinsky R. An x-ray diffraction determination of the chemical structure of colchicine. *Acta Cryst.* 1952;5:437–440.
245. Dewar MJS. Structure of colchicine. *Nature.* 1945;155:141–142.
246. Jurenka JS. Anti-inflammatory properties of curcumin, a major constituent of *Curcuma longa*: A review of preclinical and clinical research. *Altern Med Rev.* 2009 Jun;14(2):141–153. Erratum in: Altern Med Rev. 2009 Sep;14(3):277.
247. Nelson KM, Dahlin JL, Bisson J, Graham J, Pauli GF, Walters MA. The essential medicinal chemistry of curcumin. *J Med Chem.* 2017 Mar 9;60(5):1620–1637.
248. Prasad S, Aggarwal BB. Turmeric, the golden spice: From traditional medicine to modern medicine. In: Herbal Medicine: Biomolecular and Clinical Aspects (Second Edition) (eds. IFF Benzie, S Wachtel-Galor). CRC Press/Taylor & Francis, Boca Raton, FL; 2011. Chapter 13. Available from: https://www.ncbi.nlm.nih.gov/books/NBK92752/
249. Lestari ML, Indrayanto G. Curcumin. Profiles drug subst. *Excip Relat Methodol.* 2014;39:113–204.
250. Mahady GB, Pendland SL, Yun G, Lu ZZ. Turmeric (*Curcuma longa*) and curcumin inhibit the growth of *Helicobacter pylori*, a group 1 carcinogen. *Anticancer Res.* 2002;22:4179–4181.
251. Reddy RC, Vatsala PG, Keshamouni VG, Padmanaban G, Rangarajan PN. Curcumin for malaria therapy. *Biochem Biophys Res Commun.* 2005;326:472–474. doi: 10.1016/j.bbrc.2004.11.051
252. Vera-Ramirez L, Perez-Lopez P, Varela-Lopez A, Ramirez-Tortosa M, Battino M, Quiles JL. Curcumin and liver disease. *Biofactors.* 2013;39:88–100. doi: 10.1002/biof.1057
253. Wright LE, Frye JB, Gorti B, Timmermann BN, Funk JL. Bioactivity of turmeric-derived curcuminoids and related metabolites in breast cancer. *Curr Pharm Des.* 2013;19:6218–6225.
254. Anand P, Kunnumakkara AB, Newman RA, Aggarwal BB. Bioavailability of curcumin: Problems and promises. *Mol Pharm.* 2007;4:807–818.
255. Han HK. The effects of black pepper on the intestinal absorption and hepatic metabolism of drugs. *Expert Opin Drug Metab Toxicol.* 2011;7:721–729.
256. Shoba G, Joy D, Joseph T, Majeed M, Rajendran R, Srinivas PS. Influence of piperine on the pharmacokinetics of curcumin in animals and human volunteers. *Planta Med.* 1998;64:353–356.
257. Anand P, Kunnumakkara AB, Newman RA, Aggarwal BB. Bioavailability of curcumin: Problems and promises. *Mol Pharm.* 2007;4:807–818.
258. Han HK. The effects of black pepper on the intestinal absorption and hepatic metabolism of drugs. *Expert Opin Drug Metab Toxicol.* 2011;7:721–729.
259. Shoba G, Joy D, Joseph T, Majeed M, Rajendran R, Srinivas PS. Influence of piperine on the pharmacokinetics of curcumin in animals and human volunteers. *Planta Med.* 1998;64:353–356.
260. Basnet P, Skalko-Basnet N. Curcumin: An anti-inflammatory molecule from a curry spice on the path to cancer treatment. *Molecules.* 2011;16:4567–4598.
261. Lao CD, Ruffin MT 4th, Normolle D, Heath DD, Murray SI, Bailey JM, Boggs ME, Crowell J, Rock CL, Brenner DE. Dose escalation of a curcuminoid formulation. *BMC Complement Altern Med.* 2006 Mar 17;6:10.
262. Vogel A, Pelletier J. Examen chimique de la racine de Curcuma. *J Pharm.* 1815;1:289–300.
263. Daube FW. Ueber Curcumin, den Farbstoff der Curcumawurzel. *Berichte Der Deutschen Chem Gesellschaft* 1870;3:609–613.
264. Miłobdzka J, Kostanecki V, Lampe V. Zur Kenntnis des Curcumins. *Berichte Der Deutschen Chem Gesellschaft* 1910;43:2163–2170.
265. Wiegrebe W, Kramer WJ, Shamma M. The emetine alkaloids. *J Nat Prod.* 1984;47:397–408.
266. [No authors listed]. Position paper: Ipecac syrup. *J Toxicol Clin Toxicol.* 2004;42(2):133–143.
267. Kushlaf HA. Toxic myopathies. In: Clinical Neurotoxicology (ed. MR Dobbs). W.B. Saunders, 2009:159–173.
268. Scholar E. Emetine. In: xPharm: The Comprehensive Pharmacology Reference (eds. SJ Enna, DB Bylund). Elsevier, 2009:1–4.
269. Dabbous H, Shokouh-Amiri H, Zibari GB. Chapter 73 – Amebiasis and other parasitic infections. In: Jarnagin, Blumgart's Surgery of the Liver, Biliary Tract and Pancreas, 2-Volume Set (Sixth Edition) (ed. R William). Elsevier, 2017:1083–1101.e5.

270. Aronson JK. Ipecacuanha, emetine, and dehydroemetine. In: Meyler's Side Effects of Drugs (Sixteenth Edition) (ed. JK Aronson). Elsevier, 2016:311–312,

271. Mukherjee K, Biswas R, Chaudhary SK, Mukherjee PK. Chapter 18: Botanicals as medicinal food and their effects against obesity. In: Mukherjee, Evidence-Based Validation of Herbal Medicine (ed. Pulok K). Elsevier, 2015:373–403.

272. Shishodia S, Harikumar KB, Dass S, Ramawat KG, Aggarwal BB. The guggul for chronic diseases: Ancient medicine, modern targets. *Anticancer Res.* 2008 Nov–Dec;28(6A):3647–3664.

273. Deng R. Therapeutic effects of guggul and its constituent guggulsterone: Cardiovascular benefits. *Cardiovasc Drug Rev.* 2007 Winter;25(4):375–390.

274. Singh SP, Sashidhara KV. Lipid lowering agents of natural origin: An account of some promising chemotypes. *Eur J Med Chem.* 2017;140:331–348.

275. Upadhyay J, Durgapal S, Jantwal A, Kumar A, Rana M, Tiwari N. Chapter 3.2.10 – Guggulu [*Commiphora wightii* (Arn.) Bhandari.]. In: Naturally Occurring Chemicals Against Alzheimer's Disease (eds. T Belwal, SM Nabavi, SF Nabavi, AR Dehpour, S Shirooie). Academic Press, 2021:317–328.

276. Gabriel KM. Herbal, dietary supplements, and cardiovascular disease. In: Encyclopedia of Heart Diseases (ed. KM Gabriel). Academic Press, 2006:453–461.

277. Witkamp RF. Biologically active compounds in food products and their effects on obesity and diabetes. In: Comprehensive Natural Products II (eds. H-W (Ben) Liu, L Mander). Elsevier, 2010:509–545.

278. Wu J, Xia C, Meier J, Li S, Hu X, Lala DS. The hypolipidemic natural product guggulsterone acts as an antagonist of the bile acid receptor. *Mol Endocrinol.* 2002;16(7):1590–1597.

279. Srinivasan S, Vinothkumar V, Murali R. Chapter 22 – Antidiabetic efficacy of citrus fruits with special allusion to flavone glycosides. In: Bioactive Food as Dietary Interventions for Diabetes (Second Edition) (eds. RR Watson, VR Preedy). Academic Press, 2019:335–346.

280. Nagasako-Akazome Y. Chapter 58 – Safety of high and long-term intake of polyphenols, Sherma Zibadi, Polyphenols in Human Health and Disease (eds. RR Watson, VR Preedy). Academic Press, 2014:747–756.

281. Hwang SL, Shih PH, Yen GC, Chapter 80 – Citrus flavonoids and effects in dementia and age-related cognitive decline. In: Diet and Nutrition in Dementia and Cognitive Decline (eds. CR Martin, VR Preedy). Academic Press, 2015:869–878.

282. Izawa K, Amino Y, Kohmura M, Ueda Y, Kuroda M. Human–environment interactions – Taste. In: Comprehensive Natural Products II (eds. H-W (Ben) Liu, L Mander). Elsevier, 2010:631–671.

283. Benowitz NL. Pharmacology of nicotine: Addiction, smoking-induced disease, and therapeutics. *Annu Rev Pharmacol Toxicol.* 2009;49:57–71.

284. Hughes B, Nicotine. In: Encyclopedia of Toxicology (Second Edition) (ed. P Wexler). Elsevier, 2005:225–228.

285. Davis RA, Curvali M. Determination of nicotine and its metabolites in biological fluids: In vivo studies. In: Analytical Determination of Nicotine and Related Compounds and their Metabolites (eds. JW Gorrod, P Jacob). Elsevier Science, 1999:583–643.

286. Mukherjee PK. Bioactive phytocomponents and their analysis. In: Quality Control and Evaluation of Herbal Drugs (ed. PK Mukherjee). Elsevier, 2019:237–328.

287. Umit Sayin H. Psychoactive plants used during religious rituals. In: Neuropathology of Drug Addictions and Substance Misuse (ed. VR Preedy). Academic Press, 2016:17–28.

288. Basu D, Ghosh A, Sarkar S. Addictions in India. In: Neuropathology of Drug Addictions and Substance Misuse (ed. VR Preedy). Academic Press, 2016:1025–1035.

289. Fields TJ. Wiegand, opium and the constituent opiates. In: Encyclopedia of Toxicology (Third Edition) (ed. P Wexler). Academic Press, 2014:698–701.

290. Banys P. Substance abuse. In: Encyclopedia of the Neurological Sciences (Second Edition) (eds. MJ Aminoff, RB Daroff). Academic Press, 2014:337–346.

291. Zheng H, Law PY, Loh HH. Opioid Receptors Reference Module in Life Sciences. Elsevier, 2020.

292. Ferdousi M, Finn DP. Chapter 4 – Stress-induced modulation of pain: Role of the endogenous opioid system. In: Progress in Brain Research (ed. S O'Mara). Elsevier, 2018;239:121–177.

# 3 Plant Profile, Phytochemistry, and Ethnopharmacological Uses of Ashoka, Ashwagandha, and Amla

*Ahmed Al-Harrasi*
Natural and Medical Sciences Research Center, University of Nizwa

*Saurabh Bhatia*
Natural and Medical Sciences Research Center, University of Nizwa
School of Health Sciences, University of Petroleum and Energy Studies

*Sridevi Chigurupati*
Qassim University

*Tapan Behl*
Chitkara University

## 3.1 INTRODUCTION

Traditional concepts of medicine are mainly based on plants and their extracts to relieve and treat diseases. These concepts in the form of medicines have been used since prehistoric times. Utilization of herbal preparation was first promoted by Egyptians (1500 BC) and further developed by Greeks and Romans; however, later on, Ayurveda and Traditional Chinese Medicine (TCM) became more popular. The approach of preparing medicines from plants and plant extracts using traditional concepts is known as herbalism, and an expert who practises these concepts of using plants for healing is known as a herbalist. Medicines derived from plants are named phytomedicines or herbal medicines, and phytotherapy is a process of using plant or plant-based products to alleviate diseases. Human civilization was solely reliant on plant-based products to increase strength, longevity and cure different ailments before the discovery of aspirin from *Spiraea ulmaria*. This plant was already prescribed in ancient Greek medicine for the treatment of pain and fever. Also, the same plant is documented in Egyptian papyri for fever and swelling [1]. Even with the development and wide access to the most modern or synthetic medications, the utilization of traditional medicines has been revived as they are relatively considered safe and healthier than modern medications. However, the challenge associated with the plant-based products is multicomponent system, which causes hindrance in establishing the exact mechanism of action, identifying toxic elements, components interactions with each other, and physiological interactions. This factor also influences the clinical development of plant-based products. However, development of antimicrobial-resistant strains, long-term side effects or toxicity profile, and cost constrains of synthetic medication always encouraged the utilization of synthetic medicines. Evidences also revealed that Sumerians (c. 4100–1750 BCE) 5000 years ago recognized therapeutic uses of various plants such as laurel, caraway, and thyme [2]. Based on archaeological studies, it was revealed that 60,000 years ago herbal medicine was practised in Iraq and 8000 years ago in China [3, 4]. Still almost 8% of the population across the globe are relying on the folk medicine, which is consistently practised across different parts of the world. Traditional systems of medicine are classified as per their country of origin and prevalent utilization across different countries such as Ayurvedic system in India, Amachi in Tibet, Unani system in Greece, TCM in China, Homoeopathy in Germany, Sa-sang Traditional Medicine in Korean, and Kampo Traditional Medicine in Japan. Recently, WHO has also been taking more interest in traditional systems of medicine and advancing scientific validation of these medicines. In 2005, based on the first worldwide investigation on traditional and complimentary medicines, WHO issued a report on national guidelines on traditional medicines and regulation of herbal medicines. In a report published in 2019, WHO disclosed that 88% of the Member States have acknowledged their use of T&CM, which corresponds to 170 Member States [5]. Thus, a resumption of interest in these traditional systems especially when we have immediate access to synthetic drugs conveys that both the systems are equally important for human race and require more attention with scientific validation. After the arrival of modern synthetic medicine, herbal medicines have been criticized by allopathic practitioners as these traditional formulations despite their long history of effective use lack scientific evidence. However, due to the revival of interest in using folk medicines, toxicities and side effects are associated with synthetic drugs, emergence of antimicrobial-resistant strains, extraordinary expenditure in development of synthetic drug (especially during clinical trials, in research and development, formulation, and development) [6]. Since 1950, merely about 1200 new drugs have been sanctioned by the US FDA [7]. Now many pharmaceutical firms have started working on plant-based products and changed their policies in context of development and discovery of plant-based natural products [8–11]. Additionally, several allopathic practitioners prescribe plant-based products to the patients as an alternative treatment apart from prescribed main course medicine for the effective treatment of certain diseases such as cancer. Recently, it was found that complementary and alternative medicines (CAMs) and plant-based medicines are more frequently used by patients with higher levels of education and income. In spite of being familiar with this fact, the following questions arise:

- How these traditional formulations prevent or treat diseases?
- What is the scientific basis of their mechanism of action?
- What is the clarity related with toxicity profile and exact chemical composition?

Most of the communities or patients would like to continue with these formulations as they have been used from ages with proven records of their safety and therapeutic profiles.

### 3.1.1 Traditional Systems

Natural product science has emerged as a more pragmatic science relating to the learning of different types of natural products. Natural products can be utilized alternatively in the place of standard allopathic medicine, whereas complementary medications are used in conjunction with standard medical treatment. Alternative medicine is an approach to offer the therapeutic effects of medicine derived from natural resources; however, it lacks scientific evidence of its effectiveness. Thus, these are alternatives to standard medications and deprived of scientific explanation and evidence. Terms like folk medicines, herbal medicines, holistic medicines, complementary medicines (CM), integrated or integrative medicines (IM), and CAM are the part of alternative medicines. These alternative systems in diverse forms are prevalent and practised differently in various parts of the world. Basic principles and the procedures utilized among various alternative systems of medicine differ significantly from each other; however, objectives of treatment and management of diseases to maintain good health remain the same. Complementary and alternative healthcare and medical practices (CAM) are currently not considered a part of conventional medicine, however, involve diverse systems, therapeutic and well-being approaches as well as products. CAM complements mainstream medicine and also helps in diagnosis, treatment, and/or prevention of medicines. It involves a large and varied range of procedures, with both therapeutic and diagnostic approaches such as spiritual healing, reflexology, massage, iridology, homoeopathy, herbalism, chiropractic, chelation therapy, aromatherapy, and acupuncture.

### 3.1.2 ROLE OF CAM

There is criticism associated with the definition of CAM; thus, CAM doesn't have any standard definition. To differentiate CAM with alternative medicines, US National Center for Complementary and Integrative Health (NCCIH) provided separate definitions for the two terms. CAM is considered when non-mainstream practice is used together with standard medicine, whereas alternative system of medicine is considered when non-mainstream practice is employed as a replacement for standard allopathic medicine. However, these two therapies in certain cases overlap with each other as aromatherapy can occasionally be considered a complementary therapy, and in other situations, it is considered an alternative therapy. Clinical utilization or conventional medicine practice is regulated by rules that guarantee that physicians are suitably capable, skilled, trained, and well qualified and follow all the ethics of practice.

CAM contributes towards "vitalism", which suggests that all creatures are controlled by a vigorous energy [12]. CAM is defined as any procedure planned to improve or sustain human well-being which doesn't involve standard allopathic medication. Out of these procedures, some are occasionally known as holistic or traditional medicine; however, these approaches don't include all forms of CAM. Additionally, CAM consists of not only procedures based on traditional systems, but also it includes other natural therapies aimed at bringing together the mind, body, and spirit such as:

- art therapy
- biofeedback
- chiropractic medicine
- hypnosis
- prayer
- specialty diets
- therapeutic touch

The reasons of sudden increase in need for CAM are not known. However, it was assumed that it is accepted due to more stress on long-term ill health, ageing, and lifestyle-associated illness (rather than acute illness). Especially, CAM plays an important role in those cases where standard medications may not be available or less or not effective, for an instance, the use of acupuncture for chronic pain [13]. A wide acceptance of CAM is reported in Australia where it's widely used as an essential component of healthcare sector. It was found that 42% nationals use CAM treatments. During the year 2000, an investigation based on South Australia demonstrated that Australians invested nearly US$2.3 billion on alternative treatment where this investment increased up to 62% from 1993. It was also found that United States and Great Britain are following the same patterns [14–16]. Elderly population with lifestyle-based complications and those suffering from acute illness can use CAM. These days, various essential oils have been used for various therapeutic applications; thus, the therapeutic strength of essential oils can also be utilized to fight against various ailments. Apart from the recent evolution in crude drugs, various formulations based on phytochemicals as well as crude drugs, especially by using natural polymers, have been reported. Such cell- or tissue-targeted formulations have been utilized these days to treat various ailments [17–37]. In this chapter, we listed various crude drugs that have been used traditionally for a very long period. Because of their ethnomedicinal importance, these crude drugs have been evolved further in herbal drug industry [38–76].

## 3.2 ASHOKA

### 3.2.1 Vernacular Names

It is commonly called *Ashoka*, Sorrowless Tree, Asoka Tree, Ashoka Tree in India it's called by different names such as *Achenge*, *Alshth*, *Ashoka* (Hindi, Bengali, Gujarati, Marathi), *Ashokadamara* (Kannada), Ashokmu, Asogam (Tamil), *Asokam* (Malayalam), *Karkeli* (Sanskrit), *Kenkalimara*, *Oshok* (Bengali), *Sita ashok*, *Sita ashoka*, *Vanjulamu* (Telugu).

### 3.2.2 Biological Source

Ashoka bark is obtained from the *Saraca asoca* (Roxb.) De Wilde (*S. asoca*) and is one of the most important commercially viable medicinally active, perennial, evergreen tree, and ornamental plants extensively used in Ayurveda and traditional systems of medicine. Almost 20 plant species belong to this flowering Genus, *Saraca* out of which, 4 species (*S. thaipingensis, S. asoca, S. declinate*, as well as *S. indica*) are from India [77].
   **Family:** Caesalpiniacea

### 3.2.3 Geographical Distribution

*S. asoca* also known as *S. indica* is one of the most antique holy plants, extensively present across India. *S. asoca* is native to the Western Ghats of India where it is present in the hills of central and eastern Himalayas (up to an elevation of about 750 m), west coast of the subcontinent and northern plains. It's mainly found in Burma, Ceylon, Malaya, East Bengal, and Western Peninsula. Except India, it's widely cultivated in Sri Lanka where it's also traditionally used for the treatment of gynaecological disorders [78].

### 3.2.4 Chemical Composition

Extracts derived from *S. asoca* stem bark chemically contain flavonoids, lignin, phenolic compounds, tannins, and terpenoid. The previous report suggested the presence of main compounds such as (–)-epicatechin, (+)-catechin, leucocyanidin, and procyanidin B-2, 11′-deoxyprocyanidin B4, in *S. asoca*. However, these phytochemicals have not been identified in *Polyalthia longifolia*, while other chemical components, including clerodane diterpenoids and duorene alkaloids, have been found [79].
   Chemical compounds such as multiple phenolic functionalities, including (–) 3-*O*-D-glucoside, (–)-procyanidin derivatives, 11′-deoxyprocyanidin B, (–) epicatechol, amyrin apigenin-D-glucoside, catechin, catechol, ceryl alcohol cyanidin-3,5-diglucoside, epicatechin, gallic acid, kaempferol 3-*O*-β-D-glucoside, kaempferol-3-*O*-α-L-rhamnoside, leucocyanidin, leucopelargonidin, methyl- and ethyl-cholesterol derivatives, *n*-octacosanol, pelargonidin diglucoside, procyanidin B-2, quercetin as well as its D-glucoside, quercetin and β-sitosterol, were characterized from *S. asoca*. Among all, gallic acid and epicatechin are considered major phytochemicals of *S. asoca* (bark), with broad therapeutic activities.
   Five lignan glycosides, 5-methoxy-9-βxylopyranosyl-(–)-isolariciresinol, epiafzelechin-(4β→8)-epicatechin and icariside E3, lyoniside, nudiposide, procyanidin B2, schizandriside, three flavonoids, (–)-epicatechin, β-sitosterol glucoside were isolated from dried bark [80].

### 3.2.5 Common Uses

Ashoka trees, including their all parts such as flowers, leaves, roots, and seeds, mainly bark, have been traditionally used for several health benefits as a medicine in various ailments and cures. Since the bark is enriched with a high level of alkaloids, tannin, catechol, flavanoides, glucosides, sterol, and other organic calcium compounds [81], it's used as female tonic to treat various gynaecological ailments, particularly in the management of menstrual imbalances associated with abdominal pain, congestion, dysmenorrhoea, excessive bleeding, pain, and uterine spasms.

Bark is also used for vaginal discharge due to oestrogen imbalance called leucorrhoea and has stimulative and astringent effects on an innermost lining layer of the uterus and ovarian tissues [82, 83]. The bark is used to cure indigestion, intestinal inflammation, piles, sores, and menstruation imbalances. Other parts of the tree such as flowers are used for the treatment of diabetes, sexually transmitted infections such as syphilis, and dysentery. It is used as a natural blood purifier and prevents dermal allergies.

### 3.2.6 TRADITIONAL USES

The word Ashoka is a Sanskrit word that means without sorrow or sorrow-less, which itself indicates the medicinal role of a tree, i.e., signifying a role of bark in keeping a woman healthy and young [84, 85]. Since ancient time, Ashoka tree has been recognized as a sacred tree mainly by Hindu and Buddhist communities, and due to its more medicinal properties, Ashoka tree has been traditionally utilized in the gynaecological disorders.

A complete ancient Indian text on medicine called Charaka Samhita (100 AD) in the Vedanasthapan (pain killer, reducing inflammation as well as fever) category also reported several therapeutic uses of *S. indica* [86]. Not only bark, but also other parts, possesses various medicinal activities such as antihyperglycaemic, antipyretic, antibacterial, anthelmintic, activity, and so forth, which are well described in the literature [87]. Stem bark collection time of Ashoka has been already specified in the ancient Indian literature that should be in Sharada Ritu (Ashwin-Kartik).

### 3.2.7 PHARMACOLOGICAL USES

Tannins-enriched bark of Ashoka offers astringent property for ceasing excessive menstrual bleeding, bleeding haemorrhoids, bleeding ulcers, and haemorrhagic dysentery.

So far, different parts of Ashoka have been reported for diverse pharmacological activities such as analgesic, antibacterial, anticancer, antihyperglycaemic, antimutagenic, antioxidant, antipyretic, anxiolytic, central nervous system (CNS) depressant, genoprotective effect, and molluscicidal. Additionally, plant materiala have also been utilized for ammenorhoea, clots, CNS performance, genito-urinary functions, skin infections as well as pain in uterus during menstrual cycle [88].

### 3.2.8 TRADITIONAL FORMULATIONS

Ashoka bark is a major component in numerous old marketed Ayurvedic preparations like Ashokarishta, Ashokaghrita, and Ashoka Kwath; thus, it is greatly utilized by the herbal and pharmaceutical industry [89].

### 3.2.9 MECHANISM OF ACTION

It was found that aqueous and alcoholic fractions obtained from *S. asoca* (bark) contain phenolic as well as nonphenolic glycosides that are responsible for stimulatory effect on isolated human uterus [90]. Furthermore, it was revealed that fraction containing phenolic glycosides was found to be active on uterus; however, it was ineffective on CNS, cardiovascular (CVS), and smooth muscles [90].

An earlier report suggested that gallic acid is present in *S. asoca* suppressed high-fat diet-induced elevation in plasma cholesterol, triglycerides, or both, or a low high-density lipoprotein (HDL) cholesterol level, fatty liver disease, oxidative stress in rats and exhibits several other pharmacological uses such as neuroprotective, antimicrobial, and anticancer properties. It was also found that epicatechin possesses insulin-like activity and prevents cellular damage against low-density lipoprotein (LDL) (oxidized) by reducing oxidative stress and retaining NO synthase [91]. Another study also demonstrated that by suppressing pancreatic cholesterol esterase, lowering solubility of cholesterol in micelles, and binding of bile acids, both gallic acid and epicatechin showed cholesterol-lowering activity [91].

It was also found that *S. asoca* (seeds extract) showed the same mechanism of action like indomethacin by blocking the biological enzymes, which is responsible for COX (-I, -II) production thus decreases the level of prostaglandins. It's because of the presence of flavonoids to possess anti-inflammatory, analgesic, and antipyretic effects.

### 3.2.10  MARKETED PRODUCTS

Dabur Ashokarishta: Ashokarishta comprises several traditional plants such as Ashoka, Dhataki, Musta, Haritaki, and Amlaki, exhibiting anti-inflammatory and rejuvenating effects as suggested by Ayurveda. Ashoka Elixir Drops (containing J. Ashok, Helonias, Xanthoxylum, *Viburnum opulus*) by Hahnemann Scientific Laboratory is a popular homoeopathic remedy that contains Janosia Ashoka to reduce all types of uterine problems and menstrual disorders. It is also helpful in the treatment of leucorrhoea of all types of painful and irregular menses.

### 3.2.11  RECENT RESEARCH

Saracin isolated from *S. indica* seeds is more specific for binding $N$-acetylneuraminyl-$N$-acetyllactosamine [Neu5Ac-$\alpha$-(2–6)/(2-3)-D-Gal-$\beta$-(1–4)-D-GlcNAc]. Saracin has been reported for its potential in stimulating cell division, thus acting as mitogen for human lymphocytes; however, this cell division stimulatory activity of saracin could be prevented by fetuin. It was also found that Saracin could promote interleukin-2 release in vitro in human peripheral mononuclear cells. Various other reports revelaed that Saracin showed more affinity towards CD8+ than CD4+ T cells [92–94]. Also extracts obtained from *S. asoca* suppressed all transcription factors/DNA interactions. Antiarthritic activity of *S. asoca* has been assessed, and it was found that it showed promising results in reducing inflammation, pain, and oxidative stress [95].

Acetone fraction of *S. asoca* seeds has been studied for its antiarthritic property of its oral administration on Freund's adjuvant-induced arthritis in Wistar albino rats after acute and subacute toxicities. It was found that acetone fraction reduced inflammation, glucosamine, and hydroxy-proline levels at high dose. Also, radiology-based investigation showed considerable reduction in arthritis-like symptoms with mark decrease in inflammation after treatment with *S. asoca* seeds.

A recent study revealed that aqueous fractions obtained from *Bougainvillea spectabilis* and *S. asoca* show effective larvicidal activity against *Aedes aegypti* and thus can be used as efficient eco-friendly approach in the management of *A. aegypti* (vector of dengue/chikungunya) [96].

Another recent study also revealed the oestrogenic potential of standardized ethanolic fraction of *S. asoca* flowers using an ovariectomized female albino Wistar rat model. It was found that quercetin-, kaempferol-, $\beta$-sitosterol-, and luteolin-enriched fraction showed oestrogenic effects at high doses [97].

### 3.2.12  TOXICITY

The previous preclinical study demonstrated that an herbal formulation (muco-adhesive vaginal tablet) containing hydro-alcoholic extracts of dried stems of *Azadirachta indica* A. Juss, and *S. asoca* showed high therapeutic efficacy and antimicrobial action when compared to standard. This clinical investigation was performed on 363 females. It was found that prepared formulation showed no toxicity and sustained a drug-release pattern [98].

Toxicity studies based on previous reports revealed that *S. indica bark extract* has no toxicity and can be safely used against breast cancer treatment [99].

Study based on animals showed that *S. asoca* methanolic extract (0.3 and 1.2 g/kg body weight) is non-toxic for both acute and subacute oral administrations. Also, dose was tolerated, and no adverse behavioural effect was found at LD50 = 6.526 g/kg for acute administration. In comparison to control animals, haematological parameters showed a considerable decrease in platelet count (432 ± 98.3, $p < 0.05$) [100].

### 3.2.13 STATUS OF THE MEDICINAL PLANT

Overexploitation of *S. asoca* resulted in a shortage of the original raw plant materials and thus resulting in cost increase and extensive adulteration or substitution. Currently *S. asoca* figures in the Red List of "Threatened Species" by International Union for Conservation of Nature (IUCN) and is reported to be endangered.

### 3.2.14 COMMERCIAL TRADE

Based on NMPB, New Delhi, and FRLHT, Bengaluru, (2007), local requirement of *Ashoka* bark was found to be more than 1000–2000 MT annually. NMPB has revealed that nearly more than 1178 medicinal plants are used by herbal drug productions based in India and 242 medicinal plants have been used annually of more than 100 MT. *S. asoca* falls into a category of plants with an annual demand greater than 100 MT per year [101, 102].

Bark of *S. asoca* in trade also is called "Ashoka chhal" that forms a highly consumed medicinal plants used by phytopharmaceutical industry in the preparation of renowned traditional formulations such as "Ashoka arishtam" (a fermented formulation). It is projected that the overall use of "Ashoka chhal" by India's herbal industry was more than 2000 MT per year in 2005–2006 [103].

### 3.2.15 SUBSTITUENTS AND ADULTERANTS

Due to the overexploitation and increase in demand, substitution and adulteration of the Ashoka bark has increased [104]. Vast adulteration (80%) in Ashoka bark has been reported in 2016 [105]. Previous reports suggested the substitution and adulteration of *S. asoca* with common adulterants and substituents such as *Trema orientalis*, *Bauhinia variegata*, *Mesua ferrea*, *Shorea robusta*, and *P. longifolia* [106]. *P. longifolia* (Annonaceae) is considered the most common adulterant of *S. asoca* that it is also often locally named "Ashoka" due to this, and *P. longifolia* is now called "False Ashoka" in English and indigenous to Asian countries such as India and Sri Lanka and extensively grown and present also in Bhutan, China, and several other countries. Similarly, one more adulterant/substituent of *S. asoca* is called *T. orientalis* (Cannabaceae) that is present in Australia, Africa, and Asian countries. This plant has been commonly used in the treatment of diarrhoea and epilepsy [107, 108]. Another adulterant/substituent indigenous to China, Myanmar, and North Thailand, called *Bauhinia variegate*, is used in the treatment of a urological condition when blood appears in urine and menstrual periods with unusually continued bleeding with other pharmacological properties [107–111].

Recently, *Kingiodendron pinnatum* (Roxb. ex DC.) Harms belonging to the same family have been recommended as a substituent to the Ashoka bark and found as a suitable alternative to *S. asoca* in Ashokarishta and offers a technical justification for Ashokarishta in gynaecological diseases [112].

### 3.2.16 CLINICAL TRIALS

Preclinical studies on muco-adhesive vaginal tablets comprising *A. indica* and *S. asoca* as mentioned in Charaka Samhita have been performed to explore the role of herbal formulation in the treatment of excessive vaginal discharge. Preclinical data revealed antimicrobial property with no toxic effects due to the presence of phenolic compounds. Also, formulation in double-blind randomized clinical study showed sustained drug release with a high therapeutic efficacy pattern among 363 females [113].

A single-blinded randomized placebo-controlled clinical trial of a classical Ayurvedic formulation Ashokarista containing *S. asoca* has been performed to assess its role in the treatment of menorrhagia and dysmenorrhoea. It was found that the treatment of dysmenorrhoea and menorrhagia patients with Ashokarista increases haemoglobin level, reduces the erythrocyte sedimentation, increases in serum albumin content, and decreases in blood clotting time with substantial rise in both serum cholesterol and triglyceride levels [114].

Randomized, double-blinded, placebo-controlled studies showed that saracatinib was well tolerated in this osteosarcoma patient, however, was no apparent impact of the drug in this double-blinded, placebo-controlled trial on osteosarcoma. Additionally, Src suppression alone could not be enough to inhibit metastatic development in osteosarcoma [115].

## 3.3  ASHWAGANDHA

### 3.3.1  VERNACULAR NAMES

Withania root, winter cherry, or Indian ginseng in Sanskrit is called Ashvagandha, and in Hindi, it's known as Asgandha and asgand in Urdu.

### 3.3.2  BIOLOGICAL SOURCE

It's obtained from the dried roots and stem obtained from *Withania somnifera*. As per Ayurvedic pharmacopoeia, Ashwagandha comprises mature roots in dried form which are derived from *W. somnifera* (*Withania* genus has 23 sp.) [116, 117].
    **Family:** Solanaceae

### 3.3.3  GEOGRAPHICAL DISTRIBUTION

It's a perennial shrub, widely present in India, extensively grown in some regions of Rajasthan as well as of Madhya Pradesh. Roots are usually harvested during winter and further processed into short pieces or powder. Except India, it's also found in Afghanistan, Baluchistan, Sind, Sri Lanka and distributed in the Mediterranean regions. Also, it's present at high altitude (at greater than 5000 ft. in the Himalayas). *W. somnifera* is widely available in dried regions of India (western part). Apart from its natural diversity, the cultivation of *W. somnifera* is also promoted in open field across India for its further utilization as crude drugs. *W. somnifera* cultivation has been highly spotted in Bikaner and Pilani areas of Rajasthan, Rajputana, Punjab, and Manasa [118, 119].

### 3.3.4  CHEMICAL COMPOSITION

*W. somnifera* contains various phytochemicals such as steroidal lactones, saponins, and alkaloids that are responsible for providing various pharmacological properties such as antistress/adaptogenic, antitumour, tonic, immunomodulatory, anxiolytic antiarthritic properties to this plant. *W. somnifera* contains naturally occurring C28-steroidal lactones with anolides that are mainly responsible for the medicinal value of this plant. Due to the presence of these steroidal lactones, Ashwagandha crude drug is classified as phyto-steroidal drug. There are almost 12 alkaloids, 35 different analogues of withanolides, and numerous sitoindosides are present in *W. somnifera* [120, 121].

Withaferin A (1) is considered the first natural steroidal lactone isolated from the shoot of *W. somnifera*, however, available in other medicinal plants such as *Acnistus arborescens* and other plants belonging to Solanaceae. It was found that withaferine A (1) and withanolide D (2) mainly responsible for a medicinal value of this medicinal plant [122]. This plant also contains other chemical components that contribute to the medicinal value such as alkaloidal components (tropine, cuscohygrine, anahygrine, etc.) and steroidal components such as withanolide, withanone, withaferin A, withanolides with a glucose at carbon 27 and withasomniferols [122].

Withaferin A is one of the main withanolidal active principles isolated from the plant. Withaferin A present in Ashwagandha shows chemogenetic variation; thus, three chemotypes (I-III) have been reported. Biologically and structurally, withanolides have structural and biological properties that match with ginsenosides, the main components present in the ginseng [123]. Various reports of this medicinal plant (aqueous extracts) on experimental models like rabbits and rodents have revealed lipid lowering and radical scavenging effects. Most important pharmacologically active components include alkaloids (anahygrine somniferine, cuseohygrine, choline, anaferine, isopelletierine, hygrine, topine, withasomnine, withanolide D, etc.), withanolide glycosides known as withanosides IV, V, VI saponins, and steroidal lactones (withaferins, withanolides) [124]. There are some anti-stress agents also present in *W. somnifera* such as sitoindosides (sitoindosides VII-X) and acylsterylglucosides [125]. Except antistress agents, plant extracts also contain immunomodulatory compounds in aerial part such as 5-dehydroxy withanolide-R and withasomniferin-A. Withanolides exhibits various biological effects such as anti-inflammatory, immunomodulatory, etc. [126]. Withanolides, such as Withanolide A, Withaferin A, Withanolide D, Withanone, have also been reported for their antioxidant, antiproliferation, encouraging apoptosis, and preventing angiogenesis. As per previous reports, these glycowithanolides affect the cortical and striatal antioxidant enzyme functions (superoxide dismutase [SOD], catalase [CAT], and glutathione peroxidase) in a rodent model. Other than this, *W. somnifera* also contains several other components such as withaniol.

### 3.3.5 COMMON USES

Ashwagandha is commonly used to treat or prevent various ailments associated with reproductive system and is extensively used as an adaptogen/anti-stress agent to improve the body's resistance against stress. Additionally, Ashwagandha stimulates the body's immune system by developing the cell-mediated immunity. Except these uses, it's used as tonic and promotes diuresis and sexual desires. It's extensively used as tonic for kids and can be used to treat various complications such as leucoderma, constipation, sleeping disorder, arthritis, cluster of boils, skin inflammation. Ashwagandha root has also been recommended in combination for scorpion-sting and snake venom treatment. Ashwagandha (Nagori) is considered superior variety and gives promising results when used in fresh powder form.

### 3.3.6 TRADITIONAL USES

*W. somnifera* or winter cherry is a traditional plant that has been utilized in Ayurveda, the antique Hindu system of medicine, for more than 3000 years. In Ayurveda, this plant is used as a Rasayana mainly

- to promote physical and mental health;
- to enhance resilience of the body against various ailments and diverse adverse environmental conditions;
- to rejuvenate the human body system or tissue or cell exposed to any external or internal injury;
- to improve endurance;
- to build a sense of mental happiness;
- to prevent ageing.

In India, *W. somnifera* is generally named "Ashwagandha" or "Asgandh" as cited in different text. Ashwagandha is a fusion of two Sanskrit words: "Ashwa" means horse and "gandha" means smell. These two words represent one of the major macroscopic character of the medicinal plant, i.e., odour. Roots of the *W. somnifera* smell like horse and assumed to give power and strength like horse when used. Ashwagandha is the most common plant used in Ayurvedic "Rasayana" also called "Sattvic Kapha Rasayana" herb. Utilization of Ashwagandha is tracked from the Vedic age and later to Ayurveda. Ashwagandha is stated as "Asvabati" in Atharvaveda and Rigveda. Astanga Hridaya, Susruta Samhita, Charaka Samhita also mentioned the importance of Ashwagandha. Ashwagandha is well known for its beneficial effects in various diseases since ancient times. *W. somnifera* has been used for various ailments such as Daurbalya (weakness), Kashaya (skinniness), Shotha (inflammatory disorders), and Vataroga (neurological problems). Since ancient times, it has been utilized as health tonic (Rasayana) as well as Vajikarana (facial appearance, strength, potent, encourage good physique, sexually ecstatic, and strength) plant. As per Ayurveda, *W. somnifera* is recognized for the following:

- improve strength (Balya)
- libido enhancing (Kamarupini)
- nourishing (Pustida)

- sexual performance enhancer (Brusya)
- spermatogenic (Vajikari)

Various ancient scriptures of Ayurveda reveal the importance and uses of Ashwagandha. Ashwagandha appeared in Sanskrit text on Ayurveda, i.e., Charaka Samhita as distinct medicinal plant groups like Bringhaniya Mahakasaya of Sutrasthana chapter to build strength and Virechanopaga (to assist the process of purgation), Madhuraskandha (sweet taste major drugs) herb groups of Vimana Sthana chapter (Charak Samhita, 1949). In an ancient Sanskrit text on medicine and surgery, Susruta Samhita it reported in Sutrasthana Chapter in distinct medicinal plants categories such as Brihana, Vamak, and Utsadan Dravyas [127]. According to Charaka Samhita (Charak Samhita, 1949), Ashwagandha is indicated in diseases like

- glandular diseases (granthi)
- impotency (klaibya)
- neuromascular pain (vatavyadhi)
- skin diseases (visarpa)
- skin problems (kustha)
- tuberculosis (yakshma)
- urticaria (kandu)

In Sushruta Samhita, it is indicated in the treatment of diseases like [127]

- diseases of the Kapha (kaphavikara)
- dourbalya (weakness)
- gynaecological complications (sutikavikara)
- pain is main symptom (vatarakta)
- yaksma (tuberculosis)

Ashwagandha is also mentioned in Astanga Hridaya for the treatment of

- aural keloid treatment (unmantha chikitsa),
- weakness
- epilepsy
- neuromuscular pain (Ashtanga Hridayam of Vagbhata, 2005).

Ashwagandha mentioned in Bhaisajya Ratnabali for the treatment of

- neuromascular pain (vatavyadhi)
- impotency (dhawajabhanga)
- consumption (sosa)

Utilization of Ashwagandha in form of Mahanarayan taila is also reported for Vata-Vyadhi (neuromascular pain). In a historical Ayurvedic "Bhavaprakasha" book, *W. somnifera* utilization is revealed in the Guruchyadivarga chapter, helpful in vatavyadhi, weakness, granthi as well as shotha.

### 3.3.7 Pharmacological Uses

*W. somnifera* has potential to restore the strength and function of the brain and nervous system. To promote well-being, Ashwagandha also improves the function of the reproductive system. As mentioned above, it improves the body's resistance against different types of stress such as oxidative stress by presenting antioxidant effects to protect the body from cellular injury. *W. somnifera* also possesses anxiolytic, hepatoprotective effects and improves haemoglobin level and red blood cell (RBC) count [128].

### 3.3.8 TRADITIONAL FORMULATIONS

One of the most extensive uses of Ashwagandha is in the form of churna (fine sieved powder) that can be mixed with water and honey. Also, its extensive use in the form of Taila has been reported since several decades. Various traditional formulations comprising Ashwagandha is listed in Figure 3.1. Ashwagandha is a component of various traditional formulations such as

- Ashtanga Hridayam (13 preparations)
- Charaka Samhita (21 preparations)
- Sharangdhar Samhita (12 preparations)
- Susruta Samhita (13 preparations)
- Bhaishjyaratnavali (12 preparations)

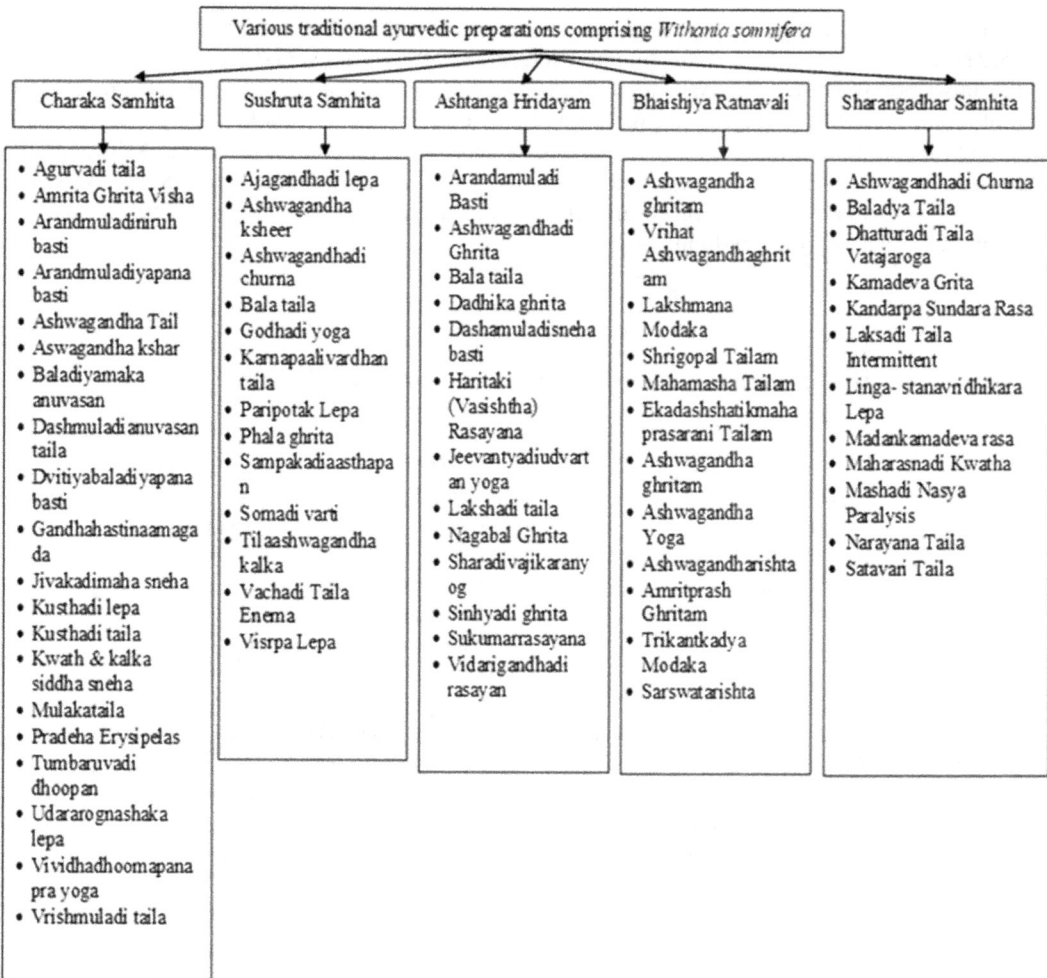

Various traditional ayurvedic preparations comprising *Withania somnifera*

| Charaka Samhita | Sushruta Samhita | Ashtanga Hridayam | Bhaishjya Ratnavali | Sharangadhar Samhita |
|---|---|---|---|---|
| • Agurvadi taila<br>• Amrita Ghrita Visha<br>• Arandmuladiniruh basti<br>• Arandmuladiyapana basti<br>• Ashwagandha Tail<br>• Aswagandha kshar<br>• Baladiyamaka anuvasan<br>• Dashmuladi anuvasan taila<br>• Dvitiyabaladiyapana basti<br>• Gandhahastinaamagada<br>• Jivakadimaha sneha<br>• Kusthadi lepa<br>• Kusthadi taila<br>• Kwath & kalka siddha sneha<br>• Mulakataila<br>• Pradeha Erysipelas<br>• Tumbaruvadi dhoopan<br>• Udararognashaka lepa<br>• Vividhadhoomapana pra yoga<br>• Vrishmuladi taila | • Ajagandhadi lepa<br>• Ashwagandha ksheer<br>• Ashwagandhadi churna<br>• Bala taila<br>• Godhadi yoga<br>• Karnapaalivardhan taila<br>• Paripotak Lepa<br>• Phala ghrita<br>• Sampakadiaasthapan<br>• Somadi varti<br>• Tilaashwagandha kalka<br>• Vachadi Taila Enema<br>• Visrpa Lepa | • Arandamuladi Basti<br>• Ashwagandhadi Ghrita<br>• Bala taila<br>• Dadhika ghrita<br>• Dashamuladisneha basti<br>• Haritaki (Vasishtha) Rasayana<br>• Jeevantyadiudvartan yoga<br>• Lakshadi taila<br>• Nagabal Ghrita<br>• Sharadivajikaranyog<br>• Sinhyadi ghrita<br>• Sukumarrasayana<br>• Vidarigandhadi rasayan | • Ashwagandha ghritam<br>• Vrihat Ashwagandhaghritam<br>• Lakshmana Modaka<br>• Shrigopal Tailam<br>• Mahamasha Tailam<br>• Ekadashshatikmaha prasarani Tailam<br>• Ashwagandha ghritam<br>• Ashwagandha Yoga<br>• Ashwagandharishta<br>• Amritprash Ghritam<br>• Trikantkadya Modaka<br>• Sarswatarishta | • Ashwagandhadi Churna<br>• Baladya Taila<br>• Dhatturadi Taila Vatajaroga<br>• Kamadeva Grita<br>• Kandarpa Sundara Rasa<br>• Laksadi Taila Intermittent<br>• Linga- stanavridhikara Lepa<br>• Madankamadeva rasa<br>• Maharasnadi Kwatha<br>• Mashadi Nasya Paralysis<br>• Narayana Taila<br>• Satavari Taila |

**FIGURE 3.1** Traditional formulations of Ashwagandha.

### 3.3.9  MECHANISM OF ACTION

Adaptogens exert their stress-protective effect by regulating homoeostasis by several mechanisms of action associated with the hypothalamic pituitary adrenal axis, also by controlling key mediators of the stress response, such as heat shock proteins, stress-activated c-Jun N-terminal protein kinase cortisol, and nitric oxide. Withanolides act against cancerous cells in different ways as mentioned below [129]:

- via the stimulation of oxidative stress
- suppression of proteasome-mediated ubiquitin degradation of nuclear factor of kappa light polypeptide gene enhancer in B-cells inhibitor
- suppression of transcription factor signal transducer and activator of transcription 3
- suppression of chaperone protein
- dysregulation of cytoskeletal and structural proteins
- suppression of angiogenesis by hypoxia-inducible factor 1

### 3.3.10  MARKETED PRODUCTS

Ashwagandha is available in the form of dry roots, powders, and liquid extracts which can further be formulated in the form of fine powder, oil, tablet, capsules, and cream (Table 3.1). One of the most popular forms of Ashwagandha is the dry form, especially powder form. Various formulations available in market are mentioned in the table. Ashwagandha (powdered) is available in the form of capsules is widely consumed as health supplement. These days, Ashwagandha is consumed across several nations such as Asia Pacific, Japan, Middle East, North America, and Western Europe. The table represents the list of various marketed formulations containing *W. somnifera* as a major ingredient.

**TABLE 3.1**
**Various Marketed Formulations Available Containing**
***W. somnifera***

| Product Name | Company |
| --- | --- |
| Adaptonyl® | Silab – France |
| Amrutha Kasthuri | Pankajakasthuri Herbals India Ltd. |
| Ashwaganda, AF – 5404 | Bio-Botanica® |
| Ashwagandha Extract | Synthite Industries |
| Ashwagandharista | Baidynath Ayurved Bhawan |
| Brento | Zandu Pharmaceutical Works Ltd. |
| Dabur Stresscom | Dabur Ltd. India |
| Himalaya Ashwagandha | The Himalaya Drug Company |
| Himalaya Ashwagandha Capsules | Himalaya Global Holdings Ltd. |
| KSM-66 Ashwagandha | Ixoreal Biotech, India |
| Lovemax | BACFO Pharmaceuticals Ltd. |
| NAT activ®, NAT healthy™ | Naturex S.A, France |
| Organic Ashwagandha Powder | Merlion International (I) Pvt. Ltd. |
| Sensoril Extract | Natreon Ltd. |
| Vigomax | Charak Pharmaceuticals Pvt. Ltd. |
| Vital Plus | Mukthi Pharma |
| Viwithan™ Extract | Vidya Herbs Inc. |
| Winter Cherries Extract HS 2951 | Grau Aromatics, GmbH & Co. KG |

### 3.3.11 Toxicity

As per previous reports, Ashwagandha root extracts have not shown any toxicity in acute and subacute toxicity studies. It was found that the administration of Ashwagandha root extracts at 2000 mg/kg once daily up to 14 days has not caused any undesirable effects or death in study. Further oral administration of Ashwagandha root extracts at dose up to 2000 (mg/kg) among rodents for 28 days has not caused any toxicity [130]. Another subacute toxicity study demonstrated that the administration of Ashwagandha extracts with Panax ginseng for 90 days showed no toxicity [131]. Further safety findings after the administration of Ashwagandha-based different extracts are also evident. These fractions that were obtained from Ashwagandha did not show any severity, toxicity, abnormal behaviour, and stress [132]. Also, treatment with extracts hasn't demonstrated any variations in usual behaviour food consumption, weight of the body, and other biochemical parameters. Histopathological changes have not been observed [133]. Nevertheless, rise in haemoglobin level, RBC number, and acid phosphatase in a rodent model and considerable drop in the weight of male organs (thymus/adrenal/spleen), only in rats, have been observed after treatment with ethanolic extracts. Administration of isolated compounds such as withaferin A, sitoinoside IX, and sitoinoside cause primarily depression, subsequently stimulation of CNS in rodents.

### 3.3.12 Recent Research

In context with COVID-19, a potential phytochemical was identified from *W. somnifera* that directly binds with the pore region of the E protein and thus blocks its channel activity and ultimately leads to the inhibition of virus replication [134]. A recent study showed that standardized root extracts of the plant containing withanolide-exerted moderate vasorelaxant effect in the endothelium intact rat aortic rings, which was lesser than acetylcholine. Standardized root extracts and withanolide increased acetylcholine-induced relaxation in the rat aortic rings with an accumulation of nitrite content, enhanced NO levels, eNOS expression, and eNOS phosphorylation in Ea.hy926 cell line [135].

### 3.3.13 Clinical Trials

A randomized, double-blinded, placebo-controlled trial, on 150 healthy subjects scoring high on non-restorative sleep measures administered with 120 mg of standardized Ashwagandha extracts (Shoden®) for six weeks, improved sleep quality by considerably increasing the sleep (non-restorative) among individuals [136].

Another clinical study over the effectiveness of folk preparation containing 1 g of Giloy Ghanvati (*Tinospora cordifolia*) and 2 g of Swasari Ras (traditional herbo-mineral formulation), 0.5 g each of Ashwagandha (*W. somnifera*), and Tulsi Ghanvati (*Ocimum sanctum*) when administered to COVID-19 positive patients via oral route to the patients for seven days showed that this treatment accelerates virological clearance, supports faster recovery, and parallelly decreases the chances of viral dissemination. Also, this treatment was found to reduce inflammation, thereby decreasing the severity of COVID-19 infection with undesirable effects [137].

Another clinical investigation demonstrated the role of Ashwagandha root fraction in offering improvement in cardiorespiratory strength as well as recovery among athletic grownups. It was found that extracts showed improvement in cardiorespiratory strength without causing any undesirable effects [138].

A randomized double-blinded, placebo-controlled trial has been performed to evaluate the efficacy of *W. somnifera* (Ashwagandha) root extract in patients with obsessive–compulsive disorder. It was found that *W. somnifera* extract could be valuable as a safe and effective adjunct to a selective serotonin reuptake inhibitor in the treatment of obsessive–compulsive disorder [139].

Findings obtained from the trial conducted in 2019 on an aqueous extract of Ashwagandha showed considerable rise in pain acceptance level as well as period against systematic pain among healthy human individuals. Treatment of schizophrenia patients demonstrated considerable decrease in Positive as well as Negative Syndrome Scale with less side effects [140]. Likewise, another trial based on the treatment of schizophrenic patients with Ashwagandha extract showed fruitful results in the anxiety as well as depression treatment [141]. Another trial finding showed that standardized *W. somnifera* extract considerably increased testosterone as well as dehydro-epiandrosterone sulphate (salivary) level among healthy subjects [142]. Similarly, treatment with *W. somnifera* extract (standardized) reduced anxiety with decrease in dehydroepiandrosterone sulphate as well as cortisol [143]. When clinically tested among insomnia and anxiety patients, *W. somnifera* extract considerably improved sleep- and anxiety-associated complications [142, 143]. It was also found that the administration of *W. somnifera* standardized extract showed improved fatigue condition after chemotherapy [144]. The most popular preparation of *W. somnifera*, KSM- 6 effectively augmented oxygen consumption among individuals and also effective in improving quality of life [145]. Another clinical study showed the administration of somnifera root extract among HIV patients with pulmonary tuberculosis considered adjuvant in combination with anti-tuberculosis drugs [146].

### 3.3.14  ADULTERANTS AND SUBSTITUENTS

Non-root parts of Ashwagandha, mainly aerial parts of *W. somnifera*, which are rich in withaferin A as well as other withanolides, are considered the major substituents. Other reported substituents such as *Mucuna pruriens*, *Trigonella foenum-graceum*, or *Senna auriculata* were also found in marketed samples of Ashwagandha [147].

### 3.3.15  ANNUAL TRADE

In the US market, the mainstream *W. somnifera*-based supplements are retailed, and it was found that sales of supplements increased from the period 2014–2017 with an inflation of 39%. Price of Ashwagandha roots was found to be US$2.46–US$3.56 in comparison to Ashwagandha leaves, which is US$0.34–US$0.82 in 2018. In India, especially in Madhya Pradesh, cultivated area extended up to 5000 ha and overall, in India overall, about 10,770 ha of land is used to cultivate *W. somnifera* with a yearly production of 8429 MT.

### 3.3.16  STATUS OF THE MEDICINAL PLANT

In 2015–2016, Ashwagandha (*W. somnifera*) was cultivated in 1491 ha.

## 3.4  ALOE VERA

### 3.4.1  VERNACULAR NAMES

Name Aloe is originated from the Arabic word "alloeh" and Hebrew, "halal" that indicates bitter shiny substance. Majority of the plants belonging to the genus Aloe are harmless and safe; however, some contain a hemlock-like toxic substance that can cause harm to human body. The scientific title for Indian aloe or *Aloe vera* is *Aloe barbadensis* as well as its scientifically also named *A. vera* [148, 149]. Nevertheless, *A. vera* is also recognized by various other names [150].

### 3.4.2  BIOLOGICAL SOURCE

*A. vera* is a juicy, cactus-like tender 1–2 ft. tall perennial, drought-resisting plant with prickly and bitter leaves, comprising a high water level (99–99.5%). Aerial part mainly contains green, thick, sharp-pointed, elongated, lance-shaped leaves, which retains high level of water in form of gel due to which plants can survive more than seven years without water [151, 152]. Except gel plants also contain solid mass (0.5–1%) that contains various bioactive compounds such as phenolic, vitamins (vitamins A, C, and E, B12, folic acid, and choline), saponins, polysaccharides (both simple/complex), lignin, organic acids, minerals (fat and water soluble), enzymes, and phenolic compounds [153]. Leaf is made up of three distinct layers starting from inner layer comprising soft and transparent, gel-like mass containing parenchymatous cells. This gel-like semisolid mass mainly contains water with other bioactive compounds such as glucomannans, vitamins, lipids, amino acids, and sterols [154, 155]. The subsequent layer called middle layer or latex contains bitter anthraquinones glycosides. The outermost layer called rind comprising 15–20 cells provides protection to semi-solid translucent mass present in the form of gel and facilitates the production of carbohydrates and proteins. There are two different types of exudates produced by this plant: one is synthesized by the parenchyma or sclerenchyma cells called latex and the other one is produced by leaf central tubular cells called internal gel or the fillet. *A. vera* gel is soft clear, tasteless, and fragrance-free [156, 157]. Latex contains components responsible for laxative effects, whereas gel is utilized for treating various skin-related problems. However, gel has not shown any laxative effects as it doesn't contain bitter principles such as anthraquinones glycosides [157]. Considering latex as well as gel, *A. vera* has two different commercial applications as a topical skin-protective agent in cosmetic products and laxative to treat various gastro intestinal tract (GIT)-related complications.

**Family**

*A. vera* belongs to the family Aloeaceae that is nowadays incorporated in the Asphodelaceae or to a generally confined family Liliaceae [158].

### 3.4.3 GEOGRAPHICAL DISTRIBUTION

*A. vera* is a succulent xerophytic plant well adapted to harsh environmental conditions where water availability is less by storing an ample amount of water in the tissue. Aloe species are extensively present among the Asian, African, and the eastern European regions and distributed nearly across the world. Aloe genera contains more than 400 species; however, only a limited number of species, including *A. arborescens*, *A. vera*, and *Aloe ferox*, have been utilized for international trade. *A. vera* is cultivated in varied climatic conditions such as temperate and subtropical regions though its native to southern and eastern Africa and later it was cultivated and used in North Africa and the Mediterranean countries [148]. *A. vera* is not adapted to freezing temperatures conditions, however, well adapted in drought-prone regions thus does not require any special attention and conditions for cultivation like other medicinal plants. Thus, this plant is reasonably attractive to the farmers.

### 3.4.4 CHEMICAL COMPOSITION

*A. vera* is well known for its chemical composition especially the presence of more than 200 chemical components that provide diverse medicinal effects to this plant (Figure 3.2). Anthraquinones (tricyclic aromatic quinines) are considered the major class of biologically active chemical components present in *A. vera*. Among various natural anthraquinone compounds, Aloe emodin and chrysophanol are considered the main chemical compounds that are accountable for various biological effects. Oxidative product of aloin such as aloesin, aloin, and Aloe-emodin are attributing to various medicinal properties to *A. vera* gel [159–161].

Aloin or anthrone C-glycoside ($C_{21}H_{22}O_9$), having a molecular weight of 418, is a yellow compound that is considered the main constituent of *A. vera* latex. There are two isomers of aloin, aloin A (also called barbaloin) and aloin B (or isobarbaloin), present as a mixture of diastereomers [162–164]. As glycosides have a common feature of containing sugar (glycon) and non-sugar (aglycon) part, aloin also a glycoside. Thus, aloin contains D-glucose as glycon part and C10 of D-glucose is linked to the 10th carbon atom position of the anthrone ring in a β-configuration

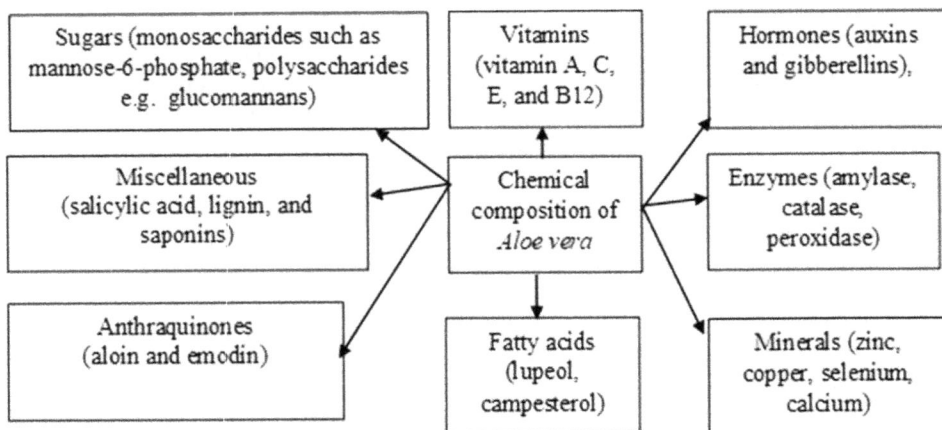

**FIGURE 3.2**  Chemical composition of *Aloe vera* [159–161].

[165]. Because of this linkage, aloin is identified as C-glycoside. This $\beta$-(1–10) C–C linkage is strong as it's unaffected by a strong acidic and alkaline environment. This linkage is also resistant against the plant enzyme, $\beta$-glycosidases, which are already known for their hydrolysing effect against gluco-oligosaccharides and plant glycosides. Even this linkage is also resistant against certain bacteria. This suggests that even the bacterial spoilage cannot easily affect the shelf life of aloin when stored under certain conditions. However, human and animal gut bacteria have the ability to break the $\beta$-C-glycosyl bond. Extent of cleavage varies among different animals because gut environment and type of bacteria depend upon the type of gut microorganism (humans > rabbits > guinea pig > rats > mice) [166, 167]. Aloin A and aloin B represent the cathartic effects of *A. vera* latex in humans and animals. However, the laxative effect of *A. vera* is not due to aloin A and aloin B, rather it's due to the metabolic products produced from metabolism of aloin A and aloin B by bacteria present in the large intestine. Aloe-emodin-9-anthrone is considered the most common metabolite that is further metabolized by oxidation to yield a free anthraquinone aloe-emodin, to a common metabolite, aloe-emodin-9-anthrone [166, 167], and subsequent oxidation to aloe-emodin, a free anthraquinone.

Processing of *A. vera* gel, latex, and other parts should be done carefully as it can affect the chemical compositions significantly. In 2015, it was found that freeze drying of exudate resulted in high-quality-dried exudates containing a high level of aloin A (54.16%) in comparison to other drying procedures such as oven drying, shade drying, and open sun drying. Another study revealed that exposure to ultraviolet (UV) light increased aloin content in gel, rind, and latex [168].

### 3.4.5  COMMON USES

*A. vera* commonly used in the treatment of various ailments such as

- Acne
- Anal fissures
- Dandruff
- Herpes sores
- Minor burns
- Psoriasis
- Seborrhoea
- Skin abrasions
- Skin injured by radiation

### 3.4.6  TOXICITY

*A. vera* is usually considered safe; nevertheless, less side effects have been evidenced so far. *A. vera* shows some side effects such as griping, flatulence, and abdominal cramps, when used for the treatment of constipation. Chronic use of Aloe-based products can lead to diarrhoea, hypokalaemia, *Pseudomelanosis coli*, kidney failure, as well as phototoxicity and hypersensitive reactions. Also, as per the previous report, the ingestion of *A. vera* extract can also result in colorectal cancer in rats. It was also found that if the aloe product is free of aloin, it can be used as a topical preparation for sunburn [169]. In 2015, *A. vera* (whole leaf extract) has been classified by the International Agency for Research on Cancer as a possible human carcinogen (Group 2B) [170, 171]. Especially, anthraquinone- and aloin-based products can lead to tumours and colon cancer. Topical application of *A. vera* can also cause irritation to skin and sometime also cause allergic reactions when applied over open wounds or injuries; however, it can safely be used over the skin especially for skin conditioning; conditionally there should not be an open wound or injury at the site of application [172].

Chronic use of aloe orally can cause diarrhoea and cramping that can further cause electrolyte imbalances. Chronic use of aloe juice or aloe latex or in high doses can result in the imbalance of electrolytes. Imbalance of electrolytes can further decrease the level of sodium that can lead to secondary hyperaldosteronism. Also, this imbalance can result in a decrease in the level (hypokalaemia) resulting in weight loss, muscular weakness, fatigue, mental problems, and kidney damages. Aloe gel, for skin or internal use, must be devoid of aloin, as it can cause irritation to GIT. It's also advised not to use aloe over skin burns or deep injuries and it was noticed that individuals those are allergic to herbs like garlic, onions, and tulips could be allergic to aloe as well. *A. vera*-based products can also show interaction with medications and other herbs. Via oral mode, *A. vera*, with furosemide or digoxin for the treatment of cardiac complications can decrease the amount of potassium in the body; thus, it should not be used with these medications. Due to the lack of safety profile, *A. vera*-based products must be avoided via oral mode by kids, lactating as well as pregnant females. It was found that the long-term utilization of *A. vera* can increase the bleeding while surgical procedures; thus, it should be stopped no less than two weeks before any surgical procedure. Topical usage of *A. vera* with steroid creams can enhance their rate of absorption of steroidal components that can further cause complications. It's also suggested that *A. vera*-based topical products containing *A. vera*, anthraquinone (mainly aloe emodin) should not be more them 50 ppm to prevent any phototoxic reactions caused by photo-oxidative damage to both RNA and DNA [173, 174]. A recent report on acute toxicity study suggested that *A. vera* soft capsule showed no mortality or behavioural changes representing the LD50 at higher than 15,000-mg/kg bodyweight. Also, subacute toxic studies revealed that *A. vera* soft capsule did not cause any considerable toxicity up to 3330 mg/kg b.wt.. In the same study, genotoxicity-based assays suggested that *A. vera* soft capsule showed no mutagenic activity as well as no report of chromosome aberrations in testes of rodent treated with 10,000 mg/kg b.wt. [175].

Recent finding also showed that hydroxyanthracene derivatives present in the latex layer of *A. vera* leaf, mostly as aloins A, B and aloe-emodin, have not shown any genotoxicity under the reported experimental conditions [176].

### 3.4.7  PHARMACOLOGICAL USES

*A. vera* has been reported for various pharmacological properties, including antitumour, anticancer, antiarthritic, antirheumatoid, and antidiabetic properties. Various reports have been evidenced on the biological effects of *A. vera* that include antimicrobial effects of the nonvolatile components of the gel. Moreover, *A. vera* is not commonly used for constipation, GIT, skin- and immunity-related complications. In context with skin complications, so far *A. vera* has been used to treat various skin problems and effective in treating topical inflammatory reaction. Numerous investigations showed inflammatory, oxidative stress, and bacterial infection reducing capabilities of *A. vera*. *A. vera* is also known for anticancer, antimicrobial, antidiabetic, anti-inflammatory, anti-obesity, antioxidant, and immune modulatory effects. Some pharmacological effects are due to the presence of aloe as phytosterols present in aloe can directly bind to cholesterol resulting in decrease in the levels of lipids in the blood and presented a hypoglycaemic activity in diabetic mice [177]. It was found that in intestine after anthraquinones metabolized into aloe-emodin-9-anthrone as well as aloe-emodin, accountable for its laxative effects, however additional medicinal properties have not associated with a particular constituent [178]. Aloe extract protects human corneal epithelial cells from $H_2O_2$-mediated injury possibly because of its antioxidant and anti-inflammatory effects, representing that eye drops comprising aloe extract could be utilized as an alternative treatment for Fuchs endothelial corneal dystrophy that is represented by the gradual worsening of corneal endothelial cells (cornea) and considered the main reason of transplantation of cornea globally [179].

### 3.4.8 TRADITIONAL USES

*A. vera* has been used in traditional medicine from last 2000 years, dating back to biblical times and considered one of the most important components in the indigenous system of medicine of various countries, including China, India, the West Indies, and Japan. *A. vera* name is obtained from "Alloeh" which is an Arabic word used for any "shining bitter substance", whereas the last name "vera" is derived from Latin word means "true". *A. vera* is called Ghrita-Kumari in Sanskrit, which means "young girl", which suggests that this plant has the capability to give the female youth and has regenerating effect on women. As per Ayurveda, *A. vera* has four out of six Ayurvedic tastes: bitter, stiff, spicy, and sweet. Taste after its ingestion is bitter, after digestion (vipaka) is sweet, and the ultimate effect is cooling. Aloe is a plant that is known for its pure bitter taste. Aloe juice has capability in restoring the imbalance among all three doshas; however, as per Ayurveda, its effect is stronger on Pita dosha in controlling heat, metabolism, and transformation in the mind and body, i.e., it also controls digestion "Agni" or fire of the body. Further, *A. vera* has been traditionally used for the treatments of various complications such as burns and wounds of ancient civilizations. So far as per previous reports, traditionally, it has been for various other complications such as seborrheic skin irritation, burns, acne, ulcers. Apart from its therapeutic effects, *A. vera* is also traditionally used as an effective insect repellent [180].

### 3.4.9 COMMON USES

#### 3.4.9.1 As Skin Moisturizing and Anti-Aging Agent

Mucopolysaccharides present in *A. vera* retain the moisture content by binding moisture into the skin to further prevent skin damage. Aloe also trigger fibroblast to produce more collagen and elastin fibres to prevent wrinkle and increase the elasticity of the epidermal skin. Gel showed skin-softening capability by possessing consistent impact on the surface epidermal cells by placing them all together. Amino acids and zinc content present in *A. vera* soften tough skin cells and showed astringent property by tightening skin pores. Moisturizing effects of *A. vera* have also been investigated over dry skin, when dry-coated *A. vera* gel gloves prevent wrinkle, reduce erythema, and maintain skin physiological conditions [181]. It also has anti-acne effect.

#### 3.4.9.2 Sun Protective and Wound Healing Effects

*A. vera* gel has been reported for its sun-protective power by preventing the skin damage caused by radiations. However, the exact mechanism of action is not explored yet, still as per the previous report application of *A. vera*, gel increased the production of a metallothionein (antioxidant protein) in the skin that further inhibits hydroxyl radical's production and averts glutathione peroxidase as well as SOD suppression. Various evidence also revealed that the application of gel decreases the secretion and production of keratinocyte-obtained cytokines, including IL-10 and thus inhibits delayed-type hypersensitivity caused by UV [182]. A recent report showed the synergistic effect of bone marrow-derived-mesenchymal stem cells when administered with Chitosan or Aloe vera gel to enhance the healing process of wound healing more than chitosan gel treatment. Thus, it was concluded that this gel can be considered efficient tactics for the treatment of burn injuries [183]. Recent new non-drug combined therapy-based study showed that *A. vera* gel when combined with ultrasound and soft mask for the treatment of mild-to-severe facial acne significantly improved acne by reducing the number of papules and the area of hyperpigmented lesions and improved skin roughness as well as local blood circulation [184].

A recent study also showed that ozonated *A. vera* oil is useful in improving the therapeutic response of full-thickness defects, resulting in rise in the number of fibroblasts and collagen thickening that, in turn, increase the rate of wound healing in Sprague-Dawley rats [185]. Recently, a clinical study performed over 40 subjects who had experienced surgical procedure showed that silicone gel comprising onion extract as well as *A. vera* is efficient as silicone gel sheets for postoperative scar prevention [186].

### 3.4.9.3   Bone Fracture Effects

A recent study based on effects of *A. vera* on penile fractures formed experimentally in a rat model showed that the local application of *A. vera* onto the penile fractures region without closing with suture reduced inflammation in rats [187].

### 3.4.10   THERAPEUTIC PROPERTIES

Based on the previous report, after topical and oral administration of *A. vera*, glucomannan and gibberellin, binds with proteins present over fibroblast, further, to increase its action as well as proliferation, thereby considerably increasing collagen production [188]. *A. vera* gel administration enhanced the synthesis of the level of collagen in wound, changed the composition of collagen (more type III) as well as enhanced the rate of crosslinking of collagen. This resulted in the improved injury reduction as well as improvement in the breaking strength of resultant scar [189]. Oral as well as topical applications also enhance the production of dermatan sulphate as well as hyaluronic acid in the wound [190].

### 3.4.10.1   Anti-Inflammatory Effects

*A. vera* also supresses the cyclooxygenase pathway as well as decreases prostaglandin E2 formation. A new phytochemical (C-glucosyl chromone) from *A. vera* gel extracts has been reported for its potential in reducing inflammation [191].

Recent clinical investigation over menopause females with complications of atrophic vaginitis symptoms showed that *A. vera* loaded cream showed similar effects like oestrogen loaded cream to treat atrophic vaginitis [192].

### 3.4.10.2   Immunomodulatory Effects

A chemical component of *A. vera* called alprogen supresses the calcium entry into mast cells to ultimately supress the leukotriene as well as histamine release [193]. Another chemical component of *A. vera* is called acemannan when administered to mice implanted with murine sarcoma cells, resulted in the stimulation of tumour necrosis factor and interleukin-1. This biological stimulation led to the cell death and deterioration of the cancerous cells [194]. The previous report also showed that low-molecular-weight components have been found to supress the release of ROS from stimulated human neutrophils [195].

### 3.4.10.3 Effects on GIT

Anthraquinones derived from *A. vera* latex are considered potent laxative. Additionally, it increases mucus production and shows laxative effects by intestinal peristalsis. A recent report also suggested that *A. vera* mitigates ulcerative colitis by increasing colon mucus barrier functions apart from its effect in reducing inflammation. Additionally, it was also reported that aloin A could be a main active constituent of *A. vera* to improve ulcerative colitis [196].

### 3.4.10.4 Antitumour, Anticancer, and Antiviral Effects

Anthraquinone present in *A. vera* shows direct or indirect effects by triggering immune system to act indirectly against virus or directly act against virus. It was found that anthraquinone effectively supresses different enveloped viruses, including influenza, herpes simplex as well as varicella zoster [197]. Recently, a polysaccharide obtained from *A. vera* has been reported for its ability to supress the benzopyrene interaction with primary rodent liver cells, so as to avert the development of complex called cancer-initiating benzopyrene-DNA adducts [198]. Recent report-based simple randomization technique showed that Curcumin as well as aloe gel have been found to be promising in improving oral submucous fibrosis; however, aloe gel was found to be more effective in burning sensation improvement without any side effects [199].

### 3.4.10.5 Antimicrobial, Antiseptic, and Amoebicidal Effects

Lupeol, cinnamonic acid, phenols, as well as salicylic acid are antimicrobial components isolated from *A. vera*, reported for its inhibitory effects against viruses, fungi, and bacteria [200]. Antidiabetic effects: Antidiabetic properties of aloe extract in rodent model could be due to their potential to normalize lipid-based biomarkers so as to improve sensitivity of insulin as well as modulate beta-cell function [201]. In vitro study at various concentrations of *A. vera* ethanol extract and honey against *Acanthamoeba* spp. cysts was investigated in comparison with chlorhexidine different incubation periods. It was found that *A. vera* ethanol extract and honey showed significant cysticidal effects on Acanthamoeba cysts [202]. Findings from the previous studies showed that ethanol extracts of *A. vera* both leaf and root can be used along with conventional antibiotics to fight against infections that are so prevalent in the skin infection (caused by *Escherichia coli*, *Shigella*, *Salmonella* spp., and *Staphylococcus aureus*) [203].

### 3.4.11 Traditional Formulations

In the 13th century, Ayurvedic Physician, Acharya Sharangdhara referred to one of the most popular formulations of *A. vera* known as Kumariasava [204]. Later in the 14th century, Acharya Shodal for the first time introduced this plant into the Nighantus. Most popular Ayurvedic preparations of *A. vera* are as follows:

- Kumāri tailam
- Kumarika vati
- Kumaripaka
- Kumaryasava
- Rajahpravartani vati

### 3.4.12 Mechanism of Action

As far as the mechanism of action is concerned *A. vera* latex components, C-glycosides aloin A and aloin B, which are usually metabolized to aloe-emodin-9-anthrone by gut bacteria present of rodents and humans. Further, aloe-emodin-9-anthrone metabolized to form aloe-emodin and rhein. It was found that *A. vera* products, acemannan, and aloin A do not represent genotoxicity. On contrary, aloe-emodin showed genotoxic effects. Thus, this demonstrates that, tumour developing ability of *A. vera* could be due to the conversion of the anthrone *C*-glycosides to aloe-emodin. This metabolites further alone or in combination with the components cause adenomas and carcinomas in the large intestine of rats.

**TABLE 3.2**
**Different Marketed Formulations of *Aloe vera***

| Product Name | Company |
| --- | --- |
| NaturSense's Organic Aloe Vera gel | Silab – France |
| Aloe Vera-based gel, cream, tonic, and juice | Himalaya, India |
| Aloe Vera-based gel, cream, tonic, oil and juice | Ayur, India |
| Aloe Vera-based gel, cream, tonic, oil, and juice | Dabur, India |
| Pura D'or's cold-pressed Healing Organic Aloe Vera Gel | Pankajakasthuri Herbals India Ltd |
| Seven minerals Aloe Vera Gel | Bio-Botanica® |
| Beauty by Earth After Sun Aloe Vera Gel | Synthite Industries |
| Aloe vera extracts and concentrates | Terry Laboratories, Inc. |
| Aloe vera products | Foodchem International Corporation, China |
| Aloe vera-based drinks, lotions, cosmetics, supplements and detergents. | Forever Living Products International, Inc. |
| Soft drinks, topical gels, capsules, powder, and other aloe vera raw materials | Aloe Laboratories, Inc. |
| Gel, whole leaf, cosmetics, and juice. | Aloecorp, Inc. USA |

## 3.4.13  MARKETED PRODUCTS

*A. vera* offers a broad range of products in cosmetics (healing, moisturizing, and skin softening), food, and medicine as mentioned in Table 3.2. Pulp (also known as gel) of this plant has wide applications in cosmetics. In cosmetics, *A. vera*-based creams and gels are the commonest form of skin care products used in the market. Quantity of *A. vera* may vary among these dosage forms. Certain cream for skin injuries contains 0.5% *A. vera* content, while some may comprise more than 70% of aloe for a skin condition like psoriasis. Among the health supplements, content of aloe is not the same. For GIT purpose (anthraquinones in *A. vera* have been reported for laxative effects), especially for a condition like constipation aloe juice (100–200 mg) or aloe extract (50 mg), administration is daily recommended. *A. vera* gel and phytosterols derived from *A. vera* have been evidenced for the type-2 diabetes mellitus treatment by controlling long-term blood glucose level. Thus, because of this one tablespoon of the gel has been recommended for daily use, however, a greater dose of aloe via oral route could be harmful [205].

## 3.4.14  INDUSTRIAL APPLICATIONS

A recent study showed that Aloe gel-based coating alone or in blend form improved the quality and shelf life of food studied for longer duration as well as decreasing post-harvesting damages [206].

- Decreases the fungal contamination due to its antifungal effects.
- Maintains quality.
- Possesses elevated acidity.
- High ascorbic acid as well as anthocyanins.
- High enzymatic activities.

A recent research also showed that *A. vera* has several applications in food industry due to its various health benefits like antifungal, anti-inflammatory, anti-microbial, antioxidant, aphrodisiac, emollient, and purgative properties. Moreover *A. vera* gel is also known for its functional and nutritional benefits [207].

### 3.4.15   ADULTERANTS AND SUBSTITUENTS

Various adulterants have been found and some are very common such as maltodextrin in *A. vera* gel, powder or adulteration of liquid preparations with water [208].

### 3.4.16   CLINICAL TRIALS

Clinical studies showed that Aloe administration can be used to treat many complications associated with GIT (Table 3.3). Further pharmacological investigations and clinical studies showed that *A. vera* could be effective in treating diabetes, hyperlipidaemic, cancer, microbial infections as well as ulcers with improved pharmacokinetic, dosage, and toxicity profile (Table 3.3).

## 3.5   AMLA

### 3.5.1   VERNACULAR NAMES

Indian gooseberry, emblic myrobalans and Malacca tree (English), amalaka (Sanskrit), and amla (Hindi).

### 3.5.2   BIOLOGICAL SOURCE

*Emblica officinalis* Gaertn. or *Phyllanthus emblica* Linn. has an important value in Ayurveda for therapeutical reasons to develop lost vigour as well as vitality.
   **Family:** Euphorbiaceae

### 3.5.3   GEOGRAPHICAL DISTRIBUTION

This medicinal plant was originally native to India but is today found growing in Pakistan, Uzbekistan, Sri Lanka, Southeast Asia China, and Malaysia [242].

### 3.5.4   CHEMICAL COMPOSITION

Amla fruit enriched with ascorbic acid contains numerous bioactive constituents, mainly polyphenolic-based compounds as mentioned in earlier reports [243–245]. Sugar-substituted phenolics such as flavone glycosides, phenolic glycosides, and flavonol glycosides, as well as tannins, such as emblicanin A, emblicanin B, phyllaemblicin B, and punigluconoin, are reported in fruit's pulp [246, 247].

   In 2016, researchers isolated two components such as new diphenyl ether derivative and bis-abolane-type sesquiterpenoid with 23 well-known components from the Amla fruits. It was also found that some components demonstrated considerable antioxidant effects. Among these potential antioxidants, one component increased $H_2O_2$-treated PC-12 cells survival [248]. A recent study (in 2020) showed that fruits of *P. emblica* contain aryltetralin-type lignan with 12 already identified polyphenolic components. Out of these, certain polyphenolic components were isolated first time from Euphorbiaceae family [249]. In 2018, two new triterpenes, the secofriedelane type secofriedelanophyllemblicine and the ursane-derived saponin ursophyllemblicoside, were isolated from the roots of the edible fruit-producing *P. emblica* out of which secofriedelanophyllemblicine showed a moderate cytotoxicity against K562 and HepG2 cancer cell lines [250].

### 3.5.5   TRADITIONAL USES

Amla relishes extraordinary privilege in Indian folklore, as the first to be created in the universe [251]. Amla is perhaps one of the most valuable plants in numerous traditional and folk systems of medicine in India. In Ayurveda, Amla is a potent rejuvenator and immunomodulator effective in delaying

**TABLE 3.3**

**Clinical Studies on *Aloe vera* Preparations**

| Aloe sp./Product/ Chemical Component | Complication/s | Trial Type/Time Period/Dose | Participants | Outcome | Ref. |
|---|---|---|---|---|---|
| Topical cream containing 0.5% aloe juice powder | Chronic anal fissure pain, wound treatment and haemorrhaging upon defection | A prospective double-blinded clinical trial; three times per day for six weeks | Aged 20–70 years | Considerable variation in fissure pain (anal) and haemorrhoids | [209] |
| Tongue protector and 0.5-ml *A. vera* | Burning mouth syndrome | Prospective, randomized, double-blinded, clinical evaluation; three times a day for three months | 75 Patients with burning mouth syndrome | Simultaneous prescription of tongue protector and aloe is effective | [210] |
| Quality-tested aloe extract | Radiation-induced skin injury | Three-Arm Randomized Phase III trial; for one, two, and four weeks | 246 Patients with breast cancer | No improvement in the symptoms or reduction in the skin reaction severity | [211] |
| Gingivitis *A. vera* | A commercially available dentifrice containing *A. vera* | A randomized controlled clinical trial; for 6, 12, and 24 weeks | 90 Patients diagnosed with chronic generalized gingivitis | Significant improvement in gingival and plaque index scores as well as microbiologic counts compared with placebo dentifrice | [212] |
| Gel based on *A. vera*; concentrations of 10, 20, 50, and 100% | Anti-irritant effect | Trial over 18 years; for 24 h | 15 Patients | Effective in the recovery from skin rash | [213] |
| Gel based on *A. vera* | Caesarean injury healing | Clinical trial (randomized) | 90 Women | – | [214] |
| Eye drops based on aloe | Ocular surface squamous neoplasia | Human-based clinical investigation | | | [215] |
| Crude gel based on *A. vera* | Scabies | An open, noncomparative study; daily for four weeks | 30 Individuals | Efficacy equivalent to benzyl benzoate | [216] |
| Cream based on aloe | Mild-to-moderate plaque psoriasis | A randomized, comparative, double-blinded; eight weeks | 87 Patients | Could be more effective than 0.1% triamcinolone acetonide | [217] |
| Aloe gel with tretinoin and vehicle | Acne vulgaris | A randomized, double-blind, prospective trial; for eight weeks | 60 Subjects with mild-to-moderate acne vulgaris | Considerable improvement in mild-to-moderate acne vulgaris | [218] |
| Aloe sterols (cycloartenol and lophenol) | Dry skin conditions | A randomized, double-blinded, placebo-controlled trial; five tablets (40 mg) each day | 48 Japanese women aged | Activation of collagen as well as hyaluronic acid formation by fibroblasts | [219] |

*(Continued)*

**TABLE 3.3 (Continued)**
**Clinical Studies on *Aloe vera* Preparations**

| *Aloe* sp./Product/ Chemical Component | Complication/s | Trial Type/Time Period/Dose | Participants | Outcome | Ref. |
|---|---|---|---|---|---|
| Aloe polysaccharides | Skin damage by radiation | Self-controlled investigation; one year | 185 Newly diagnosed nasopharyngeal carcinoma patients | None | [220] |
| Aloe polymannose multinutrient complex | Alzheimer's | An open-label trial; 4 teaspoons per day for 12 months | Adults diagnosed with Alzheimer's disease | Significant improvement | [221] |
| Aloe mouthwash | Oral lichen planus | A randomized double-blinded clinical trial; four weeks | 46 patients | Efficacy equivalent to triamcinolone acetonide | [222] |
| Aloe juice | Refractory irritable bowel syndrome | A trial; 30 ml twice daily for eight week | 33 individuals with constipation and IBD | Decrease pain and flatulence | [223] |
| Aloe gel | Ulcerative colitis | A randomized, double-blinded, placebo-controlled trial; 100 ml twice daily for four weeks | 45 valuable hospital out-patients | Decrease in the illness | [224] |
| Aloe extract cream | Psoriasis vulgaris | A placebo-controlled, study; five successive days | 60 patients | Effectively reduce psoriasis without any undesirable effects | [225] |
| Aloe cream (in combination with *Calendula officinalis* ointment) | Diaper dermatitis in children and infants | Clinical double-blind trial; ten days | 66 Newborns with dermatitis | No any undesirable effects | [226] |
| Acemannan, polysaccharide from *A. vera* Carbopol1 934P | Oral aphthous ulceration | Seven days | 180 Subjects with recurrent aphthous ulceration and 100 healthy subjects as a control group | Efficacy equivalent to triamcinolone acetonide | [227] |
| *A. vera* /olive oil cream *A. vera* | Chronic skin lesions after exposure to sulphur mustard | A randomized double-blind clinical trial; two times daily for six weeks | 67 Iranian chemical warfare injured veterans | Effectiveness is equivalent to betamethasone 0.1% | [228] |
| *A. vera* (high mol wt fraction) | Hepatic periportal fibrosis | Oral administration for 12 weeks | 40 Patients and 15 healthy volunteers | Considerable improvement in fibrosis, reduction in inflammation, activities of the liver enzymes, MDA levels, glutathione content, fibrosis markers | [229] |
| *A. vera* ointment | Wound healing of episiotomy | Clinical study for five days | 74 primiparous women | A significant activity observed | [230] |

*(Continued)*

**TABLE 3.3 (Continued)**
**Clinical Studies on *Aloe vera* Preparations**

| *Aloe* sp./Product/ Chemical Component | Complication/s | Trial Type/Time Period/Dose | Participants | Outcome | Ref. |
|---|---|---|---|---|---|
| *A. vera* juice | Acute viral hepatitis | Clinical study for six weeks | 110 male and female patients | Considerable reduction in bilirubin, alanine transaminase, aspartate transaminase, and alkaline phosphatase levels | [231] |
| *A. vera* gel | Oral sub-mucous fibrosis | Clinical study for three months | 20 study subjects with oral sub-mucous fibrosis | Decrease in the burning sensation and improvement in mouth opening | [232] |
| *A. vera* gel | Vulval lichen planus | Clinical study for eight weeks | 34 female patients | A safe and effective treatment for patients with vulval lichen planus without side effects | [233] |
| *A. vera* extracts | Hepatoprotective activity | Clinical study at a dose of 10–15 ml/kg | | Reduction in elevated sGPT in subjects | [234] |
| *A. vera* cream | Posthaemorrhoidectomy pain and wound healing | Clinical study for four weeks | 49 patients | Effective in reducing postoperative pain | [235] |
| *A. barbadensis* Aloe extract | Irritable bowel syndrome | Clinical study at dose of 250 mg | 68 grownups | Satisfactory results | [236] |
| Aloe-loaded toothpaste | Plaque and gingivitis | Clinical study for six months | 15 subjects | Considerable decrease in plaque and gingivitis | [237] |
| Aloe cream containing 0.5% extract | Genital herpes in men | Clinical study for five days | 60 Individuals | Effective in treating genital herpes | [238] |
| A dentifrice containing Aloe | Plaque and gingivitis | Clinical study for 30 days | 15 Male and 15 female | Considerable decrease in gingivitis as well as plaque | [239] |
| A commercially available lotion based on *A. vera* | Radiation-induced dermatitis | Clinical study for four weeks | 60 Patients; mean age: 52 years; 67% women | – | [240] |
| *A. vera* gel | Slight-to-moderate psoriasis vulgaris | A randomized double-blinded, placebo-controlled study; two times a day for four weeks | 41 patients with stable plaque psoriasis | Modest activity | [241] |

degenerative and ageing process. Rasayanas of Amla alone or with other herbs helps in promoting longevity, enhancing digestion, treating constipation, reducing fever and cough, alleviating asthma, strengthening the heart, improving ocular power, promoting hair growth, and improving intellect [251].

### 3.5.6 Traditional Formulations

Amla is the main component of the Ayurvedic preparation such as Jeevani malt (Chirayu Pharma), Triphala churna (Zandu), and Chyavanprash (Dabur). It is a main component of the popular multi-component Ayurvedic formulations: Triphala churna and Chavanprash. "Triphala" (Sanskrit tri = three and phala = fruits) comprises three essential curative fruits:

- *P. emblica* L. or *E. officinalis* Gaertn (Amla)
- *Terminalia chebula* (harad)
- *Terminalia belerica* (baheda or bibhitaki)

Tripahala guggulu is another Ayurvedic mixture of Triphala and Gugulu (Myrrh oleoresin), which is evidenced to have a chondroprotective property of antiarthritic effects. However, Triphala churna is one of the most popular inexpensive and non-toxic preparations widely available, a drug of choice for the treatment of several diseases, particularly those of metabolism, dental, and skin conditions, and treatment of cancer [252]. This Ayurvedic formulation has beneficial effects over the physiology of heart, skin, eyes and facilitates to postpone degenerative changes, such as cataracts. In 2012, tumour inhibitory property of Triphala and its active components by the suppression of vascular endothelial growth factor-A (VEGF) induced angiogenesis actions [253]. In Triphala treatment of diseases, vascular endothelial growth factors A–induced angiogenesis is involved [254]. Diverse polyphenolic compounds present in Amla provide broad antimicrobial spectrum activity. This Ayurvedic formulation is an effective medicine to balance all three Dosha. It is considered a good rejuvenator Rasayana, which facilitates nourishment to all tissues, or Dhatu. Triphala is a constituent of about 1500 Ayurveda formulations and can be used for numerous ailments. Another preparation of Amla (Amalaki rasayana) has been extensively used in the indigenous system of Ayurvedic medicine since long time. Another known Amla-based preparation called Nisha Amalaki formulation with *Curcuma longa* Linn and *P. emblica* Linn is called for various ailments such as diabetes in Ayurvedic texts and Nighantus [255].

### 3.5.7 Pharmacological Uses

Amla contains phenolic compounds and sugar-substituted phenolics (flavone glycosides, phenolic glycosides, and flavonol glycosides) which reduce oxidative stress as well as offer diverse biological effects, including antioxidant, antibacterial, antifungal, antidiabetic, and hepatoprotective, apoptogenic, and anticancer effects. Khandelwal et al. [256] reported its cell protective effects examined by acute cadmium toxicity. It was found that the administration of Amla extract resulted in an enhanced cell survival, decreased free radical production, and higher antioxidant levels [256]. Recent in vitro studies have also demonstrated that fruit extract of Amla relieves the immunosuppressive effects of chromium in rat lymphocytes [257]. The fruit extract of Amla is known for its analgesic, anti-mutagenic, anti-oxidative, anti-pyretic, anti-tussive, anti-ulcerogenic, cardioprotective, cytoprotective, gastroprotective, hepatoprotective, and immunomodulatory effects. Several in vitro investigations showed antiaging, antidiabetic, anti-inflammatory, antioxidant, cryoprotective, hepatoprotective, nephrotoxicity modulation properties.

### 3.5.7.1 GIT Effects

In 2002, a study revealed that the oral administration of Amla extracts considerably suppressed the development of gastric lesions by significantly decreasing pyloric-ligation-induced basal gastric secretion, titratable acidity, and gastric mucosal injury. Moreover, Amla extract presented protection against ethanol-induced depletion of stomach wall mucus and reduction in nonprotein sulphhydryl concentration. Overall findings showed that Amla extract possesses antisecretory, antiulcer, and cytoprotective effects [258].

In 2008, anti-ulcerogenic effects of fresh fruit juice of *E. officinalis* and its methnolic extract were evaluated in absolute ethanol-, indomethacin-, and histamine-induced experimental ulcers in rats. It was found that *E. officinalis* fruits offered more protections than that of ranitidine (50 mg/kg). The results of the present study suggested the novel cytoprotective effects of *E. officinalis* fruits on gastric mucosal cells.

### 3.5.7.2 Antimicrobial Effects

In a recent research, an eco-friendly approach was used to synthesize silver nanoparticles using *E. officinalis* fruit extracts as a stabilizer and a reducer. Results showed that the *E. officinalis* fruit extract is an excellent bioreductant for the synthesis of AgNPs and AgNPs synthesized using *E. officinalis* that presented considerable antibacterial against both Gram-positive and Gram-negative bacterial strains [259]. A recent study also showed that silver nanoparticles synthesized using Amla fruit extract showed significant antibacterial effects and very clear zone of inhibition against *Klebsiella pneumoniae* and *S. aureus* bacteria [260].

### 3.5.7.3 Antidiabetic Effects

Based on in silico parameters such as docking scores, drug-likeness, and pharmacophore, it was found that ellagic acid, estradiol, sesamine, kaempferol, zeatin, quercetin, and leucodelphinidin are considered potential antidiabetic compounds [261]. The most cited research about pharmacological activity of Amla is study related with anti-diabetic and antioxidant effects of Methanolic extract (75%) of *T. chebula*, *T. belerica*, and *E. officinalis*. This study suggested that the oral administration of the extracts (100 mg/kg body weight) decreased the blood sugar level in normal and in alloxan diabetic rats considerably in 4 h [262]. It's now found that methanolic extracts of *E. officinalis* fruits act as a potent α-amylase and α-glucosidase inhibitor, which showed considerable antiglycation activity and significantly inhibited the oxidation of LDL under in vitro conditions [263].

### 3.5.7.4 Obesity

*P. emblica* L. is extensively used in the Ayurveda for thousands of years to treat health complications, including disorders of the immune system, diabetes, and obesity. A recent study explored the molecular mechanisms of the fruit extracts of *P. emblica* involved in the promotion of fat cell apoptosis and alleviation of adipogenesis. Digallic acid was identified as a major component in the fruit extract. Both extract and Digallic acid showed considerable anti-lipolytic activity and considerably lowered triglyceride accumulation by downregulating adiponectin, PPARγ, cEBPα, and FABP4, respectively. It was also found that fruit extract treatment caused an expression of apoptosis-related genes mainly by an upregulation of BAX and a downregulation of BCL2 resulting in an increased caspase-3 activity. It was concluded that fruit extract negatively regulates adipogenesis by promoting fat cell apoptosis and thus can be potentially used in the management of obesity [264].

### 3.5.7.5 Immunomodulatory Effects

Treatment of arsenic exposed mice with fruit extract of *E. officinalis* reduced the levels of lipid peroxidation, reactive oxygen species (ROS) production, activity of caspase-3, apoptosis and increased cell viability, levels of antioxidant enzymes, cytochrome *c* oxidase, and mitochondrial membrane potential as compared to mice treated with arsenic alone [265].

### 3.5.7.6 Hepatoprotective Effects

Recently, hepatoprotective effect of aqueous extracts of *P. emblica* on nonalcoholic steatohepatitis has been investigated. It was found that methionine and choline-deficiency diet might cause serious hepatic steatosis and mild inflammation in mice livers; however, the administration of aqueous extract of *P. emblica* could alter the rapid progression of nonalcoholic steatohepatitis [266].

### 3.5.7.7   Antiaging Effects

A recent study showed that *P. emblica* L. branch extracts containing a variety of phenolic acids, mainly sinapic and ferulic acids, can be used as safe and effective ingredient against skin ageing [267].

### 3.5.7.8   Renoprotective Effects

Renoprotective effects of Amla fruit powder by ameliorating elevated creatinine and uric acid concentration, by normalizing various oxidative stress indicators, including malondialdehyde, nitric oxide (NO), and advanced protein oxidation product, were also increased in plasma, heart, and kidney tissues in 2K1C rats. Thus, it was found that Amla supplementation may restore plasma antioxidant abilities and inhibits oxidative stress, inflammation, and fibrosis in 2K1C rats [268].

### 3.5.7.9   Anticancer Effects

Amla extracts offer anticancer effects, and the mode of action is based on free-radical and antioxidant effect, modulation of various enzymes of inflammation and carcinogenesis, modulation of cell cycle proteins, induction of apoptosis in neoplastic cells, and prevention of metastasis as suggested by Yang and Liu [269].

### 3.5.8   Industrial Applications

Amla fruit is one of the popular botanicals having many commercial units in the healthcare, food, and cosmetics industries. The fleshy and globular fruits in both raw (green) and ripe (yellowish green) forms are widely used for both dietary and medicinal benefits. The fruits are smooth striated with an obovate-obtusely triangular six-celled nut and used for culinary purposes, especially to make murabbah, juice, pickle, chutneys, and as a vegetable in various dishes.

*Meat industry:* The previous study showed that *P. emblica* extracts from plants can act as a promising natural alternative in the development of healthy meat products. In this study, researchers evaluated the efficacy of *P. emblica*, *Eucalyptus globulus*, and *T. cordifolia* in inhibiting lipase as well as oxidative stress. The study also showed that *P. emblica* inhibited lipase and reduced oxidative stress [270].

*Poultry industry:* The protozoan parasite Eimeria causes avian intestinal tract infection that affects the poultry industry globally. A recent study showed that the development of resistance against modern coccidials inhibiting agents is a major concern that requires inexpensive, as well as eco-friendly approaches. In 2021, researchers developed an eco-friendly approach by using *P. emblica* extracts as an effective alternative anticoccidial agent. It was found that extracts considerably inhibited oocyst sporulation, reduced the oocyst infectivity, decreased the faecal oocyst excretion, and decreased the pathogenicity of *Eimeria tenella* in chickens [271]. *Antipollutant:* Wastewater containing textile dye is a major concern regarding water pollution. Recent research showed eco-friendly synthesis of MgO nanoparticles using *P. emblica* for Evans blue degradation and antibacterial activity against *Pseudomonas aeruginosa* (Gram-negative), *S. aureus* (Gram-positive), *Acinetobacter baumannii* (Gram-negative), *E. coli* (Gram-negative), and *K. pneumoniae* (Gram-negative). Results showed a 90% removal of the dye from the wastewater [272]. Also, a recent report suggested that *P. emblica* fruit extracts could be utilized as a biodegradable optimal chelate for the phytoextraction of cadmium (Cd)-contaminated soils without impacting environments adversely [273]. Further, *P. emblica* aqueous extract significantly improves the effects of chromium in rice seedlings by preventing chromium uptake, lowering oxidative stress, and modulating effects of antioxidative enzymes [274]. *As reducing agent:* Recently, the eco-friendly approach of preparing reduced graphene oxide using plant extracts has attracted more attention; thus, in one of the most recent studies, the reduction of graphene oxide is carried out using vitamin

C (478.56 mg/100 ml)–enriched *P. emblica*. Findings showed that this eco-friendly approach reduced graphene oxide effectively and considered an efficient and environment-friendly approach [275]. *Agriculture industry:* Findings obtained from a recent study suggested that AgNPs using fruit extract of *P. emblica* L. act as a growth promoter at lower concentrations by delivering a potent anti-oxidant during early seedling growth as compared to chemically synthesized AgNPs-treated wheat seedlings (two wheat varieties: HD-2967 and DBW-17) [276]. *Food industry:* The fruits, also known as the berries or myrobalans, are the most important part of the plant, being of dietary, culinary, and medicinal use. Amla fruit is an important dietary agent and is used to make murabbah, burfi, laddu, fresh juice, pickle, chutneys, and curries in India.

*Nanotechnology:* One of the most cited research studies based on the role of *E. officinalis* fruit extracts in extracellular synthesis of gold and silver nanoparticles as the reducing agent to synthesize Ag and Au nanoparticles has been reported in 2005 using transmetallation reaction [277].

### 3.5.9 Marketed Products

Amla fruit is widely consumed as raw, pickle, jam, preserve, candy, powder, and juice [278]. Vitamin C content of Amla can vary depending on the varieties and is reported to range from 206.8 to 932.1 mg/100 g. A recent study showed that Amla supplemented pan bread with its superior nutritional, and sensory qualities can be a possibility to improve consumer nutrition [278, 279].

### 3.5.10 Toxicity

The LD50 of methanolic extract of *E. officinalis* fruit was found to be approximately 1125 mg/kg (p.o.). It was also found that other parameters, including body weight and survival rate, showed no side effects to methanolic extract of *E. officinalis* fruit [280]. In another study, it was found that *Amalaki rasayana* administration stably maintained/enhanced the DNA strand break repair in elderly subjects with no undesirable effects. Moreover, individuals with distinct BMI displayed differential DNA strand break restoration potential. Thus, it was found that *Amalaki rasayana* intake presented stable DNA strand break restoration with no undesirable effects [281]. In various other studies, *P. emblica* showed protective effects against arsenic-mediated toxicity, ethanol-induced hepatic damage, anti-tuberculosis drugs–induced liver toxicity as well as CCl4 induced hepatic toxicity in different animal models [282–285].

### 3.5.11 Clinical Data

#### 3.5.11.1 Treatment for Relieving Xerostomia

Recently single-blinded, two-arm parallel randomized controlled trial enrolled with 64 patients with xerostomia to study the efficacy and safety of *P. emblica* spray, for treating postoperative xerostomia, was performed. It was concluded that *P. emblica* spray seemed to be better than warm water spray for treating postoperative xerostomia [286].

*To improve endothelial function:* Clinical investigation was conducted over 59 enrolled subjects to assess the impact of fruits of *P. emblica* extract (standardized) on inflammation, endothelial dysfunction, lipid profile, and free radicals in individuals with metabolic syndrome. It was found that *P. emblica* extracts substantially increased the function of endothelial cells, reduced free radicals, inflammation as well as normalize lipid biomarkers [287].

#### 3.5.11.2 In the Management of Gastroesophageal Reflux Disease

Gastroesophageal reflux disorder is the commonest GIT diseases caused by the reflux of stomach substances resulting in regurgitation as well as heartburning. Gastroesophageal reflux disorder is divided into erosive as well as non-erosive esophagitis. Another randomized double-blinded, placebo-controlled clinical trial demonstrated that Amla (500-mg Amla tablets) could reduce frequencies of heartburn and regurgitation and improve heartburn and regurgitation severity in patients with non-erosive reflux disease [288].

### 3.5.11.3   In the Management of Hypertension

A recent clinical study showed that *P. emblica* showed a better safety profile among patients with essential hypertension [289].

### 3.5.11.4   Improve Skin Condition

A recent clinical study showed the effect of ingestion of Amla fruit extracts as well as lingonberry on various human skin complications. It was found that skin thickness and elasticity, stratum corneum water content, and the degree of wrinkles also significantly improved in a dose-dependent manner [290].

### 3.5.11.5   To Treat Dyslipidaemia

A clinical trial was performed to evaluate the efficacy of Amla extract (composed of polyphenols, triterpenoids, oils, etc. as found in the fresh wild Amla fruit) in 98 dyslipidaemic patients. It was found that Amla extract has shown significant potential in reducing total cholesterol and triglyceride levels as well as lipid ratios, atherogenic index of the plasma, and apoB/apo A-I in dyslipidaemic persons and thus has scope to treat general as well as diabetic dyslipidaemia. A single agent to reduce cholesterol as well as TG is rare. Cholesterol reduction is achieved without concomitant reduction of Co Q10, in contrast to what is observed with statins.

### 3.5.11.6   For Oral Hygiene

An examiner-blinded and randomized clinical trial was conducted over 20 healthy adults to assess the effectiveness of chewing gum (sugar-free) comprising *P. emblica* extract in altering the oral microbiome. It was found that chewing gum comprising *P. emblica* fruit extract increased salivary flow. Additionally, this herbal gum could be an effective way to improve oral cleanliness [291].

### 3.5.12   ADULTERANTS AND SUBSTITUENTS

Phyllanthus *amarus, Phyllanthus reticulatus, Phyllanthus fraternus, Phyllanthus urinaria, Phyllanthus virgatus*, and *Phyllanthus rotundifolius* are all annual herbs, growing to a similar height (60–70 cm) with a similar pattern of branching, leaf size and phylotaxy, flowers and fruits. These species are considered an allied species to *Phyllanthus embilca*. These days, tannin-based chemical makers are frequently utilized to examine the superiority of botanical and products therefrom. Nevertheless, these markers can only evaluate the integrated similarity and/or variation of the chemical components that might not indicate the beneficial effect of the botanical. Additionally, several components can influence the overall chemical profile of any botanical. Thus, recently RAPD-SCAR–based DNA marker for Amla has been developed by the researchers and utilized it for the identification of the botanical in semi-processed and processed formulations. Owing to high tannins and its acidic nature, the isolation of DNA from *P. embilca* was difficult; thus, recently a Cetyl Trimethyl Ammonium Bromide–based method of DNA isolation has been developed from fresh and dry Amla tissues as well as its semiprocessed formulations [292]. A recent study based on metabolomics and DNA barcoding for perception of Phyllanthus species such as *P. debilis, P. myrtifolius, P. reticulatus, P. emblica, P. amarus, P. urinaria, P. lawii, P. virgatus*, as well as *P. acidus* suggested that *P. deblis* as well as *P. virgatus* are not closely associated with each other; however, they are closer to each other in terms of their metabolic fingerprints. Likewise, in the same report, authors reported that *P. myrtifolius* and *P. urinaria* are more closely associated with each other with their metabolic profiling than the genetic alignment. Based on the sequence similarity, it was also found that nine Phyllanthus species presented considerable identity with the sequences of various Phyllanthus species. Differential expressions of 14 secondary metabolites from these 9 *Phyllanthus* sp. have been found. Moreover, it has been revealed that alkaloid compound zeatin was present particularly in *P. virgatus*, and delphinidin-3-*O*-β-D-glucoside was not present in *P. myrtifolius*. It was found that *P. acidus* is the most genetically distinct among the groups and least difference found among the sequence pair between *P. emblica*, *P. urinaria*, and *P. emblica–P. reticulatus* [293].

## REFERENCES

1. Jack DB. One hundred years of aspirin. *Lancet.* 1997;350(9075):437–439.
2. Falodun A. Herbal medicine in Africa-distribution, standardization and prospects. *Res J Phytochem.* 2010;4(3):154–161. [Google Scholar].
3. Leroi Gourhan A. The flowers found with Shanidar IV, a Neanderthal burial in Iraq. *Science.* 1975;190(4214):562–564. [Google Scholar].
4. Neolithic site of the cross-lake bridge. 2013 (Chinese), http://baike.baidu.com/view/1464263.htm.
5. https://www.who.int/traditional-complementary-integrative-medicine/WhoGlobalReportOnTraditional AndComplementaryMedicine2019.pdf.
6. Pan SY, Pan S, Yu ZL, et al. New perspectives on innovative drug discovery: an overview. *J Pharm Sci.* 2010;13(3):450–471. [PubMed] [Google Scholar].
7. Munos B. Lessons from 60 years of pharmaceutical innovation. *Nat Rev Drug Discov.* 2009;8(12): 959–968. [PubMed] [Google Scholar].
8. Seidl PR. Pharmaceuticals from natural products: current trends. *An Acad Bras Cienc.* 2002;74(1): 145–150. [PubMed] [Google Scholar].
9. Li X-J, Zhang H-Y. Western-medicine-validated anti-tumor agents and traditional Chinese medicine. *Trends Mol Med.* 2008;14(1):1–2. [PubMed] [Google Scholar].
10. Corson TW, Crews CM. Molecular understanding and modern application of traditional medicines: triumphs and trials. *Cell.* 2007;130(5):769–774. [PMC free article] [PubMed] [Google Scholar].
11. Schmidt BM, Ribnicky DM, Lipsky PE, Raskin I. Revisiting the ancient concept of botanical therapeutics. *Nat Chem Biol.* 2007;3(7):360–366. [PubMed] [Google Scholar].
12. Coulter ID. Integration and paradigm clash: the practical difficulties of integrative medicine. In: The Mainstreaming of Complementary and Alternative Medicine (eds. P Tovey, J Adams, G Easthope). Routledge, London; 2004:103–122.
13. Easthope G, Gill G, Beilby J, Tranter B. Acupuncture in Australian general practice: patient characteristics. *Med J Aust.* 1999;170:259–262.
14. Bensoussan A. Complementary medicine—where lies its appeal? *Med J Aust.* 1999;170:247–248.
15. MacLennan AH, Wilson DH, Taylor AW. The escalating cost and prevalence of alternative medicine. *Prev Med.* 2002;35:166–173.
16. MacLennan AH, Wilson DH, Taylor AW. Prevalence and cost of alternative medicine in Australia. *Lancet.* 1996;347:569–573.
17. Bhatia S, Sharma K, Dahiya R, Bera T. Modern Applications of Plant Biotechnology in Pharmaceutical Sciences. Academic Press, Elsevier, Cambridge, MA; 2015:164–174.
18. Bhatia S. Nanotechnology in Drug Delivery: Fundamentals, Design, and Applications. CRC Press, Boca Raton, FL; 2016:121–127.
19. Bhatia S, Goli D. Leishmaniasis: Biology, Control and New Approaches for Its Treatment. CRC Press, Boca Raton, FL; 2016:164–173.
20. Bhatia S. Natural Polymer Drug Delivery Systems: Nanoparticles, Plants, and Algae. Springer Nature, Basingstoke, UK; 2016:117–127.
21. Bhatia S. Introduction to Pharmaceutical Biotechnology, Volume 2: Enzymes, Proteins and Bioinformatics. IOP Publishing Ltd, Bristol, UK; 2018:1.
22. Bhatia S. Introduction to Pharmaceutical Biotechnology, Volume 1: Basic Techniques and Concepts. IOP Publishing Ltd, Bristol, UK; 2018:2.
23. Bhatia S. Introduction to Pharmaceutical Biotechnology, Volume 3: Animal Tissue Culture Technology. IOP Publishing Ltd, Bristol, UK; 2019:3.
24. Bhatia S. Natural Polymer Drug Delivery Systems: Nanotechnology and Its Drug Delivery Applications. Springer International Publishing, Switzerland; 2016:1–32.

25. Bhatia S. Natural Polymer Drug Delivery Systems: Nanoparticles Types, Classification, Characterization, Fabrication Methods and Drug Delivery Applications. Springer International Publishing, Switzerland; 2016:33–93.
26. Bhatia S. Natural Polymer Drug Delivery Systems: Natural Polymers vs Synthetic Polymer. Springer International Publishing, Switzerland; 2016:95–118.
27. Bhatia S. Natural Polymer Drug Delivery Systems: Plant Derived Polymers, Properties, and Modification & Applications. Springer International Publishing, Switzerland; 2016:119–184.
28. Bhatia S. Natural Polymer Drug Delivery Systems: Marine Polysaccharides Based Nano-Materials and Its Applications. Springer International Publishing, Switzerland; 2016:185–225.
29. Bhatia S. Systems for Drug Delivery: Mammalian Polysaccharides and Its Nanomaterials. Springer International Publishing, Switzerland; 2016:1–27.
30. Bhatia S. Systems for Drug Delivery: Microbial Polysaccharides as Advance Nanomaterials. Springer International Publishing, Switzerland; 2016:29–54.
31. Bhatia S. Systems for Drug Delivery: Chitosan Based Nanomaterials and Its Applications. Springer International Publishing, Switzerland; 2016:55–117.
32. Bhatia S. Systems for Drug Delivery: Advance Polymers and Its Applications. Springer International Publishing, Switzerland; 2016:119–146.
33. Bhatia S. Systems for Drug Delivery: Advanced Application of Natural Polysaccharides. Springer International Publishing, Switzerland; 2016:147–170.
34. Bhatia S. Systems for Drug Delivery: Modern Polysaccharides and Its Current Advancements. Springer International Publishing, Switzerland; 2016:171–188.
35. Bhatia S. Systems for Drug Delivery: Toxicity of Nanodrug Delivery Systems. Springer International Publishing, Switzerland; 2016:189–197.
36. Bhatia S. Systems for Drug Delivery: Safety, Animal, and Microbial Polysaccharides. Springer Nature, Basingstoke, UK; 2016:122–127.
37. Bhatia S. Stem cell culture. In: Introduction to Pharmaceutical Biotechnology, Volume 3: Animal Tissue Culture and Biopharmaceuticals. IOP Publishing Ltd, Bristol, UK; 2019;3:1–24.
38. Bhatia S. Organ culture. In: Introduction to Pharmaceutical Biotechnology, Volume 3: Introduction to Animal Tissue Culture Science. IOP Publishing Ltd, Bristol, UK; 2019;3:1–28.
39. Bhatia S. Animal tissue culture facilities. In: Introduction to Pharmaceutical Biotechnology, Volume 3: Animal Tissue Culture and Biopharmaceuticals. IOP Publishing Ltd, Bristol, UK; 2019;3:1–32.
40. Bhatia S. Characterization of cultured cells. In: Introduction to Pharmaceutical Biotechnology, Volume 3: Animal Tissue Culture and Biopharmaceuticals. IOP Publishing Ltd, Bristol, UK; 2019:3;1–47.
41. Bhatia S. Introduction to genomics. In: Introduction to Pharmaceutical Biotechnology, Enzymes, Proteins and Bioinformatics. IOP Publishing Ltd, Bristol, UK; 2018;2:1–39.
42. Bhatia S. Bioinformatics. In: Introduction to Pharmaceutical Biotechnology, Enzymes, Proteins and Bioinformatics. IOP Publishing Ltd, Bristol, UK; 2018;3:1–16.
43. Bhatia S. Protein and enzyme engineering. In: Introduction to Pharmaceutical Biotechnology, Enzymes, Proteins and Bioinformatics. IOP Publishing Ltd, Bristol, UK; 2018;2:1–15.
44. Bhatia S. Industrial enzymes and their applications. In: Introduction to Pharmaceutical Biotechnology, Enzymes, Proteins and Bioinformatics. IOP Publishing Ltd, Bristol, UK; 2018;2:21.
45. Bhatia S. Introduction to enzymes and their applications. In: Introduction to Pharmaceutical Biotechnology, Enzymes, Proteins and Bioinformatics. IOP Publishing Ltd, Bristol, UK; 2018;2:1–29.
46. Bhatia S. Biotransformation and enzymes. In: Introduction to Pharmaceutical Biotechnology, Enzymes, Proteins and Bioinformatics. IOP Publishing Ltd, Bristol, UK; 2018;3:1–13.
47. Bhatia S. Modern DNA science and its applications. In: Introduction to Pharmaceutical Biotechnology, Volume 1: Basic Techniques and Concepts. IOP Publishing Ltd, Bristol, UK; 2018;1(3):1–70.
48. Bhatia S. Introduction to genetic engineering. In: Introduction to Pharmaceutical Biotechnology, Volume 1: Basic Techniques and Concepts. IOP Publishing Ltd, Bristol, UK; 2018;1(3):1–63.
49. Bhatia S. Applications of stem cells in disease and gene therapy. In: Introduction to Pharmaceutical Biotechnology, Volume 1: Basic Techniques and Concepts. IOP Publishing Ltd, Bristol, UK; 2018;1:1–40.
50. Bhatia S. Transgenic animals in biotechnology. In: Introduction to Pharmaceutical Biotechnology, Volume 1: Basic Techniques and Concepts. IOP Publishing Ltd, Bristol, UK; 2018;1:1–67.
51. Bhatia S. History and scope of plant biotechnology. In: Modern Applications of Plant Biotechnology in Pharmaceutical Sciences. Academic Press, 2015:1–30.
52. Bhatia S. Plant tissue culture. In: Modern Applications of Plant Biotechnology in Pharmaceutical Sciences. Academic Press, 2015:31–107.

53. Bhatia S. Laboratory organization. In: Modern Applications of Plant Biotechnology in Pharmaceutical Sciences. Academic Press, 2015:109–120.
54. Bhatia S. Concepts and techniques of plant tissue culture science. In: Modern Applications of Plant Biotechnology in Pharmaceutical Sciences. Academic Press, 2015:121–156.
55. Bhatia S. Application of plant biotechnology. In: Modern Applications of Plant Biotechnology in Pharmaceutical Sciences. Academic Press, 2015:157–207.
56. Bhatia S. Somatic embryogenesis and organogenesis. In: Modern Applications of Plant Biotechnology in Pharmaceutical Sciences. Academic Press, 2015:209–230.
57. Bhatia S. Classical and nonclassical techniques for secondary metabolite production in plant cell culture. In: Modern Applications of Plant Biotechnology in Pharmaceutical Sciences. Academic Press, 2015:231–291.
58. Bhatia S. Plant-based biotechnological products with their production host, modes of delivery systems, and stability testing. In: Modern Applications of Plant Biotechnology in Pharmaceutical Sciences. Academic Press, 2015:293–331.
59. Bhatia S. Edible vaccines. In: Modern Applications of Plant Biotechnology in Pharmaceutical Sciences. Academic Press, 2015:333–343.
60. Bhatia S. Microenvironmentation in micropropagation. In: Modern Applications of Plant Biotechnology in Pharmaceutical Sciences. Academic Press, 2015:345–360.
61. Bhatia S. Micropropagation. In: Modern Applications of Plant Biotechnology in Pharmaceutical Sciences. Academic Press, 2015:361–368.
62. Bhatia S. Laws in plant biotechnology. In: Modern Applications of Plant Biotechnology in Pharmaceutical Sciences. Academic Press, 2015:369–391.
63. Bhatia S. Technical glitches in micropropagation. In: Modern Applications of Plant Biotechnology in Pharmaceutical Sciences. Academic Press, 2015:393–404.
64. Bhatia S. Plant tissue culture-based industries. In: Modern Applications of Plant Biotechnology in Pharmaceutical Sciences. Academic Press, 2015:405–417.
65. Bhatia S, Al-Harrasi A, Behl T, Anwer MK, Ahmed MM, Mittal V, Kaushik D, Chigurupati S, Kabir MT, Sharma PB, Chaugule B, Vargas-de-la-Cruz C. Anti-migraine activity of freeze dried-latex obtained from *Calotropis gigantea* Linn. *Environ Sci Pollut Res Int.* 2021;29,18(2022):27460–27478. doi:10.1007/s11356-021-17810-x
66. Bhatia S. Natural Polymer Drug Delivery Systems: Nanotechnology and Its Drug Delivery Applications. Springer International Publishing, Switzerland; 2016:28.
67. Bhatia S. Natural Polymer Drug Delivery Systems: Nanoparticles Types, Classification, Characterization, Fabrication Methods and Drug Delivery Applications. Springer International Publishing, Switzerland; 2016:45–49.
68. Bhatia S. Natural Polymer Drug Delivery Systems: Natural Polymers vs Synthetic Polymer. Springer International Publishing, Switzerland; 2016:99–100.
69. Bhatia S. Natural Polymer Drug Delivery Systems: Plant Derived Polymers, Properties, and Modification & Applications. Springer International Publishing, Switzerland; 2016:111–122.
70. Bhatia S. Natural Polymer Drug Delivery Systems: Marine Polysaccharides Based Nano-Materials and Its Applications. Springer International Publishing, Switzerland; 2016:220–221.
71. Bhatia S. Systems for Drug Delivery: Mammalian Polysaccharides and Its Nanomaterials. Springer International Publishing, Switzerland; 2016:5.
72. Bhatia S. Protein and enzyme engineering. In: Introduction to Pharmaceutical Biotechnology, Enzymes, Proteins and Bioinformatics. IOP Publishing Ltd, Bristol, UK; 2018;2:1–7.
73. Bhatia S. Industrial enzymes and their applications. In: Introduction to Pharmaceutical Biotechnology, Enzymes, Proteins and Bioinformatics. IOP Publishing Ltd, Bristol, UK; 2018;2:27.
74. Bhatia S. Introduction to enzymes and their applications. In: Introduction to Pharmaceutical Biotechnology, Enzymes, Proteins and Bioinformatics. IOP Publishing Ltd, Bristol, UK; 2018;2:5–9.
75. Bhatia S. Biotransformation and enzymes. In: Introduction to Pharmaceutical Biotechnology, Enzymes, Proteins and Bioinformatics. IOP Publishing Ltd, Bristol, UK; 2018;3:1–07.
76. Bhatia S. Laws in plant biotechnology. In: Modern Applications of Plant Biotechnology in Pharmaceutical Sciences. Academic Press, 2015:378–380.
77. Santapau H, Henry AN. A Dictionary of the Flowering Plants in India. Reprint. New Delhi, India: National Institute of Science Publication, (CSIR), 1998. p. 152.
78. PK Warrier, VPK Nambier, PM Ganpathy. Some Important Medicinal Plants of the Western Ghats, India: A Profile. International Development Research Centre, New Delhi; 2000:343–360.

79. Hegde S, Hegde HV, Jalalpure SS, Peram MR, Pai SR, Roy S. Resolving identification issues of *Saraca asoca* from its adulterant and commercial samples using phytochemical markers. *Pharmacogn Mag.* 2017 Jul;13(Suppl 2):S266–S272.

80. Dhawan BN, Patnaik GK, Rastogi RP, Singh KK, Tandon JS. Screening of Indian plants for biological activity: part VI. *Indian J Exp Biol.* 1977 Mar;15(3):208–219. PMID: 914326.

81. Kirtikar KR, Basu BD. Indian Medicinal Plants. M/s Periodical Expert Book Agency, Delhi; 1981:883–884.

82. Nudrat SZ, Usha M. In: IA Khan, A Khanum (eds.) Medicinal and Aromatic Plants of India, Part I. Ukaaz Publication, Hyderabad; 2005:35.

83. Nadkarni KM. Indian Materia Medica, Vol 1. M/s Bombay Popular Prakashan Pvt Ltd, Mumbai; 1976:1104–1106.

84. Nayak C, Siddiqui VA, Rajpal VK, Singh HS, Pal R, Singh V, Roy RK, Prakash S, Rai Y, Das KC. *Saraca indica*: a multicentric double blind homoeopathic pathogenetic trial. *Indian J Res Homoeopathy.* 2009;3(2):32–36.

85. Pradhan P, Joseph L, Gupta V, Chulet R, Arya H, Verma R, Bajpai A. *Saraca asoca*: a review. *J Chem Pharm Res.* 2009;1(1):62–71.

86. Acharya YT. Ch. 1. Ver. 10. Varanasi: Chaukhamba Orientalia; 2011. Charaka Samhita of Agnivesha; Kalpasthana; Madanakalpa. Reprint ed; p. 653.

87. Acharya YT. Ch. 37. Ver. 5. Varanasi: Chaukhamba Sanskrita Sansthana; 2010. Sushruta Samhita of Sushruta; Sutrasthana; Bhumipravibhagiya. Reprint ed; p. 159.

88. Ketkar PM, Nayak SU, Pai SR, Joshi RK. Monitoring seasonal variation of epicatechin and gallic acid in the bark of *Saraca asoca* using reverse phase high performance liquid chromatography (RP-HPLC) method. *J Ayurveda Integr Med.* 2015 Jan-Mar;6(1):29–34. doi: 10.4103/0975-9476.146568. PMID: 25878461; PMCID: PMC4395925.

89. Ghose SC. Drugs of Hindustan, Vol. 9. M/s. Hahnemann Publishing Co. Pvt. Ltd, Calcutta; 1984:259–267.

90. Satyavati GV, Prasad DN, Sen SP, Das PK Oxytocic activity of a pure phenolic glycoside (P2) from *Saraca indica* Linn (Ashoka): a short communication. *Indian J Med Res.* 1970 May;58(5): 660–663.

91. Gupta M, Sasmal S, Mukherjee A. Therapeutic effects of acetone extract of *Saraca asoca* seeds on rats with adjuvant-induced arthritis via attenuating inflammatory responses. *ISRN Rheumatol.* 2014 Mar 4;2014:959687. doi: 10.1155/2014/959687. PMID: 24729890; PMCID: PMC3960775.

92. Ghosh S, Majumder M, Majumder S, Ganguly NK, Chatterjee BP. Saracin: a lectin from *Saraca indica* seed integument induces apoptosis in human T-lymphocytes. *Arch Biochem Biophys.* 1999 Nov 15; 371(2):163–168.

93. Lampronti I, Khan MT, Borgatti M, Bianchi N, Gambari R. Inhibitory effects of Bangladeshi medicinal plant extracts on interactions between transcription factors and target DNA sequences. *Evid Based Complement Alternat Med.* 2008 Sep; 5(3):303–312.

94. Sasmal S, Majumdar S, Gupta M, Mukherjee A, Mukherjee PK. Pharmacognostical, phytochemical and pharmacological evaluation for the antipyretic effect of the seeds of *Saraca asoca* Roxb. *Asian Pac J Trop Biomed.* 2012 Oct; 2(10):782–786.

95. Gupta M, Sasmal S, Mukherjee A. Therapeutic effects of acetone extract of *Saraca asoca* seeds on rats with adjuvant-induced arthritis via attenuating inflammatory responses. *ISRN Rheumatol.* 2014 Mar 4; 2014:959687. doi: 10.1155/2014/959687. PMID: 24729890; PMCID: PMC3960775.

96. Sharma A, Tilak R, Sisodia N. Evaluation of bioactivity of aqueous extracts of *Bougainvillea spectabilis*, *Saraca asoca*, and *Chenopodium album* against immature forms of *Aedes aegypti*. *Med J Armed Forces India.* 2019 Jul;75(3):308–311. doi: 10.1016/j.mjafi.2018.07.013. Epub 2018 Oct 15. PMID: 31388235; PMCID: PMC6676307.

97. Swar G, Shailajan S, Menon S. Activity based evaluation of a traditional Ayurvedic medicinal plant: *Saraca asoca* (Roxb.) de Wilde flowers as estrogenic agents using ovariectomized rat model. *J Ethnopharmacol.* 2017 Jan 4;195:324–333. doi: 10.1016/j.jep.2016.11.038. Epub 2016 Nov 22. PMID: 27884717.

98. Gupta M, Majumdar S, Banerjee S, Pal S, Mondal T. A double-blind randomized clinical trial of novel Ayurvedic muco-adhesive extended-release vaginal tablet (NA) for treatment of leucorrhea. *Int J Clin Obstet Gynaecol.* 2020;4(2):324–333.

99. Yadav NK, Saini KS, Hossain Z, Omer A, Sharma C, Gayen JR, Singh P, Arya KR, Singh RK. *Saraca indica* bark extract shows in vitro antioxidant, antibreast cancer activity and does not exhibit toxicological effects. *Oxid Med Cell Longev.* 2015;2015:205360.

100. Mukhopadhyay MK, Nath D. Phytochemical screening and toxicity study of *Saraca asoca* bark methanolic extract. *Int J Phytomed*. 2012 Apr. [S.l.], 3(4):498–505.

101. Singh S, Krishna TH, Kamalraj S, Kuriakose GC, Valayil JM, Jayabaskaran C. Phytomedicinal importance of *Saraca asoca* (Ashoka): an exciting past, an emerging present and a promising future. *Curr Sci*. 2015;109:1790–1801.

102. https://nmpb.nic.in/content/marketing-trade-1#market-price-of-medicinal-plants

103. https://nmpb.nic.in/sites/default/files/Projects/Chapter-10.pdf

104. Singh S, Krishna THA, Kamalraj S, Kuriakose GC, Valayil JM, Jayabaskaran C. Phytomedicinal importance of *Saraca asoca* (Ashoka): an exciting past, an emerging present and a promising future. *Curr Sci*. 2015;109(10):1790–1801. doi: 10.18520/cs/v109/i10/1790-1801.

105. Urumarudappa SKJ, Gogna N, Newmaster SG, Venkatarangaiah K, Subramanyam R, Saroja SG, Gudasalamani R, Dorai K, Ramanan US. DNA barcoding and NMR spectroscopy-based assessment of species adulteration in the raw herbal trade of *Saraca asoca* (Roxb) Willd, an important medicinal plant. *Int J Legal Med*. 2016;130(6):1457–1470. doi: 10.1007/s00414-016-1436-y

106. Hegde S, Hegde HV, Jalalpure SS, Peram MR, Pai SR, Roy S. Resolving identification issues of *Saraca asoca* from its adulterant and commercial samples using phytochemical markers. *Pharmacog Mag*. 2017;13:S266–S272. doi: 10.4103/pm.pm_417_16

107. Nadkarni AK. Indian Meteria Medica. Popular Prakashan, Mumbai; 1976.

108. Khare CP. Indian Medicinal Plants: An Illustrated Dictionary. Springer, Heidelberg; 2007.

109. Farag MA, Sakna ST, El-Fiky NM, Shabana MM, Wessjohann LA. Phytochemical, antioxidant and antidiabetic evaluation of eight *Bauhinia* L. species from Egypt using UHPLC-PDA-qTOF-MS and chemometrics. *Phytochemistry*. 2015 Nov;119:41–50.

110. Patra A, Deya AK, Kundu AB, Purushothaman KK, Saraswathy A. Shoreaphenol, a polyphenol from *Shorea robusta*. *Phytochemistry*. 1992;31(7):2561–2562.

111. Chow YL, Quon HH. Chemical constituents of the heartwood of *Mesua ferrea*. *Phytochemistry*. 1968; 7(10):1871–1874.

112. Shahid AP, Sasidharan N, Salini S, Padikkala J, Meera N, Raghavamenon AC, Babu TD. *Kingiodendron pinnatum*, a pharmacologically effective alternative for *Saraca asoca* in an Ayurvedic preparation, *Asokarishta*. *J Tradit Complement Med*. 2017 Jun 26;8(1):244–250.

113. Gupta M, Majumdar S, Banerjee S, Pal S, Mondal T. A double-blind randomized clinical trial of novel Ayurvedic muco-adhesive extended release vaginal tablet (NA) for treatment of leucorrhea. *Int J Clin Obstet Gynaecol*. 2020;4(2):324–333.

114. Akhtar Y, Alamgir M, Khan MTH, Hannan JMA, Choudhuri MSK. A single blind randomised placebo controlled clinical trial of a classical Ayurvedic formulation Ashokarista in the treatment of menorrhagia and dysmenorrhea. *Orient Pharm Exp Med*. 2007;7(4):372–378.

115. Baird K, Glod J, Steinberg SM, Reinke D, Pressey JG, Mascarenhas L, Federman N, Marina N, Chawla S, Lagmay JP, Goldberg J, Milhem M, Loeb DM, Butrynski JE, Turpin B, Staddon A, Spunt SL, Jones RL, Rodler ET, Schuetze SM, Okuno SH, Helman L. Results of a randomized, double-blinded, placebo-controlled, phase 2.5 study of saracatinib (AZD0530), in patients with recurrent osteosarcoma localized to the lung. *Sarcoma*. 2020 Apr 30;2020:7935475.

116. Anonymous. The Wealth of India (Vol. X. Sp-W). Publications and Information Directorate, Council of Scientific and Industrial Research (CSIR), New Delhi; 1982:580–585.

117. Mirjalili MH, Fakhr-Tabatabaei SM, Alizadeh H, Ghassempour A, Mirzajani F. Genetic and withaferin A analysis of Iranian natural populations of *Withania somnifera* and *W. coagulans* by RAPD and HPTLC. *Nat Prod Commun*. 2009 Mar;4(3):337–346. PMID: 19413110.

118. Bhatia A, Bharti SK, Tewari SK, Sidhu OP, Roy R. Metabolic profiling for studying chemotype variations in *Withania somnifera* (L.) Dunal fruits using GC-MS and NMR spectroscopy. *Phytochemistry*. 2013 Sep;93:105–115.

119. Anonymous. The Unani Pharmacopoeia of India, Part I, Vol. I. Deptt. of AYUSH. Ministry of Health & Family Welfare, Govt. of India, New Delhi; 2007:7–8.

120. Mishra LC, Singh BB, Dagenais S. Scientific basis for the therapeutic use of *Withania somnifera* (Ashwagandha): a review. *Altern Med Rev*. 2000;5:334–346.

121. Misra L, Mishra P, Pandey A, Sangwan RS, Sangwan NS, Tuli R. Withanolides from *Withania somnifera* roots. *Phytochemistry*. 2008;69:1000–1004.

122. Singh G, Sharma PK, Dudhe R, Singh S. Biological activities of *Withania somnifera*. *Ann Biol Res*. 2010;1(3):56–63.

123. Ganzera M, Choudhary MI, Khan IA. Quantitative HPLC analysis of withanolides in *Withania somnifera*. *Fitoterapia*. 2003;74:68–76.

124. Gupta GL, Rana AC (2007). *Withania somnifera* (Ashwagandha): a review. *Phcog Rev.* 1:129–136.
125. Bhattacharya SK, Goel RK, Kaur R, Ghosal S. Anti-stress activity of Sitoindosides VII and VIII. New Acylsterylglucosides from *Withania somnifera. Phytother Res.* 1987;1:32–37.
126. Ghosal S, Srivastava RS, Bhattacharya SK, Upadhyay SN, Jaiswal AK, Chattopadhyay U. Immunomodulatory and CNS effects of sitoindosides IX and X, two new glycowithanolides form *Withania somnifera. Phytother Res.* 1989;2:201–206.
127. Bhishagratna KKL. The Sushruta Samhita, second ed., Vol. III. Chowkhamba Sanskrit Series Office, Varanasi, India [An English translation based on the original texts]. Kaviraj Kunja Lal Bhishagratna, Calcutta; 1916;3:1907–1916.
128. Bhattacharya SK, Muruganandam AV. Adaptogenic activity of *Withania somnifera*: an experimental study using a rat model of chronic stress. *Pharmacol Biochem Behav.* 2003 Jun;75(3):547–555.
129. White PT, Subramanian C, Motiwala HF, Cohen MS. Natural withanolides in the treatment of chronic diseases. *Adv Exp Med Biol.* 2016;928:329–373.
130. Prabhu MY, Rao A, Karanth KS. Neuropharmacological activity of *Withania somnifera. Fitoterapia.* 1990;3:237–240.
131. Aphale AA, Chhibba AD, Kumbhakarna NR, Mateenuddin M, Dahat SH. Subacute toxicity study of the combination of ginseng (*Panax ginseng*) and ashwagandha (*Withania somnifera*) in rats: a safety assessment. *Indian J Physiol Pharmacol.* 1998 Apr;42(2):299–302. PMID: 10225062.
132. Ilayperuma I, Ratnasooriya WD, Weerasooriya TR. Effect of *Withania somnifera* root extract on the sexual behaviour of male rats. *Asian J Androl.* 2002 Dec;4(4):295–298.
133. Sharada AC, Solomon FE, Devi PU. Toxicity of *Withania somnifera* root extract in rats and mice. *Int J Pharmacogn.* 1993;31:205–212.
134. Abdullah Alharbi R. Structure insights of SARS-CoV-2 open state envelope protein and inhibiting through active phytochemical of ayurvedic medicinal plants from *Withania somnifera. Saudi J Biol Sci.* 2021 Jun;28(6):3594–3601.
135. Pathak P, Shukla P, Kanshana JS, Jagavelu K, Sangwan NS, Dwivedi AK, Dikshit M. Standardized root extract of *Withania somnifera* and Withanolide A exert moderate vasorelaxant effect in the rat aortic rings by enhancing nitric oxide generation. *J Ethnopharmacol.* 2021 Jun;3:114296.
136. Deshpande A, Irani N, Balkrishnan R, Benny IR. A randomized, double blind, placebo controlled study to evaluate the effects of ashwagandha (*Withania somnifera*) extract on sleep quality in healthy adults. *Sleep Med.* 2020 Aug;72:28–36.
137. Devpura G, Tomar BS, Nathiya D, Sharma A, Bhandari D, Haldar S, Balkrishna A, Varshney A. Randomized placebo-controlled pilot clinical trial on the efficacy of ayurvedic treatment regime on COVID-19 positive patients. *Phytomedicine.* 2021 Apr;84:153494.
138. Tiwari S, Gupta SK, Pathak AK. A double-blind, randomized, placebo-controlled trial on the effect of Ashwagandha (*Withania somnifera* dunal.) root extract in improving cardiorespiratory endurance and recovery in healthy athletic adults. *J Ethnopharmacol.* 2021 May 23;272:113929.
139. Jahanbakhsh SP, Manteghi AA, Emami SA, Mahyari S, Gholampour B, Mohammadpour AH, Sahebkar A. Evaluation of the efficacy of *Withania somnifera* (Ashwagandha) root extract in patients with obsessive-compulsive disorder: a randomized double-blind placebo-controlled trial. *Complement Ther Med.* 2016 Aug;27:25–29.
140. Chengappa KNR, Brar JS, Gannon JM, Schlicht PJ. Adjunctive use of a standardized extract of *Withania somnifera* (Ashwagandha) to treat symptom exacerbation in schizophrenia: a randomized, double-blind, placebo-controlled study. *J Clin Psychiatry.* 2018 Jul 10;79(5):17m11826.
141. Gannon JM, Brar J, Rai A, Chengappa KNR. Effects of a standardized extract of *Withania somnifera* (Ashwagandha) on depression and anxiety symptoms in persons with schizophrenia participating in a randomized, placebo-controlled clinical trial. *Ann Clin Psychiatry.* 2019 May;31(2):123–129.
142. Lopresti AL, Drummond PD, Smith SJ. A randomized, double-blind, placebo-controlled, crossover study examining the hormonal and vitality effects of ashwagandha (*Withania somnifera*) in aging, overweight males. *Am J Mens Health.* 2019 Mar-Apr;13(2):1557988319835985.
143. Lopresti AL, Smith SJ, Malvi H, Kodgule R. An investigation into the stress-relieving and pharmacological actions of an ashwagandha (*Withania somnifera*) extract: a randomized, double-blind, placebo-controlled study. *Medicine (Baltimore).* 2019 Sep;98(37):e17186.
144. Deshpande A, Irani N, Balakrishnan R. Study protocol and rationale for a prospective, randomized, double-blind, placebo-controlled study to evaluate the effects of Ashwagandha (*Withania somnifera*) extract on nonrestorative sleep. *Medicine (Baltimore).* 2018;97:e11299.
145. Choudhary B, Shetty A, Langade DG. Efficacy of Ashwagandha (*Withania somnifera* [L.] Dunal) in improving cardiorespiratory endurance in healthy athletic adults. *Ayu.* 2015;36(1):63.

146. Maurya SP, Das BK, Singh R, Tyagi S. Effect of *Withania somnifera* on CD38 expression on CD8+ T lymphocytes among patients of HIV infection. *Clin Immunol*. 2019;203:122–124.

147. Amritha N, Bhooma V, Parani M. Authentication of the market samples of Ashwagandha by DNA barcoding reveals that powders are significantly more adulterated than roots. *J Ethnopharmacol*. 2020 Jun 28;256:112725.

148. Sahu PK, Giri DD, Singh R, Pandey P, Gupta S, Shrivastava AK, Kumar A, Pandey KD. Therapeutic and medicinal uses of *Aloe vera*: a review. *Pharmacol Pharm*. 2013;4:599–610.

149. Surjushe A, Vasani R, Saple DG. *Aloe vera*: a short review. *Indian J Dermatol*. 2008;53(4):163–166.

150. Ahlawat KS, Khatkar BS. Processing, food applications and safety of *Aloe vera* products: a review. *J Food Sci Technol*. 2011 Oct;48(5):525–533.

151. Joseph B, Raj SJ. Pharmacognostic and phytochemical properties of *Aloe vera* Linn – an overview. *Int J Pharm Sci Rev Res*. 2010;4:106–110.

152. Manvitha K, Bidya B. *Aloe vera*: a wonder plant its history, cultivation and medicinal uses. *J Pharmacog and Phytochem*. 2014;2(5):85–88.

153. Hamman J. Composition and applications of *Aloe vera* leaf gel. *Molecules*. 2008;13:1599–1616.

154. Benítez S, Achaerandio I, Pujol M, Sepulcre F. *Aloe vera* as an alternative to traditional edible coatings used in freshcut fruits: a case of study with kiwifruit slices. *LWT – Food Sci Technol*. 2015; 61:184–193.

155. Ramachandra CT, Rao PR. Processing of *Aloe vera* leaf gel: a review. *Am J Agric Biol Sci*. 2008;3: 502–510.

156. Pal S, Sahrawat A, Prakash D. *Aloe vera*: composition, processing and medicinal properties. *Curr Discov*. 2013;2:106–122.

157. Vogler BK, Ernst E. *Aloe vera*: a systematic review of its clinical effectiveness. *Br J Gen Pract*. 1999;49:823–828.

158. Marchese A, Barbieri R, Sanches-Silva A, Daglia M, Nabavi SF, Jafari NJ, Izadi M, Ajami M, Nabavi SM. Antifungal and antibacterial activities of allicin: a review. *Trends Food Sci Technol*. 2016;52:49–56.

159. Surjushe A, Vasani R, Saple DG. *Aloe vera*: a short review. *Indian J Dermatol*. 2008;53:163–166.

160. Malik I, Zarnigar HN. *Aloe vera* – a review of its clinical effectiveness. *Int Res J Phar*. 2003;4:75–79.

161. Maan AA, Nazir A, Khan MKI, Ahmad T, Zia R, Murid M, Abrar M. The therapeutic properties and applications of *Aloe vera*: a review. *J Herb Med*. 2018;12:1–10.

162. Haynes LJ, Holdsworth DK, Russell R. C-glycosyl compounds. Part VI. Aloesin, a C-glucosylchromone from Aloe sp. *J Chem Soc*. 1970:2581–2586.

163. Groom QJ, Reynolds T. Barbaloin in aloe species. *Planta Med*. 1987 Aug; 53(4):345–348.

164. Grindlay D, Reynolds TJ. The *Aloe vera* phenomenon: a review of the properties and modern uses of the leaf parenchyma gel. *Ethnopharmacol*. 1986 Jun;16(2–3):117–151.

165. Smith T., Smith H. On aloin: the cathartic principles of aloes. *Month J Med Sci*. 1851;12:127–131.

166. Che QM, Akao T, Hattori M, Kobashi K, Namba T. Isolation of a human intestinal bacterium capable of transforming barbaloin to aloe-emodin anthrone. *Planta Med*. 1991 Feb;57(1):15–19.

167. Mapp RK, McCarthy TJ. The assessment of purgative principles in aloes. *Planta Med*. 1970 Aug;18(4):361–365.

168. Saravanan R, Gajbhiye N, Makasana JS, Ravi V. Barbaloin content of aloe (*Aloe barbadensis*) leaf exudates as affected by different drying techniques. *Indian J Agric Sci*. 2015;85.

169. Guo X, Mei N. *Aloe vera*: a review of toxicity and adverse clinical effects. *J Environ Sci Health C Environ Carcinog Ecotoxicol Rev*. 2016;34(2):77–96.

170. Grosse Y, Loomis D, Lauby-Secretan B, El Ghissassi F, Bouvard V, Benbrahim-Tallaa L, Guha N, Baan R, Mattock H, Straif K. International agency for research on cancer monograph working group. Carcinogenicity of some drugs and herbal products. *Lancet Oncol*. 2013 Aug;14(9):807–808.

171. International Agency for Research on Cancer. Aloe vera. In: Some Drugs and Herbal Products, Vol. 108. International Agency for research on Cancer, Lyon, France; 2015. http://monographs.iarc.fr/ENG/Monographs/vol108/mono108.pdf (accessed October, 2015).

172. Surjushe A, Vasani R, Saple DG. *Aloe vera*: a short review. *Indian J Dermatol*. 2008;53:163–166.

173. Boudreau MD, Beland FA. An evaluation of the biological and toxicological properties of *Aloe barbadensis* (Miller) *Aloe vera*. *J Environ Sci Health*. 2006;24:103–154.

174. Christaki EV, Panagiota C, Paneri F. *Aloe vera*: a plant for many uses. *J Food Agric Environ*. 2010;8:245–249.

175. Wu J, Zhang Y, Lv Z, Yu P, Shi W. Safety evaluation of *Aloe vera* soft capsule in acute, subacute toxicity and genotoxicity study. *PLoS One*. 2021 Mar 26;16(3):e0249356.

176. Hu J, Lloyd M, Hobbs C, Cox P, Burke K, Pearce G, Streicker MA, Gao Q, Frankos V. Absence of geno-toxicity of purified *Aloe vera* whole leaf dry juice as assessed by an in vitro mouse lymphoma tk assay and an in vivo comet assay in male F344 rats. *Toxicol Rep.* 2021 Mar 9;8:511–519.

177. Pothuraju R, Sharma RK, Onteru SK, Singh S, Hussain SA. Hypoglycemic and hypolipidemic effects of *Aloe vera* extract preparations: a review. *Phytother Res.* 2016 Feb;30(2):200–207.

178. Radha MH, Laxmipriya NP. Evaluation of biological properties and clinical effectiveness of *Aloe vera*: a systematic review. *J Tradit Complement Med.* 2014 Dec 23;5(1):21–26.

179. Ceravolo I, Mannino F, Irrera N, Squadrito F, Altavilla D, Ceravolo G, Pallio G, Minutoli L. Health potential of *Aloe vera* against oxidative stress induced corneal damage: an "in vitro" study. *Antioxidants (Basel).* 2021 Feb 20;10(2):318.

180. Shelton MS. *Aloe vera*, its chemical and therapeutic properties. *Int J Dermatol.* 1991;30:679–683.

181. West DP, Zhu YF. Evaluation of *Aloe vera* gel gloves in the treatment of dry skin associated with occupational exposure. *Am J Infect Control.* 2003;31:40–42.

182. Byeon S, Pelley R, Ullrich SE, Waller TA, Bucana CD, Strickland FM. *Aloe barbadensis* extracts reduce the production of interleukin-10 after exposure to ultraviolet radiation. *J Invest Dermtol.* 1988;110:811–817.

183. Sharifi E, Chehelgerdi M, Fatahian-Kelishadrokhi A, Yazdani-Nafchi F, Ashrafi-Dehkordi K. Comparison of therapeutic effects of encapsulated Mesenchymal stem cells in *Aloe vera* gel and Chitosan-based gel in healing of grade-II burn injuries. *Regen Ther.* 2021 Mar 21;18:30–37.

184. Zhong H, Li X, Zhang W, Shen X, Lu Y, Li H. Efficacy of a new non-drug acne therapy: *Aloe vera* gel combined with ultrasound and soft mask for the treatment of mild to severe facial acne. *Front Med (Lausanne).* 2021 May 21;8:662640.

185. Taqwim Hidayat A, Thohar Arifin M, Nur M, Muniroh M, Susilaningsih N. Ozonated *Aloe vera* oil effective increased the number of fibroblasts and collagen thickening in the healing response of full-thickness skin defects. *Int J Inflam.* 2021 Feb 9;2021:6654343.

186. Pangkanon W, Yenbutra P, Kamanamool N, Tannirandorn A, Udompataikul M. A comparison of the efficacy of silicone gel containing onion extract and *Aloe vera* to silicone gel sheets to prevent postoperative hypertrophic scars and keloids. *J Cosmet Dermatol.* 2021 Apr;20(4):1146–1153.

187. Akgül KT, Doğantekin E, Özer E, Kotanoğlu MS, Gökkurt Y, Hücümenoğlu S. Histopathological effects of *Aloe vera* on wound healing process in penile fracture model: an experimental study. *Turk J Med Sci.* 2021 Aug 30;51(4):2193–2197. doi:10.3906/sag-2102-224

188. Chithra R Sajithlal GB, Chandrakasan G. Influence of *Aloe vera* on collagen characteristics in healing dermal wounds in rats. *Mol Cell Biochem.* 1998;181:71–76.

189. Heggers J, Kucukcelebi A, Listengarten D, Stabenau J, Ko F, Broemeling LD, et al. Beneficial effect of aloe on wound healing in an excisional wound model. *J Altern Complement Med.* 1996;2:271–277.

190. Chithra P, Sajithlal G, Chandrakasan G. Influence of *Aloe vera* on the glycosaminoglycans in the matrix of healing dermal wounds in rats. *J Ethnopharmacol.* 1998;59:179–186.

191. Hutter JA, Salmon M, Stavinoha WB, Satsangi N, Williams RF, Streeper RT, et al. Anti-inflammatory C-glucosyl chromone from *Aloe barbadensis*. *J Nat Prod.* 1996;59:541–543.

192. Poordast T, Ghaedian L, Ghaedian L, Najib FS, Alipour S, Hosseinzadeh M, Vardanjani HM, Salehi A, Hosseinimehr SJ. *Aloe vera*; A new treatment for atrophic vaginitis, A randomized double-blinded controlled trial. *J Ethnopharmacol.* 2021 Apr 24;270:113760.

193. Ro JY, Lee B, Kim JY, Chung Y, Chung MH, Lee SK, et al. Inhibitory mechanism of aloe single component (Alprogen) on mediator release in guinea pig lung mast cells activated with specific antigen-antibody reactions. *J Pharmacol Exp Ther.* 2000;292:114–121.

194. Peng SY, Norman J, Curtin G, Corrier D, McDaniel HR, Busbee D. Decreased mortality of Norman murine sarcoma in mice treated with the immunomodulator, acemannon. *Mol Biother.* 1991;3:79–87.

195. Hart LA, Nibbering PH, van den Barselaar MT, van Dijk H, van den Burg AJ, Labadie RP. Effects of low molecular constituents from *Aloe vera* gel on oxidative metabolism and cytotoxic and bactericidal activities of human neutrophils. *Int J Immunopharmacol.* 1990;12:427–434.

196. Shi G, Jiang H, Zheng X, Zhang D, Jiang C, Zhang J. *Aloe vera* mitigates dextran sulfate sodium-induced rat ulcerative colitis by potentiating colon mucus barrier. *J Ethnopharmacol.* 2021 Apr 8;279:114108. doi:10.1016/j.jep.2021.114108

197. Sydiskis RJ, Owen DG, Lohr JL, Rosler KH, Blomster RN. Inactivation of enveloped viruses by anthra-quinones extracted from plants. *Antimicrob Agents Chemother.* 1991;35:2463–2466.

198. Kim HS, Kacew S, Lee BM. In vitro chemopreventive effects of plant polysaccharides (*Aloe barbadensis* Miller, *Lentinus edodes*, *Ganoderma lucidum*, and *Coriolus vesicolor*). *Carcinogenesis.* 1999;20:1637–1640.

199. Nerkar Rajbhoj A, Kulkarni TM, Shete A, Shete M, Gore R, Sapkal R. A comparative study to evaluate efficacy of curcumin and *Aloe vera* gel along with oral physiotherapy in the management of oral submucous fibrosis: a randomized clinical trial. *Asian Pac J Cancer Prev.* 2021 Feb 1;22(S1): 107–112.

200. Surjushe A, Vasani R, Saple DG. *Aloe vera*: a short review. *Indian J Dermatol.* 2008;53(4):163–166.

201. Deora N, Sunitha MM, Satyavani M, Harishankar N, Vijayalakshmi MA, Venkataraman K, Venkateshan V. Alleviation of diabetes mellitus through the restoration of β-cell function and lipid metabolism by *Aloe vera* (L.) Burm. f. extract in obesogenic WNIN/GR-Ob rats. *J Ethnopharmacol.* 2021 May 23;272:113921.

202. Kadry GM, Ismail MAM, El-Sayed NM, El-Kholy HS, El-Akkad DMH. In vitro amoebicidal effect of *Aloe vera* ethanol extract and honey against *Acanthamoeba* spp. cysts. *J Parasit Dis.* 2021 Mar;45(1):159–168.

203. Arbab S, Ullah H, Weiwei W, Wei X, Ahmad SU, Wu L, Zhang J. Comparative study of antimicrobial action of *Aloe vera* and antibiotics against different bacterial isolates from skin infection. *Vet Med Sci.* 2021 May 5;7(5):2061–2067. https://doi.org/10.1002/vms3.488

204. Hashemi SA, Madani SA, Abediankenari S. The review on properties of *Aloe vera* in healing of cutaneous wounds. *Biomed Res Int.* 2015;2015:714216.

205. Tanaka M, Misawa E, Ito Y, Habara N, Nomaguchi K, Yamada M, Toida T, Hayasawa H, Takase M, Inagaki M, Higuchi R. Identification of five phytosterols from *Aloe vera* gel as anti-diabetic compounds. *Biol Pharm Bull.* 2006 Jul;29(7):1418–1422.

206. Hasan MU, Riaz R, Malik AU, Khan AS, Anwar R, Rehman RNU, Ali S. Potential of *Aloe vera* gel coating for storage life extension and quality conservation of fruits and vegetables: an overview. *J Food Biochem.* 2021 Apr;45(4):e13640.

207. Sonawane SK, Gokhale JS, Mulla MZ, Kandu VR, Patil S. A comprehensive overview of functional and rheological properties of *Aloe vera* and its application in foods. *J Food Sci Technol.* 2021 Apr;58(4):1217–1226.

208. Pelley RP, Martini WJ, Liu DQ, et al. Multiparameter analysis of commercial "*Aloe vera*" materials and comparison to *Aloe barbadensis* Miller extracts. *Subtrop Plant Sci.* 1998;50:1–14.

209. Rahmani N, Khademloo M, Vosoughi K, Assadpour S. Effects of *Aloe vera* cream on chronic anal fissure pain, wound healing and hemorrhaging upon defection: a prospective double blind clinical trial. *Eur Rev Med Pharmacol Sci.* 2014;18(7):1078–1084.

210. López-Jornet P, Camacho-Alonso F, Molino-Pagan D. Prospective, randomized, double-blind, clinical evaluation of *Aloe vera* Barbadensis, applied in combination with a tongue protector to treat burning mouth syndrome. *J Oral Pathol Med.* 2013 Apr;42(4):295–301.

211. Hoopfer D, Holloway C, Gabos Z, Alidrisi M, Chafe S, Krause B, Lees A, Mehta N, Tankel K, Strickland F, Hanson J, King C, Ghosh S, Severin D. Three-arm randomized phase III trial: quality aloe and placebo cream versus powder as skin treatment during breast cancer radiation therapy. *Clin Breast Cancer.* 2015 Jun;15(3):181–90.e1–4.

212. Pradeep AR, Agarwal E, Naik SB. Clinical and microbiologic effects of commercially available dentifrice containing *Aloe vera*: a randomized controlled clinical trial. *J Periodontol.* 2012 Jun;83(6): 797–804.

213. Han HJ, Park CW, Lee CH, Yoo CW. A study on anti-irritant effect of *Aloe vera* gel against the irritation of sodium lauryl sulfate. *Korean J Dermatol.* 2004;42(4):413–419.

214. Molazem Z, Mohseni F, Younesi M, Keshavarzi S. *Aloe vera* gel and cesarean wound healing; a randomized controlled clinical trial. *Glob J Health Sci.* 2014 Aug 31;7(1):203–209.

215. Damani MR, Shah AR, Karp CL, Orlin SE. Treatment of ocular surface squamous neoplasia with topical *Aloe vera* drops. *Cornea.* 2015 Jan;34(1):87–89.

216. Oyelami OA, Onayemi A, Oyedeji OA, Adeyemi LA. Preliminary study of effectiveness of *Aloe vera* in scabies treatment. *Phytother Res.* 2009 Oct;23(10):1482–1484.

217. Choonhakarn C, Busaracome P, Sripanidkulchai B, Sarakarn P. A prospective, randomized clinical trial comparing topical *Aloe vera* with 0.1% triamcinolone acetonide in mild to moderate plaque psoriasis. *J Eur Acad Dermatol Venereol.* 2010 Feb;24(2):168–172.

218. Hajheydari Z, Saeedi M, Morteza-Semnani K, Soltani A. Effect of *Aloe vera* topical gel combined with tretinoin in treatment of mild and moderate acne vulgaris: a randomized, double-blind, prospective trial. *J Dermatolog Treat.* 2014 Apr;25(2):123–129.

219. Tanaka M, Misawa E, Yamauchi K, Abe F, Ishizaki C. Effects of plant sterols derived from *Aloe vera* gel on human dermal fibroblasts in vitro and on skin condition in Japanese women. *Clin Cosmet Invest Dermatol.* 2015 Feb 20;8:95–104.

220. Lu RB, Chen YN. Clinical study using aloe polysaccharides to prevent and treat skin reaction induced by radiation. *Chin J Cancer Prev Treat*. 2012;19(7):541–542.

221. Lewis JE, McDaniel HR, Agronin ME, Loewenstein DA, Riveros J, Mestre R, Martinez M, Colina N, Abreu D, Konefal J, Woolger JM, Ali KH. The effect of an aloe polymannose multinutrient complex on cognitive and immune functioning in Alzheimer's disease. *J Alzheimers Dis*. 2013;33(2):393–406.

222. Mansourian A, Momen-Heravi F, Saheb-Jamee M, Esfehani M, Khalilzadeh O, Momen-Beitollahi J. Comparison of *Aloe vera* mouthwash with triamcinolone acetonide 0.1% on oral lichen planus: a randomized double-blinded clinical trial. *Am J Med Sci*. 2011 Dec;342(6):447–451.

223. Khedmat H, Karbasi A, Amini M, Aghaei A, Taheri S. *Aloe vera* in treatment of refractory irritable bowel syndrome: trial on Iranian patients. *J Res Med Sci*. 2013 Aug;18(8):732.

224. Langmead L, Feakins RM, Goldthorpe S, Holt H, Tsironi E, De Silva A, Jewell DP, Rampton DS. Randomized, double-blind, placebo-controlled trial of oral *Aloe vera* gel for active ulcerative colitis. *Aliment Pharmacol Ther*. 2004 Apr 1;19(7):739–747.

225. Syed TA, Ahmad SA, Holt AH, Ahmad SA, Ahmad SH, Afzal M. Management of psoriasis with *Aloe vera* extract in a hydrophilic cream: a placebo-controlled, double-blind study. *Trop Med Int Health*. 1996 Aug;1(4):505–509.

226. Woźniak A, Paduch R. *Aloe vera* extract activity on human corneal cells. *Pharm Biol*. 2012 Feb;50(2):147–154.

227. Bhalang K, Thunyakitpisal P, Rungsirisatean N. Acemannan, a polysaccharide extracted from *Aloe vera*, is effective in the treatment of oral aphthous ulceration. *J Altern Complement Med*. 2013 May;19(5):429–434.

228. Panahi Y, Davoudi SM, Sahebkar A, Beiraghdar F, Dadjo Y, Feizi I, Amirchoopani G, Zamani A. Efficacy of *Aloe vera*/olive oil cream versus betamethasone cream for chronic skin lesions following sulfur mustard exposure: a randomized double-blind clinical trial. *Cutan Ocul Toxicol*. 2012 Jun;31(2):95–103.

229. Hegazy SK, El-Bedewy M, Yagi A. Antifibrotic effect of *Aloe vera* in viral infection-induced hepatic periportal fibrosis. *World J Gastroenterol*. 2012 May 7;18(17):2026–2034. doi: 10.3748/wjg.v18.i17.2026. PMID: 22563189; PMCID: PMC3342600.

230. Eghdampour F, Jahdie F, Kheyrkhah M, Taghizadeh M, Naghizadeh S, Haghani H. The effect of *Aloe vera* ointment in wound healing of episiotomy among primiparous women. *Iran J Obstet Gynecol Infertil*. 2013;15(35):25–31.

231. Bhatt S, Kumar H, Sharma M, Saxena K, Garg G, Singh G. Evaluation of hepatoprotective activity of *Aloe vera* in drug induced hepatitis. *World J Pharm Pharm Sci*. 2015;4(6):935–944.

232. Sudarshan R, Annigeri RG, Sree Vijayabala G. *Aloe vera* in the treatment for oral submucous fibrosis – a preliminary study. *J Oral Pathol Med*. 2012 Nov;41(10):755–761.

233. Rajar UD, Majeed R, Parveen N, Sheikh I, Sushel C. Efficacy of *Aloe vera* gel in the treatment of vulval lichen planus. *J Coll Physicians Surg Pak*. 2008 Oct;18(10):612–614.

234. Fan YJ, Li M, Yang WL, Qin L, Zou J. [Protective effect of extracts from *Aloe vera* L. var. chinensis (Haw.) Berg. on experimental hepatic lesions and a primary clinical study on the injection of in patients with hepatitis]. *Zhongguo Zhong Yao Za Zhi*. 1989 Dec;14(12):746–748.

235. Eshghi F, Hosseinimehr SJ, Rahmani N, Khademloo M, Norozi MS, Hojati O. Effects of *Aloe vera* cream on posthemorrhoidectomy pain and wound healing: results of a randomized, blind placebo-control study. *J Alternat Complement Med (New York, NY)*. 2010;16(6):647–650.

236. Størsrud S, Pontén I, Simrén M. A pilot study of the effect of *Aloe* barbadensis Mill. Extract (AVH2001) in patients with irritable bowel syndrome: a randomized, double-blind, placebo-controlled study. *J Gastrointest Liver Dis*. 2015;24(3):275–280.

237. Namiranian H, Serino G. The effect of a toothpaste containing *Aloe vera* on established gingivitis. *Swed Dent J*. 2012;36(4):179–185.

238. Syed TA, Afzal M, Ahmad SA, Holt AH, Ahmad SA, Ahmad SH. Management of genital herpes in men with 0.5% *Aloe vera* extract in a hydrophilic cream: a placebo-controlled double-blind study. *J Dermatol Treat*. 1997;8(2):99–102.

239. Oliveira SMA, Torres TC, Pereira SLS, Mota OML, Carlos MX. Effect of a dentifrice containing *Aloe vera* on plaque and gingivitis control: a doubleblind clinical study in humans. *J Appl Oral Sci*. 2008;16:293–296.

240. Haddad P, Amouzgar-Hashemi F, Samsami S, Chinichian S, Oghabian MA. *Aloe vera* for prevention of radiation-induced dermatitis: a self-controlled clinical trial. *Curr Oncol (Tor Ont)*. 2013;20(4): e345–e348.

241. Paulsen E, Korsholm L, Brandrup F. A double-blind, placebo-controlled study of a commercial *Aloe vera* gel in the treatment of slight to moderate psoriasis vulgaris. *J Eur Acad Dermatol Venereol.* 2005;19(3):326–331.

242. Krishnaveni M, Mirunalini S. Chemopreventive efficacy of *Phyllanthus emblica* L. (amla) fruit extraction 7,12-dimethylbenz(a)anthracene induced oral carcinogenesis – a dose-response study. *Environ Toxicol Pharmacol.* 2012;34(3):801–810.

243. Patel JR, Tripathi P, Sharma V, Chauhan NS, Dixit VK. *Phyllanthus amarus*: ethnomedicinal uses, phytochemistry and pharmacology: a review. *J Ethnopharmacol.* 2011;138 (2):286–313.

244. Variya BC, Bakrania AK, Patel SS, *Emblica officinalis* (Amla): a review for its phytochemistry, ethnomedicinal uses and medicinal potentials with respect to molecular mechanisms. *Pharmacol Res.* 2016;111:180–200.

245. Yadav SS, Singh MK, Singh PK, Kumar V. Traditional knowledge to clinical trials: a review on therapeutic actions of *Emblica officinalis. Biomed Pharmacother.* 2017;93:12921302.

246. Bhattacharya A, Chatterjee, A, Ghosal S, Bhattacharya SK. Antioxidant activity of active tannoid principles of *Emblica officinalis* (amla). *Indian J Exp Biol.* 1999;37(7):676–680.

247. Yadav SS, Singh MK, Singh PK, Kumar V. 2017. Traditional knowledge to clinical trials: a review on therapeutic actions of *Emblica officinalis. Biomed Pharmacother.* 1999;93:1292–1302.

248. Zhang Y, Zhao L, Guo X, Li C, Li H, Lou H, Ren D. Chemical constituents from *Phyllanthus emblica* and the cytoprotective effects on $H_2O_2$-induced PC12 cell injuries. *Arch Pharm Res.* 2016 Sep;39(9):1202–1211.

249. Yang F, Yaseen A, Chen B, Li F, Wang L, Hu W, Wang M. Chemical constituents from the fruits of *Phyllanthus emblica* L. *Biochem Syst Ecol.* 2020;92:104122.

250. Nguyen TA, Duong TH, Le Pogam P, Beniddir MA, Nguyen HH, Nguyen TP, Do TM, Nguyen KP. Two new triterpenoids from the roots of *Phyllanthus emblica. Fitoterapia.* 2018 Oct;130:140–144.

251. Khan KH. Roles of *Emblica officinalis* in medicine—a review. *Bot Res Int.* 2009;2:218–228.

252. Baliga MS. Triphala, Ayurvedic formulation for treating and preventing cancer: a review. *J Altern Complement Med (New York, NY).* 16(12);2010:1301–1308.

253. Gupta S, Kalaiselvan V, Srivastava S, Agrawal S, Saxena R. Evaluation of anticataract potential of Triphala in selenite-induced cataract: in vitro and in vivo studies. *J Ayurveda Integr Med.* 2010; 1(4):280–286.

254. Lu K, Chakroborty D, Sarkar C, Lu T, Xie Z, Liu Z, Basu S. Triphala and its active constituent chebulinic acid are natural inhibitors of vascular endothelial growth factor-a mediated angiogenesis. *PLoS One.* 2012;7(8):e43934.

255. Shengule S, Kumbhare K, Patil D, Mishra S, Apte K, Patwardhan B. Herb-drug interaction of Nisha Amalaki and Curcuminoids with metformin in normal and diabetic condition: a disease system approach. *Biomed Pharmacother.* 2018 May;101:591–598.

256. Khandelwal S, Shukla LJ, Shanker R. Modulation of acute cadmium toxicity by *Emblica officinalis* fruit in rat. *Indian J Exp Biol.* 2002;40(5):564–570.

257. Sai Ram M, Neetu D, Yogesh B, Anju B, Dipti P, Pauline T, Sharma SK, Sarada SK, Ilavazhagan G, Kumar D, Selvamurthy W. Cyto-protective and immunomodulating properties of Amla (*Emblica officinalis*) on lymphocytes: an in-vitro study. *J Ethnopharmaco.* 2002;81(1):5–10.

258. Rajeshkumar NV, Marie T, Ramadasan K. *Emblica officinalis* fruits afford protection against experimental gastric ulcers in rats. *Pharm Biol.* 2001;39(5)375–380.

259. Ramesh PS, Kokila T, Geetha D. Plant mediated green synthesis and antibacterial activity of silver nanoparticles using *Emblica officinalis* fruit extract. *Spectrochim Acta A Mol Biomol Spectrosc.* 2015 May 5;142:339–343.

260. Renuka R, Devi KR, Sivakami M, Thilagavathi T, Uthrakumar R, Kaviyarasu K. Biosynthesis of silver nanoparticles using *Phyllanthus emblica* fruit extract for antimicrobial application. *Biocatal Agric Biotechnol.* 24;2020:101567.

261. Sharma P, Joshi T, Joshi T, Chandra S, Tamta S. In silico screening of potential antidiabetic phytochemicals from *Phyllanthus emblica* against therapeutic targets of type 2 diabetes. *J Ethnopharmacol.* 2020 Feb 10;248:112268.

262. Sabu MC, Kuttan R. Anti-diabetic activity of medicinal plants and its relationship with their antioxidant property. *J Ethnopharmacol.* 2002 Jul;81(2):155–160.

263. Nampoothiri SV, Prathapan A, Cherian OL, Raghu KG, Venugopalan VV, Sundaresan A. In vitro antioxidant and inhibitory potential of Terminalia bellerica and *Emblica officinalis* fruits against LDL oxidation and key enzymes linked to type 2 diabetes. *Food Chem Toxicol.* 2011 Jan;49(1):125–131.

264. Balusamy SR, Veerappan K, Ranjan A, Kim YJ, Chellappan DK, Dua K, Lee J, Perumalsamy H. *Phyllanthus emblica* fruit extract attenuates lipid metabolism in 3T3-L1 adipocytes via activating apoptosis mediated cell death. *Phytomedicine.* 2020 Jan;66:153129.

265. Singh MK, Yadav SS, Gupta V, Khattri S. Immunomodulatory role of *Emblica officinalis* in arsenic induced oxidative damage and apoptosis in thymocytes of mice. *BMC Complement Altern Med.* 2013 Jul 27;13:193. doi: 10.1186/1472-6882-13-193

266. Tung YT, Huang CZ, Lin JH, Yen GC. Effect of *Phyllanthus emblica* L. fruit on methionine and choline-deficiency diet-induced nonalcoholic steatohepatitis. *J Food Drug Anal.* 2018 Oct;26(4):1245–1252.

267. Chaikul P, Kanlayavattanakul M, Somkumnerd J, Lourith N. *Phyllanthus emblica* L. (amla) branch: a safe and effective ingredient against skin aging. *J Tradit Complement Med.* 2021;11(5):390–399.

268. Rahman MM, Ferdous KU, Roy S, Nitul IA, Mamun F, Hossain MH, Subhan N, Alam MA, Haque MA. Polyphenolic compounds of amla prevent oxidative stress and fibrosis in the kidney and heart of 2K1C rats. *Food Sci Nutr.* 2020 May 20;8(7):3578–3589.

269. Yang B, Liu P. Composition and biological activities of hydrolyzable tannins of fruits of *Phyllanthus emblica. J Agric Food Chem.* 2014;62(3):529–541.

270. Chauhan P, Kumar RR, Mendiratta SK, Talukder S, Gangwar M, Sakunde DT, Meshram SK. In-vitro functional efficacy of extracts from *Phyllanthus emblica, Eucalyptus globulus, Tinospora cordifolia* as pancreatic lipase inhibitor and source of anti-oxidant in goat meat nuggets. *Food Chem.* 2021 Jun 30;348:129087.

271. Sharma UNS, Fernando DD, Wijesundara KK, Manawadu A, Pathirana I, Rajapakse RPVJ. Anticoccidial effects of *Phyllanthus emblica* (Indian gooseberry) extracts: potential for controlling avian coccidiosis. *Vet Parasitol: Reg Stud Rep.* 25;2021:100592.

272. Ananda A, Ramakrishnappa T, Archana S, et al. Green synthesis of MgO nanoparticles using *Phyllanthus emblica* for Evans blue degradation and antibacterial activity. *Mater Today Proc.* 2021. doi:10.1016/j.matpr.2021.05.340

273. Guo B, Liu C, Lin Y, Li H, Li N, Liu J, Fu Q, Tong W, Yu H. Fruit extracts from *Phyllanthus emblica* accentuate cadmium tolerance and accumulation in *Platycladus orientalis*: a new natural chelate for phytoextraction. *Environ Pollut.* 2021 Jul 1;280:116996.

274. Pandey AK, Gautam A, Pandey P, Dubey RS. Alleviation of chromium toxicity in rice seedling using *Phyllanthus emblica* aqueous extract in relation to metal uptake and modulation of antioxidative defense. *S Afr J Bot.* 121;2019:306–316.

275. Madhuri DR, Kavyashree K, Lamani AR, Jayanna HS, Nagaraju G, Mundinamani S. Reduction of graphene oxide by *Phyllanthus emblica* as a reducing agent – a green approach for supercapacitor application. *Mater Today Proc.* 2021.

276. Kannaujia R, Srivastava CM, Prasad V, Singh BN, Pandey V. Phyllanthus emblica fruit extract stabilized biogenic silver nanoparticles as a growth promoter of wheat varieties by reducing ROS toxicity. *Plant Physiol Biochem.* 2019 Sep;142:460–471.

277. Ankamwar B, Damle C, Ahmad A, Sastry M. Biosynthesis of gold and silver nanoparticles using *Emblica officinalis* fruit extract, their phase transfer and transmetallation in an organic solution. *J Nanosci Nanotechnol.* 2005 Oct;5(10):1665–1671.

278. Sidhu JS, Zafar TA. Super fruits: pomegranate, wolfberry, aronia (Chokeberry), acai, noni, and amla. In: Handbook of Fruits and Fruit Processing, Vol. 2 (eds. NK Sinha, JS Sidhu, J Barta, JSB Wu, MP Cano). Wiley-Blackwell, Oxford; 2012:667–671.

279. Alkandari D, Sarfraz H, Sidhu JS. Development of a functional food (pan bread) using amla fruit powder. *J Food Sci Technol.* 2019 Apr;56(4):2287–2295.

280. Middha SK, Goyal AK, Lokesh P, Yardi V, Mojamdar L, Keni DS, Babu D, Usha T. Toxicological evaluation of *Emblica officinalis* fruit extract and its anti-inflammatory and free radical scavenging properties. *Pharmacogn Mag.* 2015 Oct;11(Suppl 3):S427–S433.

281. Vishwanatha U, Guruprasad KP, Gopinath PM, Acharya RV, Prasanna BV, Nayak J, Ganesh R, Rao J, Shree R, Anchan S, Raghu KS, Joshi MB, Paladhi P, Varier PM, Muraleedharan K, Muraleedharan TS, Satyamoorthy K. Effect of Amalaki rasayana on DNA damage and repair in randomized aged human individuals. *J Ethnopharmacol.* 2016 Sep 15;191:387–397.

282. Chaphalkar R, Apte KG, Talekar Y, Ojha SK, Nandave M. Antioxidants of *Phyllanthus emblica* L. bark extract provide hepatoprotection against ethanol-induced hepatic damage: a comparison with silymarin. *Oxid Med Cell Longev.* 2017;2017:3876040.

283. Tasduq SA, Kaisar P, Gupta DK, Kapahi BK, Maheshwari HS, Jyotsna S, Johri RK. Protective effect of a 50% hydroalcoholic fruit extract of *Emblica officinalis* against anti-tuberculosis drugs induced liver toxicity. *Phytother Res.* 2005 Mar;19(3):193–197.

284. Sultana S, Ahmad S, Khan N, Jahangir T. Effect of *Emblica officinalis* (Gaertn) on CCl4 induced hepatic toxicity and DNA synthesis in Wistar rats. *Indian J Exp Biol.* 2005 May;43(5):430–436.
285. Sayed S, Ahsan N, Kato M, Ohgami N, Rashid A, Akhand AA. Protective effects of *Phyllanthus emblica* leaf extract on sodium arsenite-mediated adverse effects in mice. *Nagoya J Med Sci.* 2015 Feb;77(1–2):145–153.
286. He H, Wen X, Chen X, Zhang G, Huang Q, Zhang Y, Lin Y. Effects of *Phyllanthus emblica* spray interventions on xerostomia after general anesthesia for gynecologic tracheal intubation: a randomised controlled trial. *Eur J Integr Med.* 2020;33:101035.
287. Usharani P, Merugu PL, Nutalapati C. Evaluation of the effects of a standardized aqueous extract of *Phyllanthus emblica* fruits on endothelial dysfunction, oxidative stress, systemic inflammation and lipid profile in subjects with metabolic syndrome: a randomised, double blind, placebo controlled clinical study. *BMC Complement Altern Med.* 2019 May 6;19(1):97.
288. Karkon Varnosfaderani S, Hashem-Dabaghian F, Amin G, Bozorgi M, Heydarirad G, Nazem E, Nasiri Toosi M, Mosavat SH. Efficacy and safety of Amla (*Phyllanthus emblica* L.) in non-erosive reflux disease: a double-blind, randomized, placebo-controlled clinical trial. *J Integr Med.* 2018 Mar;16(2):126–131.
289. Shanmugarajan D, Girish C, Harivenkatesh N, Chanaveerappa B, Prasanna Lakshmi NC. Antihypertensive and pleiotropic effects of *Phyllanthus emblica* extract as an add-on therapy in patients with essential hypertension – a randomized double-blind placebo-controlled trial. *Phytother Res.* 2021 Jun;35(6):3275–3285.
290. Uchiyama T, Tsunenaga M, Miyanaga M, Ueda O, Ogo M. Oral intake of lingonberry and amla fruit extract improves skin conditions in healthy female subjects: a randomized, double-blind, placebo-controlled clinical trial. *Biotechnol Appl Biochem.* 2019 Sep;66(5):870–879.
291. Gao Q, Li X, Huang H, Guan Y, Mi Q, Yao J. The efficacy of a chewing gum containing *Phyllanthus emblica* fruit extract in improving oral health. *Curr Microbiol.* 2018 May;75(5):604–610.
292. Warude D, Preeti C, Joshi K, Patwardhan B. DNA isolation from fresh and dry plant samples with highly acidic tissue extracts. *Plant Mol Biol Rep.* 2003;21:467a–467f.
293. Kiran KR, Swathy PS, Paul B, Shama Prasada K, Radhakrishna Rao M, Joshi MB, Rai PS, Satyamoorthy K, Muthusamy A. Untargeted metabolomics and DNA barcoding for discrimination of Phyllanthus species. *J Ethnopharmacol.* 2021 Jun 12;273:113928.

# 4 Plant Profile, Phytochemistry, and Ethnopharmacological Uses of *Terminalia bellirica, Terminalia chebula,* and *Terminalia arjuna*

*Ahmed Al-Harrasi*
Natural and Medical Sciences Research Center, University of Nizwa

*Saurabh Bhatia*
Natural and Medical Sciences Research Center, University of Nizwa
School of Health Sciences, University of Petroleum and Energy Studies

*Mohammed F. Aldawsari*
Prince Sattam Bin Abdulaziz University

*Tapan Behl*
Chitkara University

## 4.1 INTRODUCTION

For the development of potent drugs, medicinal as well as aromatic plants have always been used. Owing to their established ethnopharmacological-based data, safety profile as well as significant therapeutic effects of these sources especially crude drugs are getting more attention. Various plant tissue culture-based approaches have been recently developed to improve or manipulate the secondary metabolite synthesis to improve the yield of the targeted phytopharmaceutical. Crude drugs of these days are in great demand because of their more positive effects on the biological system than allopathic medicine. So far, thousands of phytopharmaceuticals have been isolated from the plants along with the development of their respective analogues by using synthetic chemistry. However, certain chemical components cannot be synthesized easily because of the complexity associated with the structure. Various biotechnological approaches have been used to increase the secondary metabolite synthesis in the plants. Different approaches, such as elicitor technique, immobilization, hairy root culture, media optimization, precursor feeding, genetic manipulation, selection of elite, or superior variety, have been utilized to improve the secondary metabolite production [1–14]. Challenges faced by these phytopharmaceuticals, while using them for specific ailments, are their site-specific delivery or targeted delivery, unnecessary wastage, or metabolism of drugs before their arrival at the targeted site, and undesirable or toxic effects. Thus, to overcome these limitations, various natural polymers-based formulations have been developed in the form of nanoparticles or other forms not only to improve the therapeutic effect but also their safety profile. Various algae-based polymers such as porphyran, ulvan, fucoidan, carrageenan, and alginate have been used recently to formulate natural polymer-loaded herbal formulations.

DOI: 10.1201/9781003274124-4

Advancement in nanotechnology allows considerable improvement in the therapeutic effects as well as in their safety profile [15–28]. A class of crude drugs, known as essential oils, have been recently used to treat various respiratory ailments [29–53]. Various essential oils have been recently reported for their considerable effects against different respiratory pathogens, thus, can be utilized to treat different types of upper and lower respiratory tract infections. Just like essential oils, many other classes of unorganized crude drugs, such as mucilage's, gums, and latex, have been evidenced for treating various diseases. Organized crude drugs have been used from ancient time for various ailments. Traditional ethnopharmacological evidences are getting more importance because of their reliability and consumers' trust as well as their experience associated with the respective crude drugs. In this chapter, we have discussed about the three different species of *Terminalia* that have been traditionally utilized from decades and further gaining more attention due to their ethnopharmacological importance.

## 4.2  BAHERA

### 4.2.1  Vernacular Names

In English, it is known by *Belleric myrobalan*, whereas in Hindi, it is called Bahera; in Sanskrit, Bibhitaki; Urdu, Bahera.

### 4.2.2  Biological Sources

It is obtained from the dried ripe fruits of the plant *T. bellirica* Linn. The genus "*Terminalia*" consists of more than 200 tropical trees and shrubs, which are commonly found in the tropical areas. Until now, 39 *Terminalia* species have been studied for their chemical profile. These studies have shown the presence of phytochemicals such as terpenes, tannins, flavonoids, lignans, and simple phenols [54].
   **Family:** Combretaceae

### 4.2.3  Cultivation and Collection

The genus name "*Terminalia*" is originated from Latin literature, "terminus" or "terminalis" (ending), which means a general habit of the leaves being borne over the tips of shoots. *Terminalia* species are commonly present in the southern Asia, Himalayas, Madagascar, Australia, and the tropical and subtropical regions of Africa and considered ancient plants of various Asian countries such as Bangladesh, India, Sri Lanka, Pakistan, and Nepal, as well as South-East Asia. This old plant is indigenous to Bangladesh, Bhutan, Cambodia, China, Indonesia, Laos, Malaysia, Nepal, Pakistan, Sri Lanka, Thailand, Vietnam, and India. As far as its natural habitat is concerned, plants are often present in heavy rain forests, mixed deciduous forests, or dry deciduous dipterocarp forests, associated with teak. In Sanskrit, it's called Bibhitaki, and in India, it's called by its local name Bahera that is a large giant flowering tree found all over India apart from the dry western area up to an altitude of 900 m [29].

### 4.2.4  Chemical Composition

The fleshy fruit pulp of Bahera contains 21.4% tannin [55]. In 1989, various triterpenoids and their glucosides from *T. bellirica* such as arjungenin and its glucoside, belleric acid and its glucoside, and bellericoside have been identified and isolated [56]. In 2001, a new cardenolide, cannogenol 3-$O$-β-D-galactopyranosyl-(1→4)-$O$-α-l-rhamnopyranoside (1), was isolated form the seeds of *T. bellirica* [57, 58]. In 1992, two new pentacyclic triterpene acids, bellericagenin A and B, and their glycosides, bellericaside A and bellericaside B, were isolated from the stembark of *T. bellirica*. Bahera fruits comprise several triterpenoids (such as β-sitosterol and belleric acid) and the saponin glycosides (including bellericanin and bellericoside). In addition, they also contain polyphenolic components, including chebulagic acid, ellagic acid, ethyl gallate, gallic acid, and phyllemblin. Furthermore, the presences of lignans (anolignan B, thannilignan, termilignan, hydroxy-3′,4′-[methylenedioxy] flavan) and a fixed yellow oil [59] have also been reported. Chemically, it's different from *T. chebula*, as it doesn't contain corilagin and chebulic acid [59]. The previous report showed that gallic acid as well as octyl gallate derived from a methanolic fraction of Bahera fruit suppresses breast cancer cells (MCF-7 and MDA-MB-231). Both octyl gallate and gallic acid displayed reduced MCF-7 andMDA-MB-231 survival and caused apoptosis, with IC50 values of octyl gallate and gallic acid as 40 μM and 80 μM, respectively, without any toxic effect on normal breast cells (MCF-10A). Thus, both octyl gallate and gallic acid could act as potential anticancer agents [60]. There are many factors such as processing of plant materials that can influence the biological activity and chemical composition of *T. bellirica*. Fruits of *T. bellirica* in grilled form have been used as a promising treatment for diarrhoea. It was found that grilling considerably changes the levels of metabolites in *T. bellirica* fruits, which could be accountable for its enhanced therapeutic potential. Non-targeted metabolomics analysis demonstrated greater levels of valeric acid, malic acid, uridine, 1,2,3-tris-benzene, tartaric acid, succinic acid, oxalic acid, malonic acid, gallic acid, and 11-eicosenoic acid in grilled fruits. HPTLC results have shown a high level of gallic acid, on the other hand, less ellagic acid content in grilled powder of fruit when compared with dried powdered materials [61].

### 4.2.5 Traditional Uses

Its medicinal utilization is evidenced in various folk systems, including traditional Chinese medicine, Ayurveda, Unani, and Siddha. Utilization of *T. bellirica* is not only evidenced among these traditional systems of medicine but also their utilization is common in China, Tibet, Mongolia, and other indigenous systems in Vietnam, Burma, Malaysia, Thailand, and other Asian (southeast) nations. In Nepal, Bahera fruit is consumed as food by numerous ethnic groups. In the classical language of South Asia, Sanskrit, it is called Bibhitaki which means "fearless", as it takes away the fear of disease; the plant material, especially fruit part, prevents risks of illness. In Indian mythology, it is believed that *T. bellirica* was lived by devils and individuals those who used to lie down under tree were at risks to an attack. However, owing to its therapeutic effects, the tree is also acknowledged as Anila-ghnaka (in Sanskrit) or "wind-killing". Ayurveda has evidenced various uses of *T. bellirica* such as bitter, acrid, astringent, laxative, germicidal, and antipyretic. It's also used in the treatment of various diseases such as nausea, cough, diarrhoea, dysentery, indigestion, ocular-based complications, flatulence, inflammation of the small intestine, leprosy, hepatic complications, and tuberculosis [62].

*T. bellirica* has also been known for its ability to clean the blood as well as voice and to stimulate hair growth. Whole plants, especially fruits, are known as anthelmintic, antipyretic, astringent, hair tonic, and laxative. It's widely known for its benefits in treating several complications such as asthma, bronchitis, coughs, diarrhoea, indigestion, ocular complications, hepatitis, dysphonia, piles, and poisoning caused by scorpion-sting. In traditional practices, *T. bellirica* fruit (decoction) is administered to treat cough; however, fruit pulp is beneficial in diarrhoea, oedema, piles, and leprosy, and half-ripe fruit is employed as purgative. Fruits have also been known for their utilization in menstrual disorder in Khagrachari. The triterpenoid composition of fruits demonstrates considerable antimicrobial effects. Sweet kernels inside the hard covering of *T. bellirica* fruit show narcotic effects, whereas kernel oil demonstrated purgative effects and its extended utilization has been found to be well tolerated in mice [63]. Exudates such as gum obtained from the bark demonstrate demulcent and purgative effects.

### 4.2.6 Traditional Formulations

Triphala, a popular Ayurvedic formulation that means "three fruits" (in Sanskrit), consists of a mixture of three fruits obtained from *Emblica officinalis* (*Phyllanthus emblica*), *T. bellirica* (Bahera and Bibhitaki), and *T. chebula* (Haritaki). Due to its antioxidant and anti-inflammatory properties, Triphala is widely consumed in India. It's also a component of the most popular rasayanic tonic herb used for rasayanic treatment. Triphala has been known for its various therapeutic effects such as it's widely used in the treatment of constipation, ulcerative colitis, indigestion, and malabsorption and considered an effective formulation to clean colon and colon tonifying. It's also used in the treatment of anaemia, asthma, cough, fever, jaundice, leucorrhoea, obesity, and pyorrhoea [38, 39]. It was found that a herbal formulation called HP-1, containing *Phyllanthus niruri* and extracts of *T. bellirica*, *T. chebula*, *P. emblica*, and *Tinospora cordifolia* presented hepatoprotective effects against CCl4-induced toxicity in rats [64, 65].

*Bhasma* of biotite mica (an iron-rich mineral) is used for several chronic complications, works as *Rasayana*, and is a vital component of several herbo-mineral preparations. Three different types of *Abhraka bhasma* obtainable in the market are *sahasraputi*, *dashaputi*, and *shataputi*. Based on *indication for which the bhasma is to be used (as per Ayurvedic principle),* plants utilized in *Abhraka bhasma formulation vary.* Recent reports have shown a development of Krishna Vajra Abhraka bhasma using traditional *bhasma*-making technology that includes *Ficus benghalensis*, *Solanum virginiatum*, *Adhatoda vasica*, *T. bellirica*, and *Ricinus communis*. It was found that healing potential of this preparation could be due to the cow milk and specific herbs used during the process [66].

A polyherbal formulation comprising *Allium sativum* L., *T. bellirica*, *Curcuma aeruginosa*, and *Amomum compactum* which is used for hypertension treatment has been recently investigated to evaluate the acute and subchronic oral toxicity of the polyherbal formulation in rats. Results obtained from this study suggested that short- and long-term oral administration of the polyherbal formulation is safe to use within its dose recommendation [67].

Mathurameha, a Thai folk medicine, containing 26 herbs, including *T. bellirica*, was used by Ayurvedic (Chevagakomarapat) institute to cure diabetes for almost 30 years. New studies have confirmed the effectiveness as well as safety profile of the ingredients present in traditional formulation. It was found that aqueous fraction-based formulation has ability to reduce the blood sugar, improved biochemical parameters without causing any toxicity [68]. Another poly herbal formulation, Nawarathne Kalka which has been used in indigenous system of Sri Lanka comprises 14 different plants, including *T. bellirica*, which has been used for different conditions such as gastrointestinal tract complications, including diarrhoea, pain in abdomen, blood in stool, indigestion, dyspepsia as well as for arthritis (rheumatoid), and other inflammatory diseases. Recent studies have demonstrated that the potential of Nawarathne Kalka, as an antioxidant, causes cytotoxicity and suppresses advanced glycation end products [69]. *Recently Ayurvedic formulations containing T. bellirica such as Vilwadi Gulika and Mukkamukkatuvadi Gulika have been investigated, and it was concluded that* the active components of the selected medicines prove therapeutic potentials like anti-viral, immunomodulatory, and anti-inflammatory activities. Another formulation, *Indukantham Kwatham,* which doesn't contain *T. bellirica* and is used for recurrent fever, reduces the feeling of tiredness and improves the resistance power, was also studied. In addition, HPTLC analysis revealed the presence of many active molecules like umbelliferone, scopoletin, caffeic acid, ferulic acid, gallic acid, piperine, curcumin, berberine, and palmatine [70].

## 4.2.7 PHARMACOLOGICAL USES

Based on the various pharmacological studies, it was found that *T. bellirica* presents various therapeutic effects such as anticancer, antidiabetic, anti-hyperlipidaemic, anti-inflammatory, antimicrobial, antioxidant, hepatoprotective, immunomodulatory, and renoprotective effects. These pharmacological activities are attributed to the occurrence of therapeutically active components, including glucoside, tannins, corilagin, arjunolic acid, ellagic acid, chebulagic acid, ethyl gallate, gallic acid, and galloyl glucose. Fruits have been known for various ethno-medicinal effects such as laxative, astringent, anthelmintic, and antipyretic properties. In Ayurveda, as mentioned above, fruits have been known for various disorders like asthma, bronchitis, coughs, dyspepsia, diarrhoea, hepatitis, piles, and eye infections [71]. Apart from these traditional uses, fruits are present potential pharmacological effects such as antidiabetic, anti-inflammatory, antioxidant, analgesic, and antidiarrhoeal [72].

#### 4.2.7.1 Antidiabetic Activity

Recent studies have presented that *T. bellirica* fruit extracts possessed hypoglycaemic effect in alloxan-induced diabetic rats and regulated hepatic and renal functions. Extract derived from *T. bellirica* fruit showed antioxidant, α-amylase inhibitory, and antidiabetic effects, and thus, it could be beneficial for the management of hyperglycaemia and oxidative stress [73]. In 2011, findings from one study showed that methanolic extract of *T. bellirica* and *E. officinalis* can act as an effective α-amylase as well as an α-glucosidase inhibitor. Both extracts showed considerable antiglycation effects and supressed the low-density lipoprotein oxidation in vitro. Also, LC–MS data showed the existence of ellagic acid and ascorbic acid as the main compounds in extracts [74].

#### 4.2.7.2 Antinociceptive Activity

An earlier study suggested that *T. bellirica* (fruits) as well as *T. chebula* (fruits) possess potent analgesic response and might be used as a possible agent for the bioassay-based isolation of phytochemicals employed in the treatment of pain (chronic) [75].

#### 4.2.7.3 Antiplasmodial Effects

An earlier study has also shown in vitro and in vivo antiplasmodial effects and the cytotoxicity of *P. emblica* Linn, *T. chebula* Retz, and *T. bellirica* (Gaertn) Roxb extracts. It was found that all fractions demonstrated considerable antiplasmodial effects (both in vitro and in vivo) [76]. In 2008, a study was performed to establish scientific background for the medicinal use of *T. bellirica* in hyperactive GIT as well as pulmonary-based complications. It was found that fruit of *T. bellirica* has the ability to inhibit acetylcholine as well as $Ca^{+2}$, which presents its traditional use in GIT complications and lung disorders [77].

#### 4.2.7.4 ACE Inhibitory Activity

Earlier reports suggested that ACE inhibitory activity of extracts derived from *T. bellirica* could be due to the presence of gallic acid along with other metabolites in the extract. It was found that ethyl acetate fraction contains a maximum amount of gallic acid, which could be attributed to its ACE inhibitory activity [78].

#### 4.2.7.5 Angiogenic Activity

Earlier research suggested that ethanolic extract of *T. bellirica* leaf possessed a significant angiogenic activity in vivo. It was also demonstrated that promising angiogenic potential could be due to the existence of the phytochemicals [79].

#### 4.2.7.6 Anti-Atherogenic Effect

An earlier report suggested that *T. bellirica* extract showed suppressive effects over the oxidation of low-density lipoprotein and inflammation caused by macrophage in vitro, indicating that its in vivo utilization could prevent plaque development in arteries. It was found that in THP-1 macrophages (human acute monocytic leukaemia cell line), *T. bellirica* extract treatment caused a considerable downregulation of mRNA of TNF-α, IL-1β, and lectin-type oxidized low-density lipoprotein receptor one. In addition, *T. bellirica* extract decreased matrix metalloproteinase 9 release as well as an intracellular ROS release in instinctively immortalized monocyte-like cell line [80].

#### 4.2.7.7 Anti-Inflammatory Effects

The previous report suggested that ethyl acetate extracts of fruits obtained from *T. bellirica* inhibit inflammatory markers via decreasing nuclear factor-κB in lipopolysaccharide-treated raw 264.7 cells. This proinflammatory response was due to the inhibition of ROS as well as nitric oxide species, suppression of the synthesis of arachidonic acid metabolites, inflammatory markers as well as release of cytokines release by extract [81]. Recently, a team of researchers from Japan explored the molecular basis of inflammation as well as oxidative-stress-reducing capability of *T. bellirica* and its main polyphenolic components. It was discovered that *T. bellirica* extract and gallic acid reduce inflammation caused by LPS and showed antioxidant effects via Akt/AMPK/Nrf2 and MAPK/NF-κB pathways, which suggests that *T. bellirica* extract and gallic acid might be actively used to cure inflammation-associated complications [82]. In an earlier study, gallic acid, derived from *T. bellirica*, demonstrated considerable suppression against cyclooxygenase without suppression of 5-lipoxygenases activity. Thus, being a phytochemical with specific as well as reversible suppression of cyclooxygenase 2, gallic acid can be considered a potential agent to treat inflammation [83].

### 4.2.7.8 Antifibrotic Activity

Previous findings also suggested the antifibrotic activity of ethyl acetate fraction of *T. bellirica* fruit. It was found that its mechanism of action could be attributed to the suppression of collagen synthesis, cytokine secretion, and TGF-β1/Smad pathway. Moreover, ethyl acetate fraction enables apoptosis in hepatic stellate cells [84].

### 4.2.7.9 Antibacterial Activity

Recent findings showed that *T. bellirica* fruit could be a potential source for developing broad-spectrum antibacterial drugs against multidrug-resistant bacteria, bacteria, which are non-toxic to mammalian cells and possess high antioxidant activity. All fractions supress *Staphylococcus aureus* (methicillin-resistant), *Acinetobacter* spp. (MDR), and *Pseudomonas aeruginosa* (MDR) effectively. The sequential aqueous extracts (MIC, 4mg/ml) inhibited extended-spectrum β-lactamase-producing *Escherichia coli*. However, none of the extracts exhibited activity against multidrug-resistant *Klebsiella pneumoniae* [85].

Recently by using microwave radiations and optimized conditions, *T. bellirica* extract (fruit)-based colloidal silver NPs have been synthesized. Results showed that developed silver nanoparticles presented considerable catalytic responses in the presence of sodium borohydride by reducing 4-nitrophenol into 4-aminophenol. In Addition, prepared silver nanoparticles presented possible antibiofilm as well as antibacterial effects against pathogenic bacteria such as *Bacillus subtilis*, *S. aureus*, *E. coli*, and *P. aeruginosa* [86]. Earlier studies have also suggested that extract from *T. bellirica* showed strong activity against *Streptococcus mutans*. The extract also prevents the formation of biofilms by the bacteria.

### 4.2.8 INDUSTRIAL APPLICATIONS

### 4.2.8.1 Purification of Water

Variety of nanoparticles emerged as an effective catalyst in the photocatalytic degradation of real textile industrial water. Recent reports have shown the synthesis of *T. bellirica* fruit extract-mediated silver nanoparticles and its utilization in the photocatalytic degradation of wastewater from textile industries. It was found that silver nanoparticles synthesized by *T. bellirica* fruit ethanolic extracts showed good photocatalytic activity (which can be used in water purification) as well as antibacterial activities [87].

Biogenic Fe and NiFe nanoparticles prepared using *T. bellirica* extracts, which act as a reducing and capping agent, have been recently developed for water treatment applications. These nanoparticles presented higher adsorptive capacity for Cr(VI) and faster rate kinetics for methylene blue degradation in a catalyst, peroxygen-assisted oxidation process, when compared to the chemogenic nanoparticles [88].

### 4.2.8.2 For the Management of Wet Litter in Broiler Chickens

Recent studies have shown the assessment of polyherbal preparation (Stodi®) for the treatment of broiler chickens that excrete wet droppings. Stodi® is a mixture of five medicinal plants that are widely available in India and are used to treat diarrhoeal complications. For this, wet litter has been induced in broiler chicken by the administration of a more amount of magnesium chloride. The administration of polyherbal formulation improved these complications and, hence, could be used for the treatment of wet litter in chickens [89].

### 4.2.8.3 Mitigate Toxicity Induced by Medications and Other Chemicals

Recent reports suggested that fruit extracts of *T. bellirica* as well as ellagic acid showed antioxidant as well as hepatoprotective effect by reversing complications caused by chronic administration of diclofenac [90]. Recent studies have also shown the protective effect of *T. bellirica* against drugs-induced cardiotoxicity in Wistar Albino rats. It was found that the treatment of rats with a methanolic extract of *T. bellirica* demonstrated considerable decrease in creatine kinase-muscle/brain and malondialdehyde levels. In addition, this treatment also increased reduced glutathione, superoxide dismutase, and catalase activity. Treatment with *T. bellirica* also normalized oxidative stress markers such as like alkaline phosphatase, uric acid, alanine transferase, and aspartate transaminase. Furthermore, along with substantial restoration in histopathological findings of myocardium, treatment with methanolic extract of *T. bellirica* reduced the levels of total cholesterol and triglycerides in serum and considerably enhanced high-density lipoprotein levels [91]. Recent studies have demonstrated that the treatment of animals with an aqueous acetone extract of *T. bellirica* fruits mitigates CCl4-induced oxidative stress and hepatotoxicity in rats [92]. Another study showed hepatoprotective and antioxidant effects of phytoconstituent ellagic acid and *T. bellirica* fruit extracts in long-term aceclofenac (nonsteroidal anti-inflammatory drug)-treated rats. In vitro antioxidant assays showed that Bahera fruit fraction derived by using ethyl acetate possessed greater metal ion chelating and nitric oxide antioxidant effects in comparison to aqueous fraction. It was also found that in vivo efficacy of ellagic acid was more than *T. bellirica* fruit extracts [93].

## 4.2.9  Clinical Data

Clinical trial (double-blind) was performed in the group of aged peoples between 15 and 40 years to establish the therapeutic activity of *Gooseberry, Triphala,* Bahera, and myrobalan aqueous fraction against *S. mutans.* It was ultimately found that Triphala at 10% was found to be more effective than others [94].

## 4.2.10  Adulterants and Substituents

*T. bellirica* and *T. chebula* themselves considered adulterants of various medicinal plants in few reports. *T. bellirica* (bark), *T. chebula* (bark), and *Terminalia tomentosa* have been described as adulterants of Arjuna crude drug [95]. *T. bellirica* is also considered adulterant for *Tectona grandis* [96]. Recent report also showed that the extent of adulteration in raw herbal trade of 30 important medicinal plants in South India was studied. No adulteration in *T. bellirica* fruit samples [97] was found.

## 4.2.11  Current Status

### 4.2.11.1  Contraindications

- Use of *T. bellirica* must be avoided during pregnancy as it is not safe during pregnancy and breastfeeding.
- Since *T. bellirica* presents hypoglycaemic effects, it must be carefully administered to diabetic patients who are using antidiabetic medications.
- Owing to its hypoglycaemic effects, it may affect blood sugar control surgery; therefore, it must be administered in no less than two weeks prior to surgery.

### 4.2.11.2  Toxicity

No data found. In recent studies, in vivo toxicity associated with aqueous acetone extract of *T. bellirica* (Gaertn.) Roxb fruits was evaluated. It was found that the acute administration of extracts in female Wistar Albino rats as a single dose up to 2000-mg/kg body weight showed a normal weight of the organ (relative), biochemical profile, blood reports, and normal histopathology. These findings suggested that aqueous fraction of Bahera fruit is safe and can be used as a conventional herbal preparation for its oxidative stress reducing and additional healthiness [98].

### 4.2.11.3  Most Cited Research

In 2010, 34 polyphenolic compounds in methanol extracts of the fruits of *T. bellirica*, *T. chebula*, and *Terminalia horrida* were identified. Two ellagic-acid-based polyphenolic compounds were identified and showed oxidative-stress-reducing effects [99]. Another most cited research of 1997 demonstrated an isolation of anti-HIV-1, antimalarial, and antifungal activity termilignan, thannilignan, 7-hydroxy-3',4'-(methylenedioxy)flavan, and anolignan B [100]. In 2011, the antidiabetic activity of *T. bellirica* has been demonstrated. It was found that gallic acid in Bahera fruit extract is accountable for the of beta-cells regeneration and normalized all the biochemical profile related with diabetes mellitus [101].

## 4.3  ARJUNA BARK

### 4.3.1  VERNACULAR NAMES

Commonly known as Arjuna, Partha, Indradru, and Veeravriksha is a big, deciduous, perennial, pest, and tree devoid of any infection.

### 4.3.2  BIOLOGICAL SOURCE

*T. arjuna* (Roxb.), or named as Arjun indigenous to India, is an evergreen tree and a part of ancient Indian literature, for its effectiveness against cardiac disease as mentioned in the ancient Ayurvedic text book, Charaka Samhita by Charaka and practised by the Ayurvedic practitioners like Chakradatta and Bhavamishra. Other similar looking tree, *Terminalia phaeocarpa* Eichler is indigenous to Brazil, also named capitão. *T. phaeocarpa* is further found to be like *Terminalia argentea* Mart., also named Capitão, not endemic to Brazil. Both trees have been used as antidiabetic plants [102].

**Family:** Combretaceae

### 4.3.3  GEOGRAPHICAL DISTRIBUTION

*T. arjuna* plants are distributed throughout India mainly in Konkan region, Bengal, Madhya Pradesh, Deccan Plateau, Orissa, Punjab, South Bihar, and Uttar Pradesh. Apart from its presence in India, it's also found in the forests of Mauritius, Sri Lanka, and Burma [103].

### 4.3.4  CHEMICAL COMPOSITION

Main chemical components are ß-sitosterol, ellagic acid, and arjunic acid. *T. arjuna* stem bark chemically contains oleane-type triterpenes like arjunic acid, arjunolic acid, arjungenin, arjunetin, tannins, glycosides, and high level of ellagic acids, phytosterols, flavonoids (quercetin, kaempferol, luteolin, and pellargonidin), and minerals such as zinc, magnesium, copper, and calcium. Triterpenes such as arjunic acid, arjunolic acid, and arjungenin are present in *T. arjuna*. In addition, it also contains triterpene glycosides from the *T. arjuna* tree such as arjunetin, arjunoglucoside I-III, and arjunoside I-II. *T. arjuna* also contains polyphenolic compounds such as arjunin, arjunone, and arjunolone [104]. The plant also contains minerals. Chemical composition of *T. arjuna* bark is demonstrated in Figures 4.1 and 4.2.

    *T. arjuna* has been described to exhibit hypolipidaemic effects, and it's assumed that this is due to the presence of saponin glycosides in *T. arjuna*, and these chemical components might be accountable for its inotropic properties, whereas the flavonoids and phenolics might offer radical scavenging and vascular strengthening effects. Due to this, *T. arjuna* bark presents cardioprotective effects [105, 106].

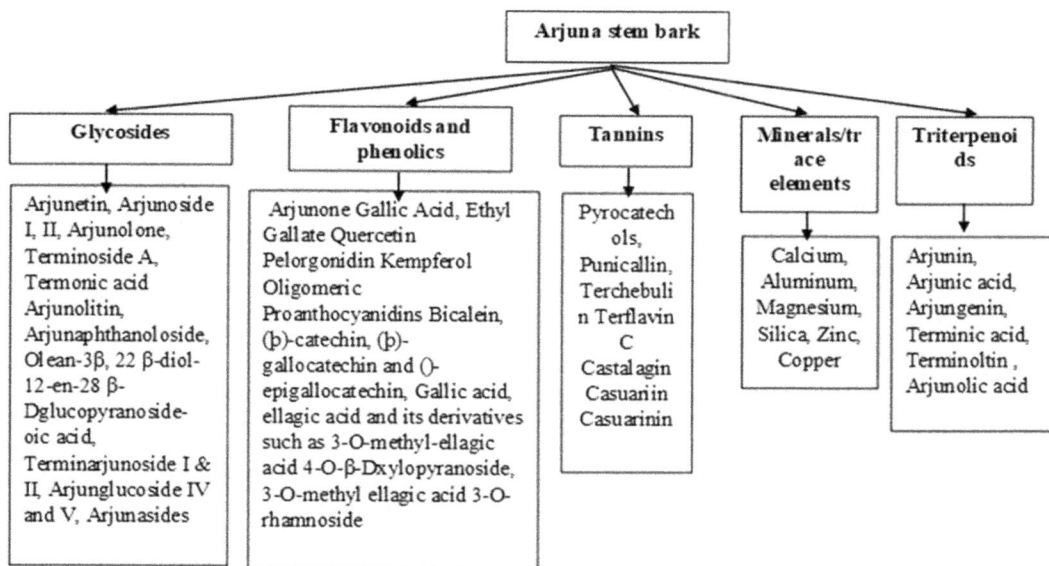

**FIGURE 4.1**  Chemical composition of *T. arjuna* bark.

### 4.3.5  TRADITIONAL USES

*T. arjuna*, commonly called Arjuna, is known for its cardioprotective effects and for other ailments to maintain good health. Fruit, leaves, and roots, especially stem bark, present excellent healing effects. Extract obtained from *T. arjuna* bark has been known for its promising outcomes in angina, i.e., "hritshool", as well as also prescribed in the atherosclerosis treatment, cardiac arrest, and hypercholesterolaemia [105, 106]. Since earlier times, Arjuna bark has been used as alcoholic

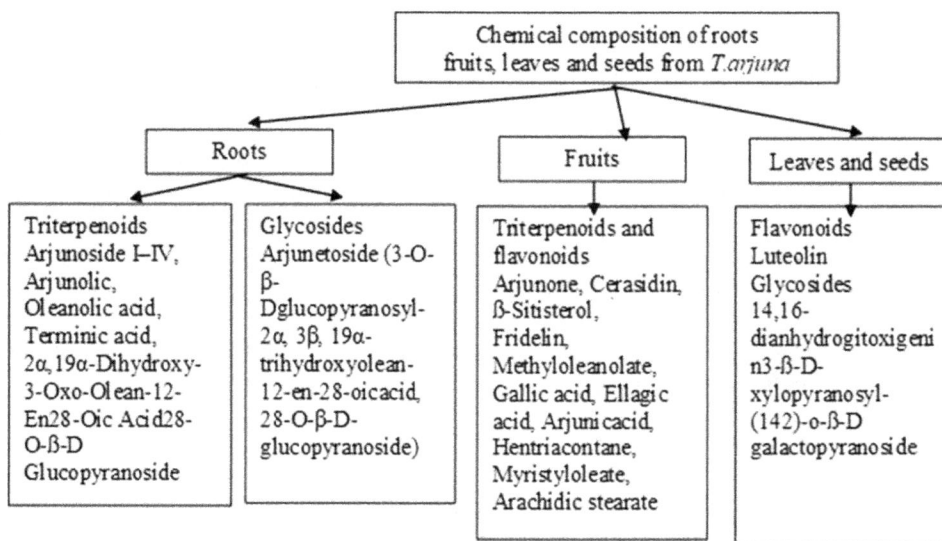

**FIGURE 4.2**  Chemical composition of roots, fruits, leaves, and seeds from *T. arjuna* bark.

decoction (asava) or administered with hot milk (kshirpak) or clarified butter (ghrita) [107, 108]. Arjuna utilization has been mentioned in the primeval Indian manuscripts of Chakradatta as well as Bhāvaprakāśa-nighaṇṭu and considered an effective treatment for heart-associated complications [109, 110, 111]. Bark of the Arjuna plant is commonly used for various therapeutic applications as mentioned in Ayurvedic terminologies sweet, acrid, astringent, febrifuge, cooling, cardiotonic, cardioprotective, and antidysenteric in nature. Arjuna stem bark has been used in different forms such as powder, decoction, hydroalcoholic extract, bark powder with ghrita (fat), or bark powder boiled in milk (kshirpak), to treat various ailments. Several Ayurvedic formulations comprising *T. arjuna*, such as Visvesaram, Trinetrarasam, Sankaravadi, Nagarjunabhram, Nagabaladi churnam, Kakubhadi churnam, Hrdayarnavarasam, Godhumakakubhatvak churnam, Godhumakakubha churnam, Cintamanirasam, Arjunatvak churnam, and Arjunatvagadi kasayam, have been used to treat cardiac problems [109].

### 4.3.6  PHARMACOLOGICAL USES

Arjuna bark offers multiple cardioprotective effects and can be considered a potential herbal drug for the treatment of coronary artery disease, hypertension, and ischaemic cardiomyopathy. Arjuna bark has been reported for its property to elevate the force of cardiac contraction, possess negative or positive inotropic/chronotropic effects on heart. The plant was found to increase the force of cardiac contraction, exert negative or positive inotropic/chronotropic effects on heart, and cause a dose-dependent reduction in heart rate and blood pressure with hypolipidaemic effect. This plant has also been reported for its cardioprotective effect by improving cardiac muscle function and then increases the pumping activity of the heart. It's extensively used as cardiotonic, especially for the patients suffering from congestive heart failure, coronary artery disease, myocardial necrosis, and ischaemia-reperfusion injury. This cardioprotective effect is attributed to its chemical components such as terpenoids, glycosides, flavonoids, and tannins [110].

### 4.3.7  TRADITIONAL FORMULATIONS

"Divya-Arjuna-Kwath" is derived from the Arjuna bark. Arjuna bark extract has been known for its potential to induce chronotropic cardiac as well as negative inotropic effects supports in recovery from cardiac arrhythmia [112]. The plant extract contains a plethora of phytochemicals and micro-nutrients that produce protective effect on cardiac morbidities. Divya Hridyamrit Vati (HAV) is a herbo-mineral formulation, comprising phytochemicals having anti-inflammatory and cardiopro-tective properties [111–117]. Arjuna Ksheera Paka, an Arjuna bark powder preparation, is known for its cardioprotective effects. Recently, inflammation-reducing potential of Arjuna Ksheera Paka (made by using cow milk) was equated with standard Arjuna hydroalcoholic extract. It was found that at early phase of inflammation, efficacy of hydroalcoholic extract was more than Arjuna Ksheera Paka; however, in the advanced inflammatory stage, Arjuna Ksheera Paka has been found to be more effective as well as equivalent to hydroalcoholic extract. Consequently, irrespective of less antioxidant effects, Arjuna Ksheera Paka possessed potential in vivo anti-inflammatory activ-ity. This could be due to the occurrence of milk solids that can behave as adjuvants to *T. arjuna* chemical components. This milk-based adjuvant can contribute to their sustained bioavailability, resulting in an increase in anti-inflammatory effects [118].

### 4.3.8  MECHANISM OF ACTION

Lipid lowering and cardioprotective effects reported in previous studies showed that treatment with *T. arjuna* improved cardiac functions by decreasing fibrosis as well as hypertrophy of the left ventricle. Moreover, *T. arjuna* considerably attenuated free radical formation and inflamma-tion via the decreasing level of cytokine in CHF rodents. There are multiple mechanisms followed by chemicals present in Arjuna extract to present cardioprotective and hypolipidaemic effects as mentioned below [119]:

- Via maintaining endogenous antioxidant enzyme activities
- Inhibiting lipid peroxidation and cytokine levels
- Via preventing the advancement of plaque formation in arteries via attenuating total cholesterol, low-density lipoprotein, triglycerides levels, and elevating the level of high-density lipoprotein
- Decrease the progression of atherosclerotic injuries
- Due to antioxidant and anti-inflammatory effects
- Via reducing oxidative stress caused by isoprenaline and decreases in the level of endogenous antioxidant enzymes such as GSH, SOD, and catalase level
- Via averting fibrosis with no rise in body weight
- Prevents the myocardium from isoproterenol-induced myocardial ischaemic reperfusion injury

Exhibits hypotensive action

- Via activation of β2-adrenergic receptor
- Or via acting directly over muscles of the cardiac tissue
- Continued decrease in blood pressure via modifying the autonomic responses

Exhibits anti-thrombotic action

- Via considerably inhibiting platelet aggregation
- Via considerably attenuating $Ca^{2+}$ release and expression of P-selectin (CD62P)

Prevents endothelial dysfunction due to the

- The presence of a high amount of bioflavonoid

Prevent heart failure

- Considerable improvement in volume fraction from left ventricular
- Considerable decreases in the size of the heart with improvement in period of exercise
- Increase in left ventricular stroke volume index and increase in left ventricular ejection fraction
- Reduction in end systolic volume as well as ventricular (left) end diastolic indices

Anti-ischaemic effects

- Via acting as an antithrombotic agent and considerably suppressing platelet adhesion as well as platelet aggregation caused by adenosine diphosphate and adrenaline
- Via reduction in angina frequency
- Via considerably dropping systolic blood pressure as well as body mass index with a slight rise in HDL cholesterol
- Improvement in left ventricular ejection fraction
- Via decreasing left ventricular mass
- Via reducing mitral valve regurgitation or leaky mitral valve
- Via improvement in the proportion of E/A representing function of the left ventricle of the heart
- Substantial improvement in the treadmill exercise

Exhibits antihypertensive effects via

- Decreasing systolic blood pressure
- Decrease in left posterior wall thickness, interventricular septum, and echocardiographic left ventricular internal diameter

### 4.3.9 MARKETED PRODUCTS

One of most popular Ayurvedic traditional formulations comprising *T. arjuna* is Arjunarishta (4-kg Arjuna tvak + 800-g Dhataki pushpa + 800-g Madhuka pushpa + 2-kg Mridvika or Draksha, i.e., dry grapes + 4-kg Guda or jaggery). It's mainly used to maintain the healthiness of heart and keep optimum blood pressure levels. In addition, as per Ayurveda, recommendations mainly suggested in Sarangadhar Samhita, Charaka Samhita, Hridaya (i.e., heart problems), Deepana (enhances stomach fire), Pachana (helps in digestion), Rochana (stimulates appetite), Anulomana (improves breathing), Mutrakrichra (i.e., dysuria), Anaha (i.e., bloating), Mutraghata (i.e., urinary obstruction), Kusthohara (treats skin disease), and Shothahara (reduces inflammation).

There are various marketed preparations available in the form of Arjun Chhal Churna (by Dabur, Himalaya, Baidyanath, and many others) which help in maintaining heart healthy and is also used in the treatment of hyperlipidaemia, hypercholesterolaemia, and polydipsia. Generally, 3–6 g (1–2 teaspoon) with water or milk is recommended.

**Dose**

*T. arjuna* bark powder and its formulations are commonly administered through oral route. An amount of 500 mg of the powdered bark of *T. arjuna* is prescribed (every 8 hours daily) for chest pain or discomfort after a heart attack along with conventional treatments, and also the same dose is recommended for congestive heart failure patients.

### 4.3.10 TOXICITY

Findings obtained from the previous study showed that oral administration of black- myrobalan fruit, Arjuna bark and its active component, 7-methyl gallic acid (250–2000 µg/ml) was found to be relatively non-toxic and further clinical trials are required for affords for identifying safe dose [120]. Recently, the toxicity of ethanol fraction *T. arjuna* bark on fresh water stinging catfish along with evaluation of changes in haematological parameters of the fishes at fatal dose was evaluated. It was found that *T. arjuna* bark extract could be considered a potent chemical substance that is poisonous to fish owing to its toxic effect on fish, mainly fish haematology [121].

After treatment, *Terminalia* can reduce the blood sugar level. Its consumption with antidiabetic drugs can cause blood sugar to go too low. Thus, its interaction must be carefully monitored.

### 4.3.11 RECENT RESEARCH

Recent research showed an effectiveness of traditional Indian formulations Divya-Arjuna-Kwath (Arjuna bark extract) and Divya Hridyamrit Vati in reducing cardiac hypertrophy caused by isoproterenol in rodents H9c2-cardiomyocytes. It was found that both preparations presented considerable biological properties in the modulation of oxidative stress caused by isoproterenol in the H9c2-cardiomyocytes and might be used as the alternative traditional formulation for treating cardiac complications [122]. Another study showed that *T. arjuna* bark powder aqueous extract prevents alcohol-induced oxidative injury to erythrocytes and offers protection to erythrocytes oxidative stress triggered by alcohol [123]. A recent study also revealed that *T. arjuna* DPP-IV (dipeptidyl peptidase-IV) inhibitory effects of *T. arjuna* were equivalent to vildagliptin. The present marketed DPP-IV inhibitors are costly antidiabetic medications and cause intolerable adverse effects; based on docking study, it was revealed that active chemical components of *T. arjuna* such as arjunetin, arjungenin, ellagic acid, and arjunic acid presented significant DPP-IV inhibitory activity when compared to synthetic DPP-IV inhibitors [124]. Recently, *T. arjuna* bark resin has been used as a gelling agent with sodium alginate to formulate pH triggered in situ gelling system using moxifloxacin HCl (MOX-HCl). It was found that formulated preparation can be alternatively used as an eye drop for its improved ocular bioavailability, better corneal permeability, and prolonged precorneal retention [125]. Due to its potential to reduce oxalate-induced morphological changes, apoptosis, and death of renal cells to further promote cell survival, *T. arjuna* bark aqueous extract has been recently reported as a natural antiurolithiatic agent against urolithiasis when tested on in vitro model system, including normal epithelial cell line [126].

## 4.3.12   CLINICAL TRIALS

In 2016, effectiveness and safety profile of extract of Arjuna bark extract (aq.) among CHF individuals was evaluated using a randomized controlled trial. It was found that an aqueous extract of Arjuna was well tolerated, however, did not improve left ventricular ejection fraction among chronic heart failure patients of New York Heart Association over 12 weeks, while there was improvement in functional capacity, antioxidant reserves, and symptom-related quality of life domains in some patients [127]. In 2013, *T. arjuna* effects among 93 individuals suffering from heart muscle disease characterized by ischaemia and idiopathic complications were evaluated. It was found that *T. arjuna* presented considerable improvement in left ventricular parameters as well as functional capacity [128]. In 1995, 12 patients with refractory chronic CHF, associated with idiopathic dilated cardiomyopathy (10 patients), previous myocardial infarction (1 patient), and peripartum cardiomyopathy (1 patient) were treated with *T. arjuna* (bark extract) as an adjuvant in a double-blind cross-over design trials to outstandingly tolerable conventional therapy (Phase I). It was found that adjuvant *T. arjuna* treatment among patients with refractory CHF, mostly related to idiopathic dilated cardiomyopathy, was found to be safe and caused long-lasting improvement in symptoms and signs of heart failure along with improvement in left ventricular ejection phase indices with definite improvement in quality of life [129]. Clinical investigation has been recently designed recently to study the long-term safety of Arjuna as an adjunct drug in chronic coronary artery disease patients, and it was found that Arjuna is safe and effective in patients with chronic coronary artery diseases. Except gastritis and constipation, no other notable adverse effects were evidenced [130].

## 4.3.13   ADULTERANTS AND SUBSTITUENTS

Practice of impacting the quality of crude drugs by misconducts such as deliberate adulteration and substitution of plant-based crude drugs, especially to compensate the overall demand in the market or to gain profit from the markets, can affect the biological potential of parent crude drugs and can also lead to the serious effects on health of end consumers. Recently, bark extracts of *myrobalan*, *Bahera* and *T. tomentosa* have also been identified as an important contaminant of *Arjuna* bark, so as to compensate great demands of parent crude drug [131, 132]. These identified adulterants of *Arjuna* bark can decrease its biological effectiveness and may cause considerable ill effects on the customers. These adulterants have been reported for their therapeutic effects; however, those therapeutic effects are not like *T. arjuna* [133], even though an addition of these adulterants can mask the pharmacological effects of *T. arjuna*. *In addition, the chemical composition of adulterants and parent plant Arjuna bark is different; thus, with acting in a* synergistic/additive/antagonistic manner, it can interfere with the active chemical components of *Arjuna*. In addition, apart from the above-mentioned adulterants, *Lagerstroemia speciosa*, *Terminalia alata*, *Terminalia catappa*, *Terminalia calamansi*, *Terminalia myriocarpa*, and *Terminalia bialata* area are being sold indiscriminately under the brand name Arjuna [134].

### 4.3.14  ANNUAL TRADE

*T. arjuna* is a high-demanded medicinal plant (trade name Arjun, Arjuna) with an estimated annual trade of 2000–5000 MT per year.

### 4.3.15  STATUS OF THE MEDICINAL PLANT

Recently, *T. arjuna* is categorized as NT (near to threatened) in the list of threat status of 242 plant species with high commercial demand (>100 MT/year) in India. As per this report, the status of *T. arjuna* in several Indian states is as follows: Karnataka (NT 1997), Madhya Pradesh (NT 2003), Kerala (NT 1997), Chhattisgarh (NT 2003), and Maharashtra (NT 2001). In 2007, *T. arjuna* has become vulnerable in Rajasthan (VU 2007) [135].

## 4.4  *TERMINALIA CHEBULA* RETZ.

### 4.4.1  VERNACULAR NAMES

Black or chebulic myrobalan, gallnut, and Fructus chebulae are some of its vernacular names.

### 4.4.2  BIOLOGICAL SOURCE

Myrobalan is the mature dried fruits of *T. chebula* Retz. (*T. chebula*) (Figure 4.3). There are only two different forms of *Haritaki*: big and small. The small type also known as *Jangi haritaki* is the *Chetaki* variety (cited in the earlier prototypes), and undeveloped, immature fruits smaller in size usually show purgative effects. The big variety, also known as *Vijaya*, is available at all places and utilized in the formulations of Ayurvedic proprietary medicine, mainly for revitalization, purificatory, and other purposes. Although the seven varieties of *T. chebula* are described by Hooker, only two varieties are found and the others are their sub-varieties, namely, *Terminalia chebula* var. *chebula*.

**Family**
Combretaceae

### 4.4.3  CULTIVATION AND COLLECTION

It is frequently grown on a commercial basis for the tannin in its fruit and for its therapeutic applications, particularly in India. It is an around 1–2-in. fruit with five ribs over the external layer. The unripe fruit is greenish in colour, whereas the ripe one in yellowish-grey in colour. Harvesting of fruits is done in between January and April, and the development of fruit begins between November and January.

**FIGURE 4.3**   Dried fruits of *T. chebula*.

### 4.4.4 CHEMICAL COMPOSITION

Myrobalan fruits are enriched with tannins (hydrolysable), and they have been extensively used to treat numerous chronic disorders. Myrobalan fruits contain about 150 phytochemicals that offer broad therapeutic activity. In 1991, 2α-hydroxymicromeric acid and two already identified constituents, 2α-hydroxyursolic acid and maslinic acid, have been obtained from myrobalan leaves [135]. Furthermore, in 1993, chebuloside I and II were obtained from the myrobalan stem bark and revealed as beta-D-galactopyran [136]. The previous study suggested that an active chemical component of *T. chebula*, i.e., chebulagic acid acts as COX-2 and 5-LOX (key enzymes involved in inflammation and carcinogenesis) dual inhibitor and stimulates apoptosis in COLO-205 cells. It also showed anti-proliferative activity against HCT-15, COLO-205, MDA-MB-231, DU-145, and K562 cell lines [137]. Chemical investigation on outgrowth (gall) of *T. chebula* revealed the presence phenolic compounds, including two phenolic carboxylic acids, hydrolysable tannins, eight triterpenoids, comprising four oleanane-type triterpene acids, and four of their glucosides [138]. A recent study suggested that *T. bellirica*, *E. officinalis*, and *T. chebula* contain K as a major element, whereas Co, Cr, and Na in *E. officinalis*, Fe, K, and Mn in *T. bellirica* and Cl and Zn in *T. chebula* were found to be the highest. This composition was determined by instrumental neutron activation analysis [139]. A recent study showed that two different tannins (hydrolysable), chebumeinin (A, B) with eight identified components have been isolated from the fruits of *T. chebula*. Out of these, some compounds were found to be active against hepatitis C virus [140]. Another study revealed the presence of three different polyhydroxytriterpenoid derivatives in myrobalan fruits with the presence of 14 already reported components [141]. Another report showed the presence of nine hydrolysable tannins, including four earlier undetermined and six artefacts, derived together with 39 compounds, from the of myrobalan fruits [142]. Recently, an antioxidant polysaccharide (neutral) called TFP-a or amylopectin containing polysaccharide enriched with α-Glc-rich polysaccharide has been also identified [143]. A recent study showed that myrobalan fresh fruits var. *tomentella* contain five new lignans, including three tetrahydrofuran and two furofuran derivatives, i.e., termitomenins A-E (1–5), as well as 10 known ones [144]. A recent study showed the isolation of two new lignan glucosides with a furofuran skeleton, termitomenins F (1) and G (2) from leaves as well as branches of myrobalan var. *tomentella*. In addition, 19 known compounds containing 5 lignan glucosides, 6 tannins (hydrolysable), and 8 phenolics (simple) have also been found. Particularly, it was found that the known tannins (hydrolysable) displayed potent α-glucosidase suppressive effects [145]. It was recently found that hydrolysable tannins and simple phenol are major chemical components in *T. chebula* fruit and present potent inhibitory effects on α-glucosidase than quercetin. In this study, apart from known hydrolysable tannin and phenolic compounds, a different tannin (hydrolysable), 2,3-(*S*)-HHDP-6-*O*-galloyl-D-glucose as well as one oleanane pentacyclic triterpene (arjungenin), was separated and identified.

### 4.4.5 TRADITIONAL USES

In Ayurvedic text, even though the plant is cited in *Brihatrayee* (three most important reference books of Ayurveda Charaka-Susruta-Vagbhatt), its varieties are not depicted by them. On the basis of the site from where the fruit is collected, and its macroscopic characteristics such as colour and shape, *Nighantus* indicated the various types of Haritaki, i.e., Chetaki, Abhaya, Jivanti, Rohini, Putana, Amrita, and Vijaya. Moreover, two different forms of Chetaki, especially white and black, are portrayed by Bhavamishra. The black form or Jangi Haritaki (black small-sized fruit of Haritaki) is described by Bhavamishra, which is available in all places. Later, its powder was used in the form of medicine for laxative purpose. The white form, also called the golden or big variety, which is six angula, is a large variety of Haritaki available in the market, which is used in the manufacture of preparations like Abhayarishta, Agastya haritakee, and Vyaghri haritakee avaleha [146, 147].

In Tibet, Chetaki is called the "King of Medicines" due to its amazing capabilities to cure various ailments and listed as an important medicinal plant in "Ayurvedic Materia Medica" due to its diverse medicinal value. In Persian medicine, *T. chebula* is one of the best choices for the treatment of haemorrhoids. Also, the fruit of *T. chebula* has been used as a contraceptive for male since ancient time; however, its pharmacological evidence related with hypotesticular properties hasn't been revealed (discussed in the pharmacology section). *T. chebula* has been commonly utilized for the treatment of memory damage, reducing the inflammation as well as ageing. *T. chebula* fruits have been since ages present in the Ayurvedic medicine and Iranian traditional medicine systems, mainly for neurologic complications and inflammation. It is widely utilized in Ayurvedic, homoeopathic, and Unani traditional systems. It was recorded initially in Ayurvedic texts because of its broad potency in curing various ailments. *T. chebula* is a famous Ayurvedic medicine introduced into China in the Sui and Tang Dynasties and has been documented and utilized therapeutically as Fructus chebulae, along with its variety *tomentella* in the Chinese Pharmacopoeia. A recent study demonstrated that hydrolysed tannins as well as lignans are the primary constituents of *T. chebula* fruits, accountable for the folk-based therapy of cough as well as sore throat. Furthermore, phenolic (simple) as well as tannins (hydrolysable) showed potent hypoglycaemic effects which could be linked to the ethno-botanical utilization of fruits of *T. chebula* on diabetes in medicine used in Tibet [146, 147].

### 4.4.6 Traditional Formulations

The dried fruit of *T. chebula* is an important traditional medicine used for intestinal and hepatic detoxification. Gurigumu-7 which is made of *T. chebula* has been used often for hepatic complications in Tibet as well as Mongolia. It was found that treatment with Gurigumu-7 may perhaps enhance the phenolic compounds (of *T. chebula*) absorption via reducing the metabolism of bacteria present in intestine [148]. In Ayurvedic classics, Mahatriphaladya Ghrita formulation (containing *T. chebula* as a component) can be used for the treatment of Timira (blurred vision) [149].

Earlier studies showed that homoeopathic preparations of *T. chebula* (MT, 3X, 6C, 30C) could target MCF7 and MDAMB231 breast cancer cell lines and decreased growth kinetics of breast cancer cells in a dose- and time-dependent manner. In comparison to mother tincture, other potencies (3X, 6C, 30C) were non-toxic to normal cells. It was found that mother tincture comprising nanoclusters and 6C contained nanoparticles of ~20-nm size. This study evidenced an anticancer property of homoeopathic preparations of *T. chebula* against breast cancer and showed their nanoparticulate nature [150].

### 4.4.7 Pharmacological Uses

#### 4.4.7.1 Inhibitory Activity on Acetylcholinesterase by Molecular Docking

A recent study has shown the development of procedure based on ultrafiltration and ultra-performance liquid chromatography-quadrupole time-of-flight mass spectrometry to quickly screen and identify specific acetylcholinesterase inhibitors from *T. chebula* fruits. It was found that chebulanin, chebulagic acid, and corilagin were accountable for acetylcholinesterase inhibitory effects, and chebulagic acid, corilagin, chebulanin, and ellagic acid could fit well into the binding pocket of the model.

#### 4.4.7.2 In Vivo Hepatoprotective Effects

*T. chebula* fruit is one of the most used herbal drugs in traditional medicine, including those for liver complications. In this context, among all bioactive components of *T. chebula* fruits, hepatoprotective effects of chebulinic acid are well reported. It was found that pretreatment with chebulinic acid could prevent *tert*-butyl hydrogen peroxide caused injury in hepatocytes (L-02) via preventing the reactive oxygen species generation, decreasing levels of lactate dehydrogenase, and augmenting expression of heme oxygenase-1 and NAD(P)H:quinone oxidoreductase via MAPK/Nrf2 pathway. Animal studies demonstrated that chebulinic acid provides considerable protection against hepatic damage caused by CCl4 in rodents via decreasing malondialdehyde, aspartate transaminase, and alanine transaminase levels, increasing SOD activity, normalized hepatic histopathology with a stimulation of the Nrf2/HO-1 signalling pathway. It was also reported that during DPPH radical scavenging activity, chebulinic acid is metabolized to chebulic acid isomers. Furthermore, in a zebrafish (transgenic) model with a hepatic expression of DsRed RFP, chebulinic acid reduced the hepatic injury caused by acetaminophen [151].

### 4.4.7.3 Antimicrobial Effects

A recent study showed the effect of ethanolic extract of *T. chebula* against pathogenic aerobic bacteria, *P. aeruginosa* via interfering with normal quorum-sensing mechanism that is an effective strategy for attenuating its virulence. Furthermore, phytochemicals present in the ethanolic extract of *T. chebula* successfully represses the Las R followed by an inhibition of quorum-sensing-mediated virulence factors production [152]. Recently methyl gallate isolated from *T. chebula* showed in vivo fluid accumulation-inhibitory, anticolonization and anti-inflammatory, and in vitro biofilm-inhibitory effects against multidrug-resistant *Vibrio cholerae*. It was found that methyl gallate was considered an effective against for the treatment of severe secretory and inflammatory diarrhoeal disease induced by multidrug-resistant *V. cholerae* [153].

### 4.4.7.4 Antiurolithiatic Property

A recent study has shown that an aqueous extract of *T. chebula* fruits exhibited antiurolithiatic effects by decreasing the high levels of oxalate and phosphate in urine and kidney tissue homogenate. Also, treatment with extracts prevented the increase of serum levels creatinine, uric acid, and blood urea nitrogen. Histopathological findings showed that extract decreased histological alterations and maintained the normal architecture of kidney tissue [154].

### 4.4.7.5 Antioestrogenic or Antiandrogenic Effects

Based on antiandrogenic effects, different components have been identified from myrobalan mainly triterpenoids as well as ellagitannin. These phytochemicals showed antiandrogenic antioestrogenic properties. It was suggested that these compounds could cause disruption of endocrine which restricts the prolonged utilization of *T. chebula*. A recent study based on molecular docking showed the screening of potential antiandrogenic components from myrobalan. Phytochemical class, including ellagitannin and triterpenoids, has been selected and docked to the three binding sites (viz., ABS, BF3, and AF2 sites) in LBD of rodent as well as Human AR. Although both triterpenoids and ellagitannin metabolites presented similar affinity to the binding sites, urolithins and its analogues presented the paramount binding, which suggests that it functions as a potent antiandrogen [155].

### 4.4.7.6 Antidepressant-Like and Anxiolytic-Like Effects

Recently, myrobalan fruits (ethanolic extract) have been reported for antidepressant and anxiolytic-like activities in rodents. Findings from this study showed that both doses (100 or 200 mg/kg, p.o.) cause considerable decrease in period of immobility. Moreover, the treatment of mouse with 200-mg/kg dose caused considerable decrease in MAO-A activity. Findings from this study demonstrated the anxiolytic as well as antidepressant-like effects of myrobalan fruits (ethanolic extract) *in* rodents [156].

### 4.4.7.7 *T. Chebula* Nanoparticles in the Treatment of Cancer

A recent study has described the in vitro cytotoxicity, mitochondrial membrane potential, and reactive oxygen species activity of eco-friendly *T. chebula*-extract-loaded nickel oxide nanoparticles. It was found that green synthesized nickel oxide nanoparticles presented the toxicity to breast cancerous cells in a dose-dependent manner demonstrating considerable cell viability, reactive oxygen species activity, and liberation of mitochondrial membrane potential [157]. Biogenic synthesis of silver palladium bimetallic nanoparticles using aqueous fruit extract of *T. chebula* has been recently investigated. This study revealed that silver palladium bimetallic nanoparticles presented potent anticancer potential against lung cancer cell lines A549 via stimulating oxidative stress mediated cell death in A549 cells. Moreover, silver palladium bimetallic nanoparticles possessed antimicrobial effects against Gram-positive and negative bacteria [158].

### 4.4.7.8 Hypotesticular and Anti-Spermatogenic Effects

A recent study has shown the development of male herbal contraceptive from extract of *T. chebula*. Treatment with different extracts of *T. chebula* presented a considerable reduction in the effects of androgen-metabolizing enzymes (17β-HSD 3β-HSD, Δ5) and decrease in testosterone level in serum. Considerable upregulation of testicular Bax gene and downregulation of Bcl-2 gene suggested the hypotesticular effects. Findings from flow-cytometric revealed a substantial reduction in sperm viability and sperm mitochondrial status in extract-treated samples. Among all, ethyl acetate extract presented considerable results with no toxicity and reduced spermiological, testicular genomic sensors and improved sperm apoptotic sensors that resulted in male contraception [159]. The previous study also revealed anti-fertility effect of aqueous-ethanolic (1:1) extract of the fruit of *T. chebula* as an approach towards herbal contraception. It was found that extract possessed a substantial anti-spermatogenic effect in male rats [160].

### 4.4.7.9 Neuroprotective Effects

Recent study showed that *T. chebula* extract reduces inflammation in microglial cells via reducing expression levels of pro-inflammatory factors (TNF-α, IL-1β, IL-6, PGE-2, COX-2) as well as nitric oxide, increasing urea and downregulating the expression of nitric oxide synthesis. Thus, it can be used as a possible anti-inflammatory agent in CNS inflammatory complications [161].

### 4.4.7.10 Mitigate Toxicity of Medications/Chemicals

Recent study revealed that *T. chebula* due to its antioxidant, anti-inflammatory, and neuroprotective properties ameliorated quinolinic-acid-mediated cytotoxicity by mitigating oxidative damage (in PC12 and OLN-93 cells). Quinolinic acid pathologic accumulation caused neuroinflammatory and demyelinating complications, including multiple sclerosis via excessive free radicals generation. *T. chebula* exhibits neuroprotection and oligoprotection via alleviating oxidative stress parameters [162].

### 4.4.7.11 α-Glucosidase Inhibitory Effects

Recent report showed that components derived from *T. chebula* extract such as 4-*O*-(2″,4″-di-*O*-galloyl-α-l-rhamnosyl) ellagic acid as well as 1,2,3,6-tetra-*O*-galloyl-4-*O*-cinnamoyl-β-D-glucose presented considerable α-glucosidase suppressive effects [163]. Another study showed that isolated compounds, 23-*O*-galloylarjunolic acid (11, IC50 21.7 μM) and 23-*O*-galloylarjunolic acid 28-*O*-β-D-glucopyranosyl ester, showed suppressive activity against α-glucosidase enzyme (Baker's yeast) [164].

### 4.4.7.12 Anti-Alzheimer's Effects via Antioxidant, Anticholinesterase, and Antiamyloidogenic Activities

Neurological complications such as Alzheimer's disease which are clinically described by the loss of cognitive function are considered one of the most horrible illnesses. β-amyloid peptide on accumulation triggers oxidative stress and facilitates neuronal injury influencing events of regular life, including learning and memory. Recent in vitro and in silico studies showed that *T. chebula* and *T. arjuna* as well its bioactive component 7-Methyl gallic acid showed potent antioxidant, anticholinesterase, and antiamyloidogenic effects [165].

### 4.4.7.13 Antianaphylactic Action

Previous studies suggested strong antianaphylactic effects of water-soluble fraction of *T. chebula* by showing a significant inhibition of histamine release and increasing effect on anti-dinitrophenyl IgE-induced tumour necrosis factor-$\alpha$ production from rat peritoneal mast cells [166].

### 4.4.7.14 Anti-Dengue

A recent study has revealed that the seed extract of *T. chebula* containing 1,2,3-benzenetriol as well as tridecanoic acid can be considered potential agents against dengue without any affecting beneficial insects (non-target) [167].

### 4.4.7.15 Anticaries Agent

Previous studies revealed that aqueous extract from *T. chebula* (mouthrinsing with a 10% solution) inhibited the growth, salivary bacterial count, glycolysis of salivary bacteria, sucrose-induced adherence, and glucan-induced aggregation of *S. mutans* [168].

### 4.4.7.16 Pain Management

TRPV1 is a protein that sense injurious heat. TRPV1 agonists have been used to treat pain. Recent study demonstrated that *T. chebula* and *Alchemilla xanthochlora* potentiate TRPV1 channel [169].

## 4.4.8 INDUSTRIAL APPLICATIONS

### 4.4.8.1 Nutraceutical Value

Earlier studies have revealed that *T. chebula* and *P. emblica* must be utilized as a natural source of antioxidant/nutraceuticals, especially as supplementing dietary foods due to the presence of high flavonoids and phenolic compounds (gallic acid, catechin, chlorogenic acid, caffeic acid, and *p*-coumaric acid). In 1991, it was also found that *T. chebula* is very beneficial and could be an essential source of dietary supplement in vitamin C, energy, protein, and mineral nutrients [170, 171]. Another study revealed that the supplementation of polyphenol rich *T. chebula* extract at different concentrations (1.06 g/kg and 3.18 g/kg of body weight) increased overall MUFA and PUFA content in muscle, while SFA was reduced, thereby enhancing the desaturation index. Additionally, the activity of $\Delta9$-desaturase increased causing improvement of total CLA content in muscle [172].

### 4.4.8.2 As Bio-Mordant for Natural Dye

*T. chebula* (harde powder) is a bio-mordant for *Kigelia africana* natural dye. Recently a bio-mordant *T. chebula* was utilized as a mordant for natural dyeing of wool using *K. africana* natural dye. It was found that dyed wool materials presented good antioxidant, antibacterial, and UV protective effects [173].

### 4.4.8.3 Corrosive Inhibitor

Considering the application in weight reduction process and electrochemical measurements, recently an extract of *T. chebula* has been utilized to supress corrosion for steel (low carbon) using sulphuric acid as corrosive media and showed fruitful results [174].

### 4.4.8.4 For Biodiesel Production

Recently, the constituents of *T. chebula* plants, including leaves and seeds, have been efficiently utilized as a green source for the synthesis of copper oxide nanoparticles and production of biodiesel, respectively. Copper oxide nanoparticles have been fabricated via solution combustion route using *T. chebula* leaves extract as a reducing-cum-fuel agent. Particularly, the synthesized copper oxide nanoparticles are employed as a heterogeneous catalyst in the biodiesel production [175].

#### 4.4.8.5 Supercapacitor Performance

Cobalt is a transition metal that has a beneficial effect on human health. Cobalt oxide (antiferromagnetic p-type semiconductor) NPs have been explored as an agent for biomedical and other applications. A recent study showed that carbon-supported cobalt oxide nanoparticles (Co3O4@C NPs) have been synthesized by simple thermolysis of cobalt(II) acetate tetrahydrate and *T. chebula* fruits. It was found that high specific capacitance of cobalt oxide nanoparticles could be ascribed to the combination of both double layer capacitance and pseudocapacitance [176].

#### 4.4.8.6 Radioprotective Effect

Previous studies suggested that *T. chebula* can protect from γ-irradiation-induced oxidative stress and may be considered a probable radioprotector. This could be due to the antioxidant effects possessed by the major phenolic compounds such as total phenolics, flavonoids, and triterpenoids contents [177].

### 4.4.9 Marketed Products

Haritaki powder (raw harad powder), harad choti (small), and harad badi (large) are some of the well-marketed products.

### 4.4.10 Toxicity

A recent study has revealed that tannin (hydrolysable)-enriched fraction that has been derived from myrobalan fruit hydroalcoholic extract hasn't presented any toxicity or death at 5000 mg/kg/p.o. till next 14 days. However, a continuous dose of 1000 mg/kg demonstrated considerable decrease in body weight. It was found that the latter dose increases serum urea, glucose, and AST level substantially, suggesting mild disturbances in liver and kidney functions. Histopathological findings showed that the latter causes a mild granulomatous inflammation in the liver. Another study revealed that the administration of methanolic extracts of *T. chebula* fruit as well its bioactive component 7-methyl gallic acid didn't show cytotoxic, genotoxic, haemolytic, and mutagenic effects at limited dose of 2000 μg/ml. Furthermore, acute and subacute toxicity investigations revealed no considerable change in body weight, behaviour, haematology, biochemical parameters, organ weight, and histopathology [178].

### 4.4.11 Clinical Data

A double-blind randomized placebo-controlled clinical trial was performed to check effects of *T. chebula* Retz. capsules in the treatment of haemorrhoids over the 104 patients. It was found that *T. chebula* may be effective on haemorrhoids and potentially supplement for the treatment of haemorrhoids [179]. In 2014, 23 patients detected with Madhumeha were registered in the clinical investigation to examine the effects of Varadi Kwatha comprising *Holarrhena antidysenterica*, Triphala, *Coscinium fenestratum*, *Cyperus rotundus*, and *Pterocarpus marsupium*.

All patients showed a good improvement in cardinal with significant decrease in mean fasting blood sugar and mean postprandial blood sugar [180]. Recently, a double-blind placebo-controlled randomized clinical trial was conducted to study the effect of Arogyavardhini and lifestyle modification in the management of metabolic syndrome. Apart from other ingredients, Arogyavardhini contains myrobalan, Bahera as well as Amla. It was found that Arogyavardhini compound, along with lifestyle modification, was found to be more effective than lifestyle modification alone in the management of metabolic syndrome [181]. In 2014, another randomized clinical trial was performed to evaluate and compare the efficacy of Triphala mouthwash containing Amalaki, Haritaki, and Bahera with 0.2% chlorhexidine in hospitalized patients with periodontal diseases. It was concluded that cost-effective and well-tolerated Triphala mouth-wash (herbal) was found to be a potent plaque-reducing agent as chlorhexidine (0.2%) in reducing plaque accumulation and gingival inflammation without any side effects [182]. Recently, a triple-blind randomized controlled clinical trial was performed in 60 patients to study the comparative effect of Triphala mouthwash, *Aloe vera* mouthwash, and chlorhexidine mouthwash on gingivitis. It was found that Triphala to be superior in reducing plaque, gingival inflammation, and bleeding compared to that of *A. vera* [183].

Abhraloha is a time-tested Ayurvedic haematinic formulation. Abhraloha tablets contain natural haematinic ingredients such as Loha bhasma (processed iron) and Abhraka bhasma (processed mica). Abhraloha containing Amalaki, Haritaki, Bibhitaki, Shunthi, Maricha, Pippali, Chitraka, Musta, Vidanga, and Shatavari has been reported for its effect in increasing haemoglobin levels and cured iron deficiency anaemia. Recently, a randomized controlled, parallel-group, assessor-blind clinical trial was performed to compare its effect with ferrous ascorbate. It was found that Abhraloha acquires haematinic activity and improves all the blood parameters and associated with fewer adverse effects compared to oral iron therapy [184]. In 2014, a clinical study was performed to evaluate anti-cataract effect of Triphaladi Ghana Vati (containing *T. chebula* as one of the components) in immature cataract. It was found that tested formulation can decrease and control the advancement of immature cataract, and combined therapy (Elaneer Kuzhambu Anjana for local application) was found more effective [185].

*Virechana Karma (therapeutic purgation) is a part of Panchakarma process (Ayurvedic purification methods) using various traditional formulations (including Triphala) have been used and it is assumed to affect* microbiota and help in the management of the obesity. Recently, a clinical investigation showed that Virechana Karma (therapeutic purgation) over the gut flora in context with overweightness. It was found that *Virechana* is useful in obesity management because of the decrease in *E. coli* as well as its efficiency over the gut flora dysbiosis [186].

### 4.4.12 Adulterants and Substituents

*Tholymis citrina* and *Tradescantia pallida* have been used to make different types of *Haritaki* and thus considered an adulterant as well as substitute of *Haritaki* in various areas.

### 4.4.13 Most Cited Research

The most cited research so far is based on the antimicrobial screening of some Indian medicinal plants, including alcoholic extracts of *T. chebula*. It was found that extract was effective against tested bacteria and showed no cellular toxicity against fresh sheep erythrocytes [187].

### 4.4.14 Patents

US20130266676A1 claimed an optimized aqueous extraction method for *T. chebula* to increase the levels of bioactive hydrolysable tannoids such as chebulagic acid, chebulinic acid, and other low-molecular-weight hydrolysable tannoids. The procedure offers an extract comprising a hydrolysable tannoid blend as an amorphous dry powder.

US20150174184A1 claimed that the use of *T. chebula* extract can be used for the treatment of osteoarthritis. It was claimed that *T. chebula* extract containing optimized and/or enriched hydrolysable tannoid blend [chebulagic acid (8–25%), around by weight chebulinic acid (15–30%) and hydrolysable tannoids (10–40%)] can be used to ease pain and inflammation in an individual affected with osteoarthritis.

US9750778B2 claimed that extracts of *T. chebula* and *T. bellirica* or combinations can be used for the treatment of uricaemia, hyperuricaemia, and gout in a human subject.

## REFERENCES

1. Bhatia S. Chapter 1: History and scope of plant biotechnology. In: Modern Applications of Plant Biotechnology in Pharmaceutical Sciences. Academic Press, 2015:1–30.
2. Bhatia S. Chapter 2: Plant tissue culture. In: Modern Applications of Plant Biotechnology in Pharmaceutical Sciences. Academic Press, 2015:31–107.
3. Bhatia S. Chapter 3: Laboratory organization. In: Modern Applications of Plant Biotechnology in Pharmaceutical Sciences. Academic Press, 2015:109–120.
4. Bhatia S. Chapter 4: Concepts and techniques of plant tissue culture science. In: Modern Applications of Plant Biotechnology in Pharmaceutical Sciences. Academic Press, 2015:121–156.
5. Bhatia S. Chapter 5: Application of plant biotechnology. In: Modern Applications of Plant Biotechnology in Pharmaceutical Sciences. Academic Press, 2015:157–207.
6. Bhatia S. Chapter 6: Somatic embryogenesis and organogenesis. In: Modern Applications of Plant Biotechnology in Pharmaceutical Sciences. Academic Press, 2015:209–230.
7. Bhatia S. Chapter 7: Classical and nonclassical techniques for secondary metabolite production in plant cell culture. In: Modern Applications of Plant Biotechnology in Pharmaceutical Sciences. Academic Press, 2015:231–291.
8. Bhatia S. Chapter 8: Plant-based biotechnological products with their production host. In: Modes of Delivery Systems, and Stability Testing, Modern Applications of Plant Biotechnology in Pharmaceutical Sciences. Academic Press, 2015:293–331.
9. Bhatia S. Chapter 9: Edible vaccines. In: Modern Applications of Plant Biotechnology in Pharmaceutical Sciences. Academic Press, 2015:333–343.
10. Bhatia S. Chapter 10: Microenvironmentation in micropropagation. In: Modern Applications of Plant Biotechnology in Pharmaceutical Sciences. Academic Press, 2015:345–360.
11. Bhatia S. Chapter 11: Micropropagation. In: Modern Applications of Plant Biotechnology in Pharmaceutical Sciences. Academic Press, 2015:361–368.
12. Bhatia S. Chapter 12: Laws in plant biotechnology. In: Modern Applications of Plant Biotechnology in Pharmaceutical Sciences. Academic Press, 2015:369–391.
13. Bhatia S. Chapter 13: Technical glitches in micropropagation. In: Modern Applications of Plant Biotechnology in Pharmaceutical Sciences. Academic Press, 2015:393–404.
14. Bhatia S. Chapter 14: Plant tissue culture-based industries. In: Modern Applications of Plant Biotechnology in Pharmaceutical Sciences. Academic Press, 2015:405–417.
15. Bhatia S. Natural Polymer Drug Delivery Systems: Nanotechnology and Its Drug Delivery Applications. Springer International Publishing, Switzerland; 2016:1–32.
16. Bhatia S. Natural Polymer Drug Delivery Systems: Nanoparticles Types, Classification, Characterization, Fabrication Methods and Drug Delivery Applications. Springer International Publishing, Switzerland; 2016:33–93.
17. Bhatia S. Natural Polymer Drug Delivery Systems: Natural Polymers vs Synthetic Polymer. Springer International Publishing, Switzerland; 2016:95–118.
18. Bhatia S. Natural Polymer Drug Delivery Systems: Plant Derived Polymers, Properties, and Modification & Applications. Springer International Publishing, Switzerland; 2016:119–184.
19. Bhatia S. Natural Polymer Drug Delivery Systems: Marine Polysaccharides Based Nano-Materials and Its Applications. Springer International Publishing, Switzerland; 2016:185–225.
20. Bhatia S. Systems for Drug Delivery: Mammalian Polysaccharides and Its Nanomaterials. Springer International Publishing, Switzerland; 2016:1–27.
21. Bhatia S. Systems for Drug Delivery: Microbial Polysaccharides as Advance Nanomaterials. Springer International Publishing, Switzerland; 2016:29–54.
22. Bhatia S. Systems for Drug Delivery: Chitosan Based Nanomaterials and Its Applications. Springer International Publishing, Switzerland; 2016:55–117.
23. Bhatia S. Systems for Drug Delivery: Advance Polymers and Its Applications. Springer International Publishing, Switzerland; 2016:119–146.
24. Bhatia S. Systems for Drug Delivery: Advanced Application of Natural Polysaccharides. Springer International Publishing, Switzerland; 2016:147–170.
25. Bhatia S. Systems for Drug Delivery: Modern Polysaccharides and Its Current Advancements. Springer International Publishing, Switzerland; 2016:171–188.
26. Bhatia S. Systems for Drug Delivery: Toxicity of Nanodrug Delivery Systems. Springer International Publishing, Switzerland; 2016:189–197.

27. Bhatia S, Sharma K, Bera T. Structural characterization and pharmaceutical properties of porphyran. *Asian J Pharm.* 2015;9:93–101.

28. Bhatia S, Sharma A, Sharma K, Kavale M, Chaugule BB, Dhalwal K, Namdeo AG, Mahadik KR. Novel algal polysaccharides from marine source: porphyran. *Pharmacogn Rev.* 2008;4:271–276.

29. Al-Harrasi A, Bhatia S. Chapter 1: Olfactory aromatherapy versus COVID-19: systematic review. In: Role of Essential Oils in the Management of COVID-19. CRC Press, 2022:1–4.

30. Al-Harrasi A, Bhatia S. Chapter 2: Epidemiology respiratory infections: types, transmission and risks associated with co-infections. In: Role of Essential Oils in the Management of COVID-19. CRC Press, 2022:6–15.

31. Al-Harrasi A, Bhatia S. Chapter 3: Origin, morphology, genome organization, growth. In: Replication, and Pathogenesis of SARS-CoV-2, Role of Essential Oils in the Management of COVID-19. CRC Press, 2022:17–41.

32. Al-Harrasi A, Bhatia S. Chapter 4: Epidemiology, clinical manifestations, diagnostic approaches and preventive measures for COVID-19. In: Role of Essential Oils in the Management of COVID-19. CRC Press, 2022:43–59.

33. Al-Harrasi A, Bhatia S. Chapter 5: Role of drug repurposing and natural Products. In: Role of Essential Oils in the Management of COVID-19. CRC Press, 2022:59–93.

34. Al-Harrasi A, Bhatia S. Chapter 6: Host immune response vs COVID 19. In: Role of Essential Oils in the Management of COVID-19. CRC Press, 2022:93–103.

35. Al-Harrasi A, Bhatia S. Chapter 7: COVID-19 effect on different organ systems. In: Role of Essential Oils in the Management of COVID-19. CRC Press, 2022:103–125.

36. Al-Harrasi A, Bhatia S. Chapter 8: Essential oil chemistry vs aromatherapy. In: Role of Essential Oils in the Management of COVID-19. CRC Press, 2022:125–135.

37. Al-Harrasi A, Bhatia S. Chapter 9: Effect of methods of extraction on chemical composition of essential oils. In: Role of Essential Oils in the Management of COVID-19. CRC Press, 2022:135–149.

38. Al-Harrasi A, Bhatia S. Chapter 10: Essential oil stability. In: Role of Essential Oils in the Management of COVID-19. CRC Press, 2022:159–169.

39. Al-Harrasi A, Bhatia S. Chapter 11: Quality control of EOs. In: Role of Essential Oils in the Management of COVID-19. CRC Press, 2022:159–179.

40. Al-Harrasi A, Bhatia S. Chapter 12: Mechanism of action of essential oils as well as its components. In: Role of Essential Oils in the Management of COVID-19. CRC Press, 2022:179–205.

41. Al-Harrasi A, Bhatia S. Chapter 13: Antimicrobial activity of essential oils in the vapor phase. In: Role of Essential Oils in the Management of COVID-19. CRC Press, 2022:169–179.

42. Al-Harrasi A, Bhatia S. Chapter 14: Antibacterial interactions between EOs and currently employed antibiotics. In: Role of Essential Oils in the Management of COVID-19. CRC Press, 2022:179–205.

43. Al-Harrasi A, Bhatia S. Chapter 15: Antibacterial mechanism of action of essential oils. In: Role of Essential Oils in the Management of COVID-19. CRC Press, 2022:205–223.

44. Al-Harrasi A, Bhatia S. Chapter 16: Anti-inflammatory, antioxidant and immunomodulatory effects of EOs. In: Role of Essential Oils in the Management of COVID-19. CRC Press, 2022:223–233.

45. Al-Harrasi A, Bhatia S. Chapter 17: Anticancer properties of essential oils. In: Role of Essential Oils in the Management of COVID-19. CRC Press, 2022:233–253.

46. Al-Harrasi A, Bhatia S. Chapter 18: Effects of essential oils on CNS. In: Role of Essential Oils in the Management of COVID-19. CRC Press, 2022:253–263.

47. Al-Harrasi A, Bhatia S. Chapter 19: Olfactory aromatherapy vs human psychophysiological activity. In: Role of Essential Oils in the Management of COVID-19. CRC Press, 2022:263–293.

48. Al-Harrasi A, Bhatia S. Chapter 20: Essential oils in the treatment of respiratory tract infections. In: Role of Essential Oils in the Management of COVID-19. CRC Press, 2022:293–313.

49. Al-Harrasi A, Bhatia S. Chapter 21: Cardioprotective effect of EOs. In: Role of Essential Oils in the Management of COVID-19. CRC Press, 2022:313–325.

50. Al-Harrasi A, Bhatia S. Chapter 22: Pharmacokinetics of essentials oils. In: Role of Essential Oils in the Management of COVID-19. CRC Press, 2022:325–355.

51. Al-Harrasi A, Bhatia S. Chapter 23: Hepatoprotective and nephroprotective effects of EOs. In: Role of Essential Oils in the Management of COVID-19. CRC Press, 2022:355–367.

52. Al-Harrasi A, Bhatia S. Chapter 24: Dispensing methods of essential oils. In: Role of Essential Oils in the Management of COVID-19. CRC Press, 2022:367–375.

53. Al-Harrasi A, Bhatia S. Chapter 25: Toxicity associated with essential oils. In: Role of Essential Oils in the Management of COVID-19. CRC Press, 2022:375–385.

54. Hazra B, Sarkar R, Biswas S, Mandal N. Comparative study of the antioxidant and reactive oxygen species scavenging properties in the extracts of the fruits of *Terminalia chebula*, *Terminalia* bellirica and *Emblica officinalis*. *BMC Complement Altern Med*. 2010 May 13;10:20.

55. Mahato SB, Nandy AK, Kundu AP. Pentacyclic triterpenoid sapogenols and their glycosides from *Terminalia* bellirica. *Tetrahedron*. 1992;48(12):2483–2494.

56. Nandy AK, Podder G, Sahu NP, Mahato SB. Triterpenoids and their glucosides from *Terminalia* bellirica. *Phytochemistry*. 1989;28(10):2769–2772.

57. Bhatia S, Al-Harrasi A, Behl T, Anwer MK, Ahmed MM, Mittal V, Kaushik D, Chigurupati S, Kabir MT, Sharma PB, Chaugule B, Vargas-de-la-Cruz C. Anti-migraine activity of freeze dried-latex obtained from *Calotropis Gigantea* Linn. *Environ Sci Pollut Res Int*. 2021; 29,18(2022): 27460–27478. doi:10.1007/s11356-021-17810-x

58. Yadava RN, Rathore K. A new cardenolide from the seeds of *Terminalia* bellirica. *Fitoterapia*. 2001;72(3):310–312.

59. Kapoor LD. CRC Handbook of Ayurvedic medicinal plants. CRC Press, Boca Raton, FL; 1990;321:11.

60. Sales MS, Roy A, Antony L, Banu SK, Jeyaraman S, Manikkam R. Octyl gallate and gallic acid isolated from *Terminalia* bellirica regulates normal cell cycle in human breast cancer cell lines. *Biomed Pharmacother*. 2018 Jul;103:1577–1584.

61. Deb A, Choudhury G, Barua S, Das B, Anindita C, Choudhury DG. Pharmacological activities of Baheda (*Terminalia* bellirica): a review. *J Pharmacogn Phytochem*. 2016;5(1):2278–4136.

62. Chalise JP, Acharya K, Gurung N, Bhusal RP, Gurung R, Skalko-Basnet N, Basnet P. Antioxidant activity and polyphenol content in edible wild fruits from Nepal. *Int J Food Sci Nutr*. 2010 Jun; 61(4):425–32.

63. Mallik P, Das P., Karon B, Das S. A review on phytochemistry and pharmacological activity of *Terminalia* bellirica. *Int J Drug Formulation Res*. 2012;3(6):1–5.

64. Tasaduq SA, Singh K, Sethi S, Sharma SC, Bedi KL, Singh J, Jaggi BS, Johri RK. Hepatocurative and antioxidant profile of HP-1, a polyherbal phytomedicine. *Hum Exp Toxicol*. 2003 Dec;22(12):639–645.

65. Mallik P, Das P, Karon B, Das S. A review on phytochemistry and pharmacological activity of *Terminalia* bellirica. *Int J Drug Formulation Res*. 2012;3(6):1–5.

66. Wele A, De S, Dalvi M, Devi N, Pandit V. Nanoparticles of biotite mica as Krishna Vajra Abhraka Bhasma: synthesis and characterization. *J Ayurveda Integr Med*. 2021;12(2):269–282.

67. Sholikhah EN, Mustofa M, Nugrahaningsih DAA, Yuliani FS, Purwono S, Sugiyono S, Widyarini S, Ngatidjan N, Jumina J, Santosa D, Koketsu M. Acute and subchronic oral toxicity study of polyherbal formulation containing *Allium sativum* L., *Terminalia bellirica* (Gaertn.) Roxb., *Curcuma aeruginosa* Roxb., and *Amomum compactum* Sol. ex. Maton in rats. *Biomed Res Int*. 2020 Mar 24; 2020:8609364.

68. Chayarop K, Peungvicha P, Temsiririrkkul R, Wongkrajang Y, Chuakul W, Rojsanga P. Hypoglycaemic activity of Mathurameha, a Thai traditional herbal formula aqueous extract, and its effect on biochemical profiles of streptozotocin-nicotinamide-induced diabetic rats. *BMC Complement Altern Med*. 2017 Jun 29;17(1):343.

69. Fernando CD, Karunaratne DT, Gunasinghe SD, Cooray MC, Kanchana P, Udawatte C, Perera PK. Inhibitory action on the production of advanced glycation end products (AGEs) and suppression of free radicals in vitro by a Sri Lankan polyherbal formulation Nawarathne Kalka. *BMC Complement Altern Med*. 2016 Jul 8;16:197.

70. C T S, M D, P R R, K M, E M A, Balachandran I. Chemical profiling of selected Ayurveda formulations recommended for COVID-19. *Beni Suef Univ J Basic Appl Sci*. 2021;10(1):2.

71. Gupta A, Kumar R, Bhattacharyya P, Bishayee A, Pandey AK. *Terminalia bellirica* (Gaertn.) roxb. (Bahera) in health and disease: a systematic and comprehensive review. *Phytomedicine*. 2020;77:153278.

72. Jayesh K, Helen LR, Vysakh A, Binil E, Latha MS. Ethyl acetate fraction of *Terminalia bellirica* (Gaertn.) Roxb. fruits inhibits proinflammatory mediators via down regulating nuclear factor-κB in LPS stimulated Raw 264.7 cells. *Biomed Pharmacother*. 2017 Nov; 95:1654–1660.

73. Gupta A, Kumar R, Pandey AK. Antioxidant and antidiabetic activities of *Terminalia bellirica* fruit in alloxan induced diabetic rats. *S Afr J Bot*. 2020;130:308–315.

74. Nampoothiri SV, Prathapan A, Cherian OL, Raghu KG, Venugopalan V V, Sundaresan A. In vitro antioxidant and inhibitory potential of *Terminalia bellirica* and *Emblica officinalis* fruits against LDL oxidation and key enzymes linked to type 2 diabetes. *Food Chem Toxicol*. 2011;49(1):125–131.

75. Kaur S, Jaggi RK. Antinociceptive activity of chronic administration of different extracts of *Terminalia bellirica* Roxb. and *Terminalia chebula* Retz. fruits. *Indian J Exp Biol*. 2010;48(9):925–930.

76. Pinmai K, Hiriote W, Soonthornchareonnon N, Jongsakul K, Sireeratawong S, Tor-Udom S. In vitro and in vivo antiplasmodial activity and cytotoxicity of water extracts of *Phyllanthus emblica*, *Terminalia chebula*, and *Terminalia bellirica*. *J Med Assoc Thai*. 2010;93 Suppl 7:S120–S126.

77. Gilani AH, Khan AU, Ali T, Ajmal S. Mechanisms underlying the antispasmodic and bronchodilatory properties of *Terminalia bellirica* fruit. *J Ethnopharmacol*. 2008;116(3):528–538.

78. Chaudhary SK, Mukherjee PK, Nema NK, et al. ACE inhibition activity of standardized extract and fractions of *Terminalia bellirica*. *Orient Pharm Exp Med*. 2012;12:273–277.

79. Prabhu VV, Chidambaranathan N, Gopal V. Evaluation and quantification of angiogenesis activity of *Terminalia bellirica* Roxb, by mice sponge implantation method. *J Young Pharm*. 2012;4(1):22–27.

80. Tanaka M, Kishimoto Y, Saita E, Suzuki-Sugihara N, Kamiya T, Taguchi C, Iida K, Kondo K. *Terminalia bellirica* extract inhibits low-density lipoprotein oxidation and macrophage inflammatory response in vitro. *Antioxidants (Basel)*. 2016 Jun 14;5(2):20.

81. Jayesh K, Helen LR, Vysakh A, Binil E, Latha MS. Ethyl acetate fraction of *T. bellirica* (Gaertn.) Roxb. fruits inhibit proinflammatory mediators via down regulating nuclear factor-κB in LPS stimulated Raw 264.7 cells. *Biomed Pharmacother*. 2017;95:1654–1660.

82. Tanaka M, Kishimoto Y, Sasaki M, Sato A, Kamiya T, Kondo K, Iida K. *Terminalia bellirica* (Gaertn.) Roxb. extract and gallic acid attenuate LPS-induced inflammation and oxidative stress via MAPK/NF-κB and Akt/AMPK/Nrf2 pathways. *Oxid Med Cell Longev*. 2018 Nov 8;2018:9364364.

83. Reddy TC, Aparoy P, Babu NK, Kumar KA, Kalangi SK, Reddanna P. Kinetics and docking studies of a COX-2 inhibitor isolated from *Terminalia bellirica* fruits. *Protein Pept Lett*. 2010; 17(10):1251–1257.

84. Chen YX, Tong J, Ge LL, Ma BX, He JS, Wang YW. Ethyl acetate fraction of *Terminalia bellirica* fruit inhibits rat hepatic stellate cell proliferation and induces apoptosis. *Ind Crops and Products*. 2015;76:364–373.

85. Dharmaratne MPJ, Manoraj A, Thevanesam V, Ekanayake A, Kumar NS, Liyanapathirana V, Abeyratne E, Bandara BMR. *Terminalia bellirica* fruit extracts: in-vitro antibacterial activity against selected multidrug-resistant bacteria, radical scavenging activity and cytotoxicity study on BHK-21 cells. *BMC Complement Altern Med*. 2018 Dec 7;18(1):325.

86. Patil S, Chaudhari G, Paradeshi J, Mahajan R, Chaudhari BL. Instant green synthesis of silver-based herbo-metallic colloidal nanosuspension in *Terminalia bellirica* fruit aqueous extract for catalytic and antibacterial applications. *3 Biotech*. 2017 May;7(1):36.

87. Sharma R. Synthesis of *Terminalia bellirica* fruit extract mediated silver nanoparticles and application in photocatalytic degradation of wastewater from textile industries. *Mater Today Proc*. 2021; 44(1):1995–1998.

88. Gunarani GI, Raman AB, Kumar JD, Natarajan S, Jegadeesan GB. Biogenic synthesis of Fe and NiFe nanoparticles using *Terminalia bellirica* extracts for water treatment applications. *Mater Lett*. 2019;247:90–94.

89. Marimuthu S, Balasubramanian B, Selvam R, D'Souza P. Evaluation of a polyherbal formulation for the management of wet litter in broiler chickens: implications on performance parameters, cecal moisture level, and footpad lesions. *J Adv Vet Anim Res*. 2019 Oct 30;6(4):536–543.

90. Gupta A, Kumar R, Ganguly R, Singh AK, Rana HK, Pandey AK. Antioxidant, anti-inflammatory and hepatoprotective activities of *Terminalia bellirica* and its bioactive component ellagic acid against diclofenac induced oxidative stress and hepatotoxicity. *Toxicol Rep*. 2021;8:44–52.

91. Chaudhary R, Singh R, Verma R, Kumar P, Kumar N, Singh L, Kumar SS. Investigation on protective effect of *Terminalia bellirica* (Roxb.) against drugs induced cardiotoxicity in Wistar albino rats. *J Ethnopharmacol*. 2020;261:113080.

92. Kuriakose J, Lal Raisa H, A V, Eldhose B, M S L. *Terminalia bellirica* (Gaertn.) Roxb. fruit mitigates CCl4 induced oxidative stress and hepatotoxicity in rats. *Biomed Pharmacother*. 2017;93:327–333.

93. Gupta A, Pandey AK. Aceclofenac-induced hepatotoxicity: an ameliorative effect of *Terminalia bellirica* fruit and ellagic acid. *World J Hepatol*. 2020 Nov 27;12(11):949–964.

94. Saxena S, Lakshminarayan N, Gudli S, Kumar M. Anti-bacterial efficacy of *Terminalia chebula*, *Terminalia bellirica*, *Embilica officinalis* and triphala on salivary *Streptococcus mutans* count – a linear randomized cross over trial. *J Clin Diagn Res*. 2017 Feb;11(2):ZC47–ZC51.

95. Sharma S, Shrivastava N. DNA-based simultaneous identification of three terminalia species targeting adulteration. *Pharmacogn Mag*. 2016;12(Suppl 3):S379–S383.

96. Kannangara S, Karunarathne S, Ranaweera L, Ananda K, Ranathunga D, Jayarathne H, Weebadde C, Sooriyapathirana S. Assessment of the applicability of wood anatomy and DNA barcoding to detect the timber adulterations in Sri Lanka. *Sci Rep*. 2020 Mar 9;10(1):4352.

97. Santhosh Kumar JU, Krishna V, Seethapathy GS, Ganesan R, Ravikanth G, Shaanker RU. Assessment of adulteration in raw herbal trade of important medicinal plants of India using DNA barcoding. *3 Biotech.* 2018 Mar;8(3):135.

98. Jayesh K, Helen LR, Vysakh A, Binil E, Latha MS. In vivo toxicity evaluation of aqueous acetone extract of *Terminalia* bellirica (Gaertn.) Roxb. fruit. *Regul Toxicol Pharmacol.* 2017;86:349–355.

99. Pfundstein B, El Desouky SK, Hull WE, Haubner R, Erben G, Owen RW. Polyphenolic compounds in the fruits of Egyptian medicinal plants (*Terminalia bellirica*, *Terminalia chebula* and *Terminalia horrida*): characterization, quantitation and determination of antioxidant capacities. *Phytochemistry.* 2010;71(10):1132–1148.

100. Valsaraj R, Pushpangadan P, Smitt UW, et al. New anti-HIV-1, antimalarial, and antifungal compounds from *Terminalia bellirica*. *J Nat Prod.* 1997;60(7):739–742.

101. Latha RC, Daisy P. Insulin-secretagogue, antihyperlipidemic and other protective effects of gallic acid isolated from *Terminalia bellirica* Roxb. in streptozotocin-induced diabetic rats. *Chem Biol Interact.* 2011;189(1–2):112–118.

102. Gomes JHS, Mbiakop UC, Oliveira RL, Stehmann JR, Pádua RM, Cortes SF, Braga FC. Polyphenol-rich extract and fractions of *Terminalia phaeocarpa* Eichler possess hypoglycemic effect, reduce the release of cytokines, and inhibit lipase, α-glucosidase, and α-amilase enzymes. *J Ethnopharmacol.* 2021 May 10;271:113847.

103. Chopra RN, Chopra IC, Handa KL, Kapur LD. *Terminalia arjuna* W&A (Combretaceae). In: Chopra's Indigenous Drugs of India, 1st ed. (eds. RN Chopra, IC Chopra, KL Handa, LD Kapur). UNDhur & Sons, Calcutta, India; 1958:421–424.

104. Saha A, Pawar VM, Jayaraman S. Characterisation of polyphenols in Terminalia arjuna bark extract. *Indian J Pharm Sci.* 2012 Jul;74(4):339–347.

105. Maulik SK, Talwar KK. Therapeutic potential of Terminalia arjuna in cardiovascular disorders. *Am J Cardiovasc Drugs.* 2012;12:157–163.

106. Dwivedi S. Terminalia arjuna Wight & Arn - a useful drug for cardiovascular disorders. *J Ethnopharmaco.* 2007;114:114–129.

107. Nadkarni AK, Nadkarni KM. Indian Materia Medica, 1st ed. Popular Book Depot, Bombay, Indiap; 1954:1198.

108. Warrier PK, Nambiar VPK, Ramankutty C. *Terminalia arjuna.* In: Indian Medicinal Plants – A Compendium of 500 Species, 1st ed., Vol. 5 (eds. PK Warrier, VPK Nambiar, C Ramankutty). Orient Longman Limited, Madras, India; 1996:253–257.

109. Charaka Samhita. Ayurveda Deepika Commentary by Chakrapani Datta: Shareerasthana 6/12. Chaukhamba Sanskrit Sansthan, Varanasi, India; 1994.

110. Akinmoladun AC, Olaleye MT, Farombi EO. Cardiotoxicity and cardioprotective effects of African medicinal plants. In: Toxicological Survey of African Medicinal Plants (ed. V Kuete). Elsevier, 2014:395–421.

111. Atanasov AG, Zotchev SB, Dirsch VM, International Natural Product Sciences Taskforce, Supuran CT. Natural products in drug discovery: advances and opportunities. *Nat Rev Drug Discov.* 2021 Mar;20(3):200–216.

112. Oberoi L, Akiyama T, Lee KH, Liu SJ. The aqueous extract, not organic extracts, of Terminalia arjuna bark exerts cardiotonic effect on adult ventricular myocytes. *Phytomedicine.* 2011 Feb 15;18(4):259–265.

113. Balkrishna A, Gohel V, Singh R, Joshi M, Varshney Y, Srivastava J, Bhattacharya K, Varshney A. Tri-herbal medicine Divya Sarva-Kalp-Kwath (Livogrit) regulates fatty acid-induced steatosis in human HepG2 cells through inhibition of intracellular triglycerides and extracellular glycerol levels. *Molecules.* 2020 Oct 21;25(20):4849. doi: 10.3390/molecules25204849. PMID: 33096687; PMCID: PMC7587968.

114. Bhāvaprakāśa-nighaṇṭu, 16th Century A.D. By: Bhāvamiśra, in: (NIIMH), N.I.o.I.M.H. (ed.), Bhāvaprakāśa-nighaṇṭu NIIMH, India.

115. Chakrapanidatta. Chakradatta with Vaidyaprabha, 2nd ed. Chaukambha Sanskrit Sansthan, Varanasi, India; 1994.

116. Campisi A, Acquaviva R, Raciti G, Duro A, Rizzo M, Santagati NA. Antioxidant activities of *Solanum nigrum* L. leaf extracts determined in in vitro cellular models. *Foods.* 2019 Feb 8;8(2):63.

117. A P, Varghese MV, S A, P SR, Mathew AK, Nair A, Nair RH, K G R. Polyphenol rich ethanolic extract from *Boerhavia diffusa* L. mitigates angiotensin II induced cardiac hypertrophy and fibrosis in rats. *Biomed Pharmacother.* 2017 Mar;87:427–436. doi: 10.1016/j.biopha.2016.12.114. Epub 2017 Jan 6. PMID: 28068633.

118. Dube N, Nimgulkar C, Bharatraj DK. Validation of therapeutic anti-inflammatory potential of Arjuna Ksheera Paka – a traditional Ayurvedic formulation of *Terminalia arjuna*. *J Tradit Complement Med.* 2016 Dec 29;7(4):414–420. doi: 10.1016/j.jtcme.2016.11.006. PMID: 29034188; PMCID: PMC5634724.

119. Kapoor D, Vijayvergiya R, Dhawan V. Terminalia arjuna in coronary artery disease: ethnopharmacology, pre-clinical, clinical & safety evaluation. *J Ethnopharmacol*. 2014 Sep 11;155(2):1029–1045.

120. Suganthy N, Muniasamy S, Archunan G. Safety assessment of methanolic extract of *Terminalia chebula* fruit, Terminalia arjuna bark and its bioactive constituent 7-methyl gallic acid: in vitro and in vivo studies. *Regul Toxicol Pharmacol*. 2018 Feb;92:347–357.

121. Suely A, Zabed H, Ahmed AB, Mohamad J, Nasiruddin M, Sahu JN, Ganesan P. Toxicological and hematological effect of Terminalia arjuna bark extract on a freshwater catfish, *Heteropneustes fossilis*. *Fish Physiol Biochem*. 2016 Apr;42(2):431–444.

122. Balkrishna A, Rustagi Y, Tomer M, Pokhrel S, Bhattacharya K, Varshney A. Divya-Arjuna-Kwath (Terminalia arjuna) and Divya-HridyAmrit-Vati ameliorate isoproterenol-induced hypertrophy in murine cardiomyocytes through modulation of oxidative stress. *Phytomed Plus*. 2021;1(4):100074.

123. Hebbani AV, Vaddi DR, Dd PP, NCh V. Protective effect of Terminalia arjuna against alcohol induced oxidative damage of rat erythrocyte membranes. *J Ayurveda Integr Med*. 2021 Apr-Jun;12(2):330–339.

124. Mohanty IR, Borde M, Kumar C S, Maheshwari U. Dipeptidyl peptidase IV Inhibitory activity of Terminalia arjuna attributes to its cardioprotective effects in experimental diabetes: in silico, in vitro and in vivo analyses. *Phytomedicine*. 2019 Apr;57:158–165.

125. Noreen S, Ghumman SA, Batool F, Ijaz B, Basharat M, Noureen S, Kausar T, Iqbal S. Terminalia arjuna gum/alginate in situ gel system with prolonged retention time for ophthalmic drug delivery. *Int J Biol Macromol*. 2020 Jun 1;152:1056–1067.

126. Mittal A, Tandon S, Singla SK, Tandon C. Modulation of lithiatic injury to renal epithelial cells by aqueous extract of Terminalia arjuna. *J Herb Med*. 2018;13:63–70.

127. Maulik SK, Wilson V, Seth S, Bhargava B, Dua P, Ramakrishnan S, Katiyar CK. Clinical efficacy of water extract of stem bark of Terminalia arjuna (Roxb. ex DC.) Wight & Arn. in patients of chronic heart failure: a double-blind, randomized controlled trial. *Phytomedicine*. 2016 Oct 15;23(11):1211–1219.

128. Bhawania G, Kumar A, Murthy SN, Kumari N, Swamie CG. A retrospective study of effect of Terminalia arjuna and evidence based standard therapy on echocardiographic parameters in patients of dilated cardiomyopathy. *J Pharm Res*. 2013;6(5):493–498.

129. Bharani A, Ganguly A, Bhargava KD. Salutary effect of Terminalia arjuna in patients with severe refractory heart failure. *Int J Cardiol*. 1995 May;49(3):191–199.

130. Dwivedi S, Chopra D, Bhandari B. Role of Terminalia arjuna Wight and Arn. in the treatment of chronic coronary artery disease from pharmacovigilance point of view. *Ayu*. 2019 Apr-Jun;40(2):104–108.

131. Sarin YK. Illustrated Manual of Herbal Drugs Used in Ayurveda. CSIR and ICMR, New Delhi; 1996.

132. Gupta AK, Tondon N, Sharma M. (eds.) Quality standards of Indian medicinal plants. In: Anonymus. Terminalia arjuna (Roxb) Wight & Arn. Indian Council of Medical Research, New Delhi; 2005: 243–252.

133. Gupta PC. Biological and pharmacological properties of *Terminalia chebula* Retz. (Haritaki) – an overview. *Int J Pharm Sci*. 2012;4:62–68.

134. Gowthami R, Sharma N, Pandey R, Agrawal A. Status and consolidated list of threatened medicinal plants of India. *Genet Resour Crop Evol*. 2021 May 25;68(6):2235–2263. doi: 10.1007/s10722-021-01199-0. Epub 2021 May 25. PMID: 34054223; PMCID: PMC8148398.

135. Singh C. 2α-Hydroxymicromeric acid, a pentacyclic triterpene from *Terminalia chebula*. *Phytochemistry*. 1990;29(7):2348–2350.

136. Kundu AP, Mahato SB. Triterpenoids and their glycosides from *Terminalia chebula*. *Phytochemistry*. 1993;32(4):999–1002.

137. Reddy DB, Reddy TC, Jyotsna G, Sharan S, Priya N, Lakshmipathi V, Reddanna P. Chebulagic acid, a COX-LOX dual inhibitor isolated from the fruits of *Terminalia chebula* Retz., induces apoptosis in COLO-205 cell line. *J Ethnopharmacol*. 2009 Jul 30;124(3):506–512.

138. Manosroi A, Jantrawut P, Ogihara E, et al. Biological activities of phenolic compounds and triterpenoids from the galls of *Terminalia chebula*. *Chem Biodivers*. 2013;10(8):1448–1463.

139. Waheed S, Fatima I. Instrumental neutron activation analysis of *Emblica officinalis*, *Terminalia bellirica* and *Terminalia chebula* for trace element efficacy and safety. *Appl Radiat Isot*. 2013; 77:139–144.

140. Ajala OS, Jukov A, Ma CM. Hepatitis C virus inhibitory hydrolysable tannins from the fruits of *Terminalia chebula*. *Fitoterapia*. 2014 Dec;99:117–123. doi: 10.1016/j.fitote.2014.09.014. Epub 2014 Sep 28. PMID: 25261266.

141. Lee DY, Yang H, Kim HW, Sung SH. New polyhydroxytriterpenoid derivatives from fruits of *Terminalia chebula* Retz. and their α-glucosidase and α-amylase inhibitory activity. *Bioorg Med Chem Lett*. 2017 Jan;27(1) 34–39.

142. Lee DY, Kim HW, Yang H, Sung SH. Hydrolyzable tannins from the fruits of *Terminalia chebula* Retz and their α-glucosidase inhibitory activities. *Phytochemistry.* 2017 May;137:109–116.

143. Jeong HK, Lee D, Kim HP, Baek SH. Structure analysis and antioxidant activities of an amylopectin-type polysaccharide isolated from dried fruits of *Terminalia chebula. Carbohydr Polym.* 2019 May 1; 211:100–108.

144. Zhang XR, Zhu HT, Wang D, Yang Z, Yang CR, Zhang YJ. Termitomenins A-E: five new lignans from *Terminalia chebula* var. tomentella (Kurz) C. B. Clarke. *Fitoterapia.* 2020 Jun;143:104571.

145. Yin J, Zhu HT, Zhang M, Wang D, Yang CR, Zhang YJ. Termitomenins F and G, two new lignan glucosides from *Terminalia chebula* var. tomentella (Kurz) C. B. Clarke. *Nat Prod Bioprospect.* 2021. 10.1007/s13659-021-00314-z. [published online ahead of print, 2021 Jun 10].

146. Zhang XR, Qiao YJ, Zhu HT, Kong QH, Wang D, Yang CR, Zhang YJ. Multiple in vitro biological effects of phenolic compounds from *Terminalia chebula* var. tomentella. *J Ethnopharmacol.* 2021 Jul 15;275:114135.

147. Ratha KK, Joshi GC. Haritaki (*Chebulic myrobalan*) and its varieties. *Ayu.* 2013 Jul;34(3):331–334.

148. Gao J, Ajala OS, Wang CY, Xu HY, Yao JH, Zhang HP, Jukov A, Ma CM. Comparison of pharmaco-kinetic profiles of *Terminalia phenolics* after intragastric administration of the aqueous extracts of the fruit of *Terminalia chebula* and a Mongolian compound medicine-Gurigumu-7. *J Ethnopharmacol.* 2016 Jun 5;185:300–309.

149. Gupta DP, Rajagopala M, Dhiman KS. A clinical study on Akshitarpana and combination of Akshitarpana with Nasya therapy in Timira with special reference to myopia. *Ayu.* 2010 Oct;31(4):473–477.

150. Wani K, Shah N, Prabhune A, Jadhav A, Ranjekar P, Kaul-Ghanekar R. Evaluating the anticancer activity and nanoparticulate nature of homeopathic preparations of *Terminalia chebula. Homeopathy.* 2016 Nov;105(4):318–326. doi: 10.1016/j.homp.2016.02.004. Epub 2016 Apr 12. PMID: 27914571.

151. Feng XH, Xu HY, Wang JY, Duan S, Wang YC, Ma CM. In vivo hepatoprotective activity and the underlying mechanism of chebulinic acid from *Terminalia chebula* fruit. *Phytomedicine.* 2021 Mar;83:153479. doi: 10.1016/j.phymed.2021.153479. Epub 2021 Jan 23. PMID: 33561764.

152. Karthick Raja Namasivayam S, Angel J, Bharani RSA, Nachiyar CV. *Terminalia chebula* and *Ficus racemosa* principles mediated repression of novel drug target Las R – the transcriptional regulator and its controlled virulence factors produced by multiple drug resistant *Pseudomonas aeruginosa* – biocompatible formulation against drug resistant bacteria. *Microb Pathog.* 2020 Nov;148:104412.

153. Bag PK, Roy N, Acharyya S, Saha DR, Koley H, Sarkar P, Bhowmik P. In vivo fluid accumulation-inhibitory, anticolonization and anti-inflammatory and in vitro biofilm-inhibitory activities of methyl gallate isolated from *Terminalia chebula* against fluoroquinolones resistant *Vibrio cholerae. Microb Pathog.* 2019 Mar;128:41–46. doi: 10.1016/j.micpath.2018.12.037. Epub 2018 Dec 19. PMID: 30578837.

154. Pawar AT, Gaikwad GD, Metkari KS, Tijore KA, Ghodasara JV, Kuchekar BS. Effect of *Terminalia chebula* fruit extract on ethylene glycol induced urolithiasis in rats. *Biomed Aging Pathol.* 2012;3:99–103.

155. Joseph R, Binitha RN. Screening of potential antiandrogenic phytoconstituents and secondary metabolites of *Terminalia chebula* by docking studies, *Mater Today Proc.* 2020;25:316–320.

156. Mani V, Sajid S, Rabbani SI, Alqasir AS, Alharbi HA, Alshumaym A. Anxiolytic-like and antidepressant-like effects of ethanol extract of *Terminalia chebula* in mice. *J Tradit Complement Med.* 2021 May 1;11(6):493–502.

157. Ibraheem F, Aziz MH, Fatima M, Shaheen F, Ali SM, Huang Q. In vitro cytotoxicity, MMP and ROS activity of green synthesized nickel oxide nanoparticles using extract of *Terminalia chebula* against MCF-7 cells. *Mater Lett.* 2019;234;129–133.

158. Sivamaruthi BS, Ramkumar VS, Archunan G, Chaiyasut C, Suganthy N. Biogenic synthesis of silver palladium bimetallic nanoparticles from fruit extract of *Terminalia chebula* – in vitro evaluation of anticancer and antimicrobial activity. *J Drug Deliv Sci Technol.* 2019;51:139–151.

159. Ghosh P, Gupta P, Tripathy A, Das B, Ghosh D. Evaluation of hypotesticular activities of different solvent fractions of hydro-methanolic extract of the fruit of *Terminalia chebula* in Wistar strain adult albino rat: genomic and flow cytometric approaches. *J Appl Biomed.* 2018;16(4):394–400.

160. Ghosh A, Jana K, Pakhira BP, Tripathy A, Ghosh D. Anti-fertility effect of aqueous-ethanolic (1:1) extract of the fruit of *Terminalia chebula*: rising approach towards herbal contraception. *Asian Pac J Reprod.* 2015;4:201–207.

161. Rahimi VB, Askari VR, Shirazinia R, Soheili-Far S, Askari N, Rahmanian-Devin P, Sanei-Far Z, Mousavi SH, Ghodsi R. Protective effects of hydro-ethanolic extract of *Terminalia chebula* on primary microglia cells and their polarization (M1/M2 balance). *Mult Scler Relat Disord.* 2018 Oct;25:5–13. doi: 10.1016/j.msard.2018.07.015. Epub 2018 Jul 11. PMID: 30014878.

162. Sadeghnia HR, Jamshidi R, Afshari AR, Mollazadeh H, Forouzanfar F, Rakhshandeh H. *Terminalia chebula* attenuates quinolinate-induced oxidative PC12 and OLN-93 cell death. *Mult Scler Relat Disord.* 2017 May;14:60–67. doi: 10.1016/j.msard.2017.03.012. Epub 2017 Apr 5. PMID: 28619434.

163. Lee DY, Kim HW, Yang H, Sung SH. Hydrolyzable tannins from the fruits of *Terminalia chebula* Retz and their α-glucosidase inhibitory activities. *Phytochemistry.* 2017 May;137:109–116. doi: 10.1016/j. phytochem.2017.02.006. Epub 2017 Feb 15. PMID: 28213992.

164. Lee DY, Yang H, Kim HW, Sung SH. New polyhydroxytriterpenoid derivatives from fruits of *Terminalia chebula* Retz. and their α-glucosidase and α-amylase inhibitory activity. *Bioorg Med Chem Lett.* 2017 Jan 1;27(1):34–39. doi: 10.1016/j.bmcl.2016.11.039. Epub 2016 Nov 16. PMID: 27890380.

165. Pugazhendhi A, Shafreen RB, Devi KP, Suganthy N. Assessment of antioxidant, anticholinesterase and antiamyloidogenic effect of *Terminalia chebula*, Terminalia arjuna and its bioactive constituent 7-Methyl gallic acid – an in vitro and in silico studies. *J Mol Liq.* 2018;257:69–81.

166. Shin TY, Jeong HJ, Kim DK, Kim SH, Lee JK, Kim DK, Chae BS, Kim JH, Kang HW, Lee CM, Lee KC, Park ST, Lee EJ, Lim JP, Kim HM, Lee YM. Inhibitory action of water soluble fraction of *Terminalia chebula* on systemic and local anaphylaxis. *J Ethnopharmacol.* 2001 Feb;74(2):133–140. doi: 10.1016/s0378-8741(00)00360-3. PMID: 11167031.

167. Thanigaivel A, Vasantha-Srinivasan P, Senthil-Nathan S, Edwin ES, Ponsankar A, Chellappandian M, Selin-Rani S, Lija-Escaline J, Kalaivani K. Impact of *Terminalia chebula* Retz. against *Aedes aegypti* L. and non-target aquatic predatory insects. *Ecotoxicol Environ Saf.* 2017 Mar;137:210–217. doi: 10.1016/j.ecoenv.2016.11.004. Epub 2016 Dec 19. PMID: 27940415.

168. Jagtap AG, Karkera SG. Potential of the aqueous extract of *Terminalia chebula* as an anticaries agent. *J Ethnopharmacol.* 1999 Dec 15;68(1–3):299–306. doi: 10.1016/s0378-8741(99)00058-6. PMID: 10624892.

169. Herbrechter R, Beltrán LR, Ziemba PM, Titt S, Lashuk K, Gottemeyer A, Levermann J, Hoffmann KM, Beltrán M, Hatt H, Störtkuhl KF, Werner M, Gisselmann G. Effect of 158 herbal remedies on human TRPV1 and the two-pore domain potassium channels KCNK2, 3 and 9. *J Tradit Complement Med.* 2020 Apr 27;10(5):446–453.

170. Bhatt ID, Rawat S, Badhani A, Rawal RS. Nutraceutical potential of selected wild edible fruits of the Indian Himalayan region. *Food Chem.* 2017;215;84–91.

171. Barthakur NN, Arnold NP. Nutritive value of the chebulic myrobalan (*Terminalia chebula* Retz.) and its potential as a food source. *Food Chem.* 1991;40(2):213–219.

172. Rana MS, Tyagi A, Hossain SA, Tyagi AK. Effect of tanniniferous *Terminalia chebula* extract on rumen biohydrogenation, Δ9-desaturase activity, CLA content and fatty acid composition in longissimus dorsi muscle of kids. *Meat Sci.* 2012;90(3):558–563.

173. Singh A, Sheikh J. Cleaner functional dyeing of wool using *Kigelia africana* natural dye and *Terminalia chebula* bio-mordant. *Sustainable Chem Pharm.* 2020;17:100286.

174. Saxena A, KiKamal T, Saxena KK, Chambyal S, Sharma A. Electrochemical studies and surface examination of low carbon steel by applying the extract of *Terminalia chebula. Mater Today Proc.* 2020;26:1360–1367.

175. Yatish KV, Prakash RM, Ningaraju C, Sakar M, Balakrishna RG, Lalithamba HS. *Terminalia chebula* as a novel green source for the synthesis of copper oxide nanoparticles and as feedstock for biodiesel production and its application on diesel engine. *Energy B.* 2021;215:119165.

176. Edison TNJI, Atchudan R, Sethuraman MG, Lee YR. Supercapacitor performance of carbon supported $Co_3O_4$ nanoparticles synthesized using *Terminalia chebula* fruit. *J Taiwan Inst Chem Eng.* 2016;68:489–495.

177. Dixit D, Dixit AK, Lad H, Gupta D, Bhatnagar D. Radioprotective effect of *Terminalia chebula* Retzius extract against γ-irradiation-induced oxidative stress. *Biomed Aging Patholo.* 2013;3(2):83–88.

178. Suganthy N, Muniasamy S, Archunan G. Safety assessment of methanolic extract of *Terminalia chebula* fruit, Terminalia arjuna bark and its bioactive constituent 7-methyl gallic acid: in vitro and in vivo studies. *Regul Toxicol Pharmacol.* 2018 Feb;92:347–357.

179. Andarkhor P, Sadeghi A, Khodadoost M, Kamalinejad M, Gachkar L, Abdi S, Zargaran A. Effects of *Terminalia chebula* Retz. in treatment of hemorrhoids: a double-blind randomized placebo-controlled clinical trial. *Eur J Integr Med.* 2019;30:100935.

180. Guddoye G, Dwibedy BK, Singh OP. Clinical evaluation of Varadi Kwatha in the management of Madhumeha (type-2 diabetes mellitus) – an attempt to provide evidence based data to the classical therapeutic claims. *J Res Educ Indian Med.* 2014;20(1):37–44.

181. Padhar BC, Dave AR, Goyal M. Clinical study of Arogyavardhini compound and lifestyle modification in management of metabolic syndrome: a double-blind placebo controlled randomized clinical trial. *Ayu.* 2019 Jul-Sep;40(3):171–178.

182. Naiktari RS, Gaonkar P, Gurav AN, Khiste SV. A randomized clinical trial to evaluate and compare the efficacy of triphala mouthwash with 0.2% chlorhexidine in hospitalized patients with periodontal diseases. *J Periodontal Implant Sci.* 2014 Jun;44(3):134–140.

183. Penmetsa GS, B V, Bhupathi AP, Rani P S, B V S, M V R. Comparative evaluation of Triphala, *Aloe vera*, and chlorhexidine mouthwash on gingivitis: a randomized controlled clinical trial. *Contemp Clin Dent.* 2019 Apr-Jun;10(2):333–337.

184. Gajbhiye S, Koli PG, Harit M, Chitrakar M, Bavane V, Chawda M. An evaluation of the efficacy, safety, and tolerability of abhraloha compared with oral ferrous ascorbate on iron deficiency anemia in women: a randomized controlled, parallel-group, assessor-blind clinical trial. *Cureus.* 2021 Apr 7;13(4):e14348.

185. Bhati H, Manjusha R. Clinical study on evaluation of anti-cataract effect of Triphaladi Ghana Vati and Elaneer Kuzhambu Anjana in Timira (immature cataract). *Ayu.* 2015 Jul-Sep;36(3):283–289.

186. Chaturvedi A, Nath G, Yadav VB, Antiwal M, Shakya N, Swathi C, Singh JP. A clinical study on Virechana Karma (therapeutic purgation) over the gut flora with special reference to obesity. *Ayu.* 2019 Jul-Sep;40(3):179–184.

187. Ahmad I, Mehmood Z, Mohammad F. Screening of some Indian medicinal plants for their antimicrobial properties. *J Ethnopharmacol.* 1998;62(2):183–193. doi: 10.1016/s0378-8741(98)00055-5

# 5 Plant Profile, Phytochemistry, and Ethnopharmacological Uses of *Swertia chirayita, Tribulus terrestris,* and *Plumbago zeylanica*

*Ahmed Al-Harrasi*
Natural and Medical Sciences Research Center, University of Nizwa

*Saurabh Bhatia*
Natural and Medical Sciences Research Center, University of Nizwa
School of Health Sciences, University of Petroleum and Energy Studies

*Sridevi Chigurupati*
Qassim University

*Tapan Behl*
Chitkara University

*Deepak Kaushik*
M. D. University

## 5.1 INTRODUCTION

Utilization of traditional drugs, their quality, safety, efficacy, and consistency are important concerns, as compromise in any of these parameters can considerably impact the phytochemistry of the respective drugs. Various traditional drugs have been utilized since immemorial time. Some of the important crude drugs along with their chemical as well as pharmacological profiles and marketed formulations are listed in Table 5.1. Herbal preparations are evaluated for various quality parameters to safeguard their safety profile. Maintaining phytochemical consistency among herbal preparations that contain complex mixtures of different secondary metabolites is critical; however, it is important for their efficacy. Following stringent managements of processing as well as preparation of extracts to produce authentic and consistent herbal preparation is required to maintain phytochemical consistency as well as the efficacy. Thus, consistent phytochemical composition is required to maintain safety as well as quality profile in order to further establish its uniform therapeutic efficacy. Herbal preparations are always derived from different sources, sometimes by using one or more than one plant extract in order to make composition more complex as even single-plant extract may contain different phytochemicals. Thus, adherence to stick guidelines is required to maintain consistency among all these products. As these all-herbal preparations play an important role in health-care industry, thus it is important to assess the quality as well as safety profile of these all products. Owing to the complex multicomponent

DOI: 10.1201/9781003274124-5

**TABLE 5.1**

**Traditional Crude Drugs, Their Chemical Profiles, Pharmacological Uses as well as Marketed Formulations**

| Traditional Name | Common Vernacular Name | Botanical Source | Morphology | Chemical Nature | Pharmacology, Raditional and Common Uses | Marketed Formulations | Ingredients |
|---|---|---|---|---|---|---|---|
| Amla, amalki | Indian gooseberry, temperature (cooling) | *Emblica officinalis* previously *Phyllanthus emblica* (Euphorbiaceae) | Sweet, sour, pungent, bitter, astringent, fruits are depressed, globose called as rasnaya | 20 times more vitamin C than orange (625 mg–1814 mg/100 g of fruits) and it is not denatured by boiling | Chakra and Sushruta almost described the direct or indirect role of amla in every disease<br>Powerful antioxidant<br>Cooling herb (anti-inflammatory)<br>It balances all three doshas (vata, pitta, and kapha)<br>Hair tonic<br>Rejuvenating and antiageing property | Chyawanprash<br>Trifla<br>Brahmi amla hair oil<br>Amalaki oil<br>Amalaki ghrit<br>Amalaki churna (AC)<br>Dry powder (10–20 g)<br>Extract (5–10 ml) | AC + mishri (gastric problem)<br>A + bahera + harad (habitual constipation)<br>AC + jamun churna (diabetes)<br>A + honey + lime juice (rejuvenation and antiageing)<br>AC + milk/water (rasnaya) |
| Kankari (thorny plant) | Yellow berried nightshade | *Solanum xanthocarpum* (Solanaceae) | Berries: globous, glabrous, yellow<br>Seeds: glabrous<br>Stems: aculeate<br>Leaves: ovate<br>Flowers: cymose inflorescence | Solanine, solanidine (roots) | Antiasthmetic<br>Aperients<br>Astringent<br>Digestive<br>Febrifuge<br>Throat infections<br>Cough and bronchitis<br>Urinary tract infections | Fruit pulp: skin infection<br>Flowers: promote digestion<br>Whole died plant: indigestion and stomach problems<br>1 g kantakari Kantakari + black pepper + honey (asthma)<br>Fruit decoction gargle (gums 7 teeth problems) seeds smoke (gems from mouth) | *Dashamoola ghrit*<br>*Dashamoola arishta*<br>*Dashamoola taila*<br>*Dashamoola kwath*<br>Decoction: 30 ml 3 times a day<br>Powder: 2–3 times a day with warm water |
| Shatavari (hundred husbands) | Indian asparagus, temperature (cooling) | *Asparagus racemosus* (Liliaceae) | Leaves: green, needle-like cladodes, Roots: clusters or fascile. Whole plant is sweet or bitter | Root contains four steroidal saponnins; Shatavarin (1–4), called as phyto-oestrogen triterpenoid sapponin | Rejuvenative to female reproductive track, and blood and to build the body<br>Galactogouge<br>Aphrodisiac<br>Used in dyspepsia dysmenorrheal and leucorrhoea disorders<br>Reduces vatta and pitta increase Kappha | *Shatavari ghrita*: AR juice butter + honey + long pepper: aphrodisiac property<br>*Phala ghrita*: AR juice + butter + cow milk + cures barrenness and genital disorders of females<br>*Narayana taila*: diseased of nerves and joints<br>*Vishnu taila*: AR juice + cow milk + seasamum oil + juice of: nervous disorders<br>*Prameha mihira taila*: urinary disease |

*(Continued)*

**TABLE 5.1 (Continued)**

**Traditional Crude Drugs, Their Chemical Profiles, Pharmacological Uses as well as Marketed Formulations**

| Traditional Name | Common Vernacular Name | Botanical Source | Morphology | Chemical Nature | Pharmacology, Raditional and Common Uses | Marketed Formulations | Ingredients |
|---|---|---|---|---|---|---|---|
| Anantmul, Indian ipecacuahna (climber) | "Tylos" meaning "knot" and "phoros" meaning "bearing" | *Tylophora indica, Tylophora asthmatica* (Apocynaceae) | Leaves: ovate oblong, chordate, Flowers: pale yellow, purple within, in lateral Cymes Fruit: a follicle; Roots: fleshy, Latex: watery | Phenanthroindo-lizidine alkaloid (tylophorine, tylophorinine, tylophorinidine, and septidine) | In Ayurveda, the plant has been used in treatment of asthma. The alkaloid of other traditional uses are in curing jaundice and inflammation | It has antitumour, immunomodulatory, antioxidant, antiasthmatic, muscle relaxant | Tylophora powder form, 400–500 mg given once daily to asthmatic patients for six days to cure asthma Yograj guggul Rasanasaptak quath Rasnadi churna Rasnadi quath Rasnadi lepa Rasnadivati |
| Bhilava, bhallataka | Marking nut tree, Semecarpus (simeion + carpus + anacardium = means marking + nut + cardium) = heart shaped marking nut) | *Semecarpus anacardium* Linn (Anacardiaceae) *Bhallatak* literary means sharp like spear. Temperature (hot) | Ushna (hot in potency), its mere skin contact may cause boils. To reduce its hot potency and toxicity, it is subjected to purification procedure | Bhilavanol A, Bhilavanol B, and other phenolic compounds | *Deepaniya* (promotes digestion); *Kushtaghna* (useful in skin diseases); *Mutra sangrahaniya* (cause urine retention); *Bhallatak* is classified as toxic plant in Ayurveda and (coconut water and oil used as antidote) | Used for dyeing, and promoting hair growth in folk medicine. Amrut Bhallatakavaleha[a] (General tonic and vitalizer) Bhallatakasava[a] (Neuralgia and asthma) Suran vatak[b] (Piles and anorectal diseases) Sanjeevani Vati[b] (Dysentery and diarrhoea) Bhallatak Parpati[a] (Rheumatic diseases) Narasimha Churna[a] (General restorative) | |
| Kalajira (seed of blessing; heal every disease except death) | Black seed or black cumin, kalaunji | *Nigella sativa* Linn (Ranunculaceae) | Seeds are small dicotyledonous, trigonus, angular, regulose-tubercular, black externally and white inside; aromatic and bitter taste | Thymoquinone, carvacrol, 4-terpineol, t-anethol, sesquiterpene longifolene (1%–8%) α-pinene, thymol, isoquinoline alkaloids (nigellicimine), pyrazol alkaloids or indazole alkaloids (nigellidine, nigellicine) | Diuretic, emmenagogue, galactagogue, anthelmintic, thermogenic, carminative, anodyne, deodorant, digestive, constipating, sudorific, febrifuge, expectorant, purgative, abortifacient. They are used in ascites, cough, jaundice, piles, skin diseases, anorexia, dyspepsia, flatulence, diarrhoea, dysentery, intrinsic haemorrhage and amenorrhea. Seed oil is a local anaesthetic | Stimulate the body's energy and helps recovery from fatigue and dispiritedness Flavouring, to improve digestion and produce warmth | Crushed seeds + vinegar + skin disorders Nigella + fenugreek + Commiphora Pancha jiraka paka (ajowan + carum + cumin + anise + fwnugreek + coriander = puerperal desease Roasted nigella = pain |

(Continued)

**TABLE 5.1 (Continued)**
Traditional Crude Drugs, Their Chemical Profiles, Pharmacological Uses as well as Marketed Formulations

| Traditional Name | Common Vernacular Name | Botanical Source | Morphology | Chemical Nature | Pharmacology, Raditional and Common Uses | Marketed Formulations | Ingredients |
|---|---|---|---|---|---|---|---|
| Bach | Sweet flag (longetivity, memory, and intellect) | *Acorus calamus* Acoraceae (Acoron + calamos = the pupil of the eye + reed like appearance) | Rhizomatous herb having oil (oxygenated sesquiterpenes), Sympoidal distichously alternate leaf | Sesquiterpenes, flavonoids, α- and β-asarone, xanthone, steroids | For mental disorders it is registered in *Pakistani Materia Medica* Some *in vitro* and *in vivo* studies have shown *Acorus calamus* oil induce malignant tumours, due to β-asarone Sweet flag powder + lukewarm salt-water: induces vomiting and relieves phlegm, while easing coughs and asthma In epilepsy, the powders of sweet flag, brahmi, and jatamamsi work well, when given with honey | Mental disorders (polygala root + sweet flag), chakra used leaves stalks in powder form or decoction in internal prescription Sushruta used juice of the plant for promoting intellect, longetivity, and memory In rheumatism, rheumatic fever and inflamed joints, the paste applied externally alleviates the pain and swelling For digestive ailments such as flatulence, loss of appetite, abdominal dull pain, and worms | Sarasvata choorna: epilepsy, hysteria, and as a brain tonic Granule Asabi (unani preparation): excellent nervine tonic which improves memory, reception as well as the speech. As it stimulates the uterine contractions, so it is used to augment the labour pains. It is also salutary in dysmenorrhoeal |
| Rasna (sugandha) | Alpinia + galanga (Italian botanist who catalogued and described exotic plants + mild ginger) | *Alpinia galangal* (greater galangal) *Alpinia officinarum* (minor galangal) (Zingiberaceae) | Pungent, hot, and spicy taste with an aromatic ginger like odour. In north Indai pluche lanceolata used as rasana | Contain volatile oils (alpha pinene, cinole, linalool, and sesquiterpenes lactones (galangol and galangin) | First reference was found in Shodal Nighantu and the first compound was Kulinjanaavaleha 3 g rhizome powder + milk used in impotence, herb was included in a number of sex tonics In Arabic medicine herb was administered as sex tonic In Siddha all compounds of Sahasrayoga are based on *Alpinia galangal* In north India *Pluchia lanceolata* is used as Raasanaa | Rhizome (carminative), stomachic, cicularory stimulant, diaphoretic, anti inflammatory. In South India its used for rheumatism, intermittent fever, dyspepsia, and respiratory disorders Drug is irritant in higher doses | Kulanjan churna Kulanjan taila Kulanjan rhizome |

(Continued)

**TABLE 5.1 *(Continued)***

**Traditional Crude Drugs, Their Chemical Profiles, Pharmacological Uses as well as Marketed Formulations**

| Traditional Name | Common Vernacular Name | Botanical Source | Morphology | Chemical Nature | Pharmacology, Raditional and Common Uses | Marketed Formulations | Ingredients |
|---|---|---|---|---|---|---|---|
| Punarnava (spreading hogweed) | Rejuvenative drug (plant was named in honour of *Hermann Boerhaave,* a famous Dutch physician of the 18th century) | *Boerhaavia diffusa* (nyctaginaceae) is known as Raktapunarnava (bitter, acrid, cool, and light) it is constipating. it overcomes the morbid kapha and pitta. *Trianthema portulacastrum* (aizoaceae) known as Svetapunarnavba (hot, bitter, and dry) It alleviates morbid kapha and pitta Rajanigantu a blue variety is also mentioned | Plant is mentioned in the Atharvaveda with the name "Punarnava", because the top of the plant dries up during the summer season and regenerates again during the rainy season | Two plants are recognized with the help of flower colour | Boeravinone (A–F), Punarnavoside, Punarnavine | Diuretic, stomachic, in jaundice, adaptogenic, immunodulatory, antioxidant aphrodisiac and nootropic activities *Charaka Samhita* (swedopaga, anuvasanupaga, kaashara, vayasthapana, kushtha) *Sushruta Samhita* (pittaj ashmari, shotha, mooshaka vish, alarka visha, with dhatura seed powder in jvara) *Ashtanga Sangraha* (kaashara mahakshaya and vayasthapana mahakshaya, madatya) *Chakradatta Samhita* Ayurveda Sara Sangrah Adarsha Nighantu Sharangdhar Samhita | Ayurveda Sara Sangrah Punarnavasava, Punarnavashtaka kwatha, Punarnavadi mandur Sukumara ghrita Shothaghna lepa, 20-30 g of the drug is to be used for decoction therapeutically in Shotha (inflammation) Pandu (anaemia) Punarnavasavam, kumaryasavam, dhanvanatram lkulambu, cyavanaprasam, vastyamayantaka ghrtam Punarnava arka, Punarnava Kshaar, Punarnava arishta Punarnavadi taila, Punarnavashtaka kwatha Punarnavadi kwatha |

*(Continued)*

**TABLE 5.1 (Continued)**

**Traditional Crude Drugs, Their Chemical Profiles, Pharmacological Uses as well as Marketed Formulations**

| Traditional Name | Common Vernacular Name | Botanical Source | Morphology | Chemical Nature | Pharmacology, Raditional and Common Uses | Marketed Formulations | Ingredients |
|---|---|---|---|---|---|---|---|
| Chitrak | Leadwort | *White flowered Chitrak* (*Plumbago zeylanica*, India) *Red flowered Chitrak* (*Plumbago rosea*, Sikkim) *Blue flowered chitrak* (*Plumbago capensis*, Africa) Plumbaginaceae | Roots are light yellow coloured when fresh, reddish brown when dry Care: roots are powerful poison, in malyagiri hills of Orrisa. Chitrak decoction is used for nigh blindness (50 g/100 ml water) | Skin disease (leprosy) Roots are appetizer, abortifacient, dyspepsia, anasarca, infusion (black fever and influenza), leaves are eaten for abortion, whole plant is powerful irritant and is used by malingerers to produce facial dermatitis | Plumbagin with higher percentages in the root | *Red flowered Chitrak more effective than P. zeylinica* *Red flowered Chitrak more effective than P. zeylinica and P. rosea* | Chittrakadi gutika Chittarak haritaki Chittarak ghrit Chittrakadi di churna Sudarshan churna Yograj guggul Agnitundi vati Punarnava mandur |
| Prickly chaff plant | Apamarga | *Achyranthes aspera* (Amaranthaceae) | Dedoublement of stamens, dormancy imposed by covering structures and introse type of anther in this species have been reported | – | Gynaecological disorders diuretic in the treatment of dropsy dermatological disorders Used in vamana therapy to alleviate Haemorrhoids Hiccoughs Abdominal disorders | – | Apamarga ksara taila (Ear diseases) Apamarga ksara (caustics for external use) Samshodana taila For cleansing in wounds Apamarga varti Excessive menstrual flow |

<sup>a</sup> Bhallatak as a major ingredient.
<sup>b</sup> Bhallatak as a minor ingredient.

nature, herbal preparations often possess the capability to target at multiple sites of the pathogen and sometimes host to treat or prevent the disease. Usually, to maintain quality standards, pharmacopoeia quality monographs (chemical markers-based information) to ensure the consistency, have been followed. However, an idea of quality marker, also known as Q marker, can be implemented by assessing the property-effect-components of herbal preparations. Owing to the utilization of herbal preparations at global scale, it is important to safeguard its safety profile and assess its quality as several adverse incidents such as adulteration and drug spoilage can lead to compromise in the quality of the raw materials or the finished products. Quality of the herbal preparations can be categorized into internal as well as external factors. External concerns mainly include drug contamination (due to external factors such as admixture with toxic metals, pesticides residues, and infected by microbes), misidentification, and adulteration, whereas internal factors include complexity and non-uniformity associated with the components of the herbal preparations. Both can significantly impact the quality of herbal preparations [1, 2]. Thus, stick adherence to the standard guidelines such as Good Agricultural and Collection Practices, Good Manufacturing Practices to ensure the quality of the drugs is required. Another aspect is improving the quality of medicinal plants by using various biotechnological approaches such as genetic engineering, plant tissue culture techniques to significantly improve the content of active component. Further various formulations can also be prepared to improve the therapeutic efficacy of the crude drug, especially by using the natural polymers [2–30].

## 5.2 CHIRATA

### 5.2.1 VERNACULAR NAMES

Chiretta, Chiravata, Kiratatikta, Qasabuzzarirah, kairata

### 5.2.2 BIOLOGICAL SOURCE

*Swertia chirata*
   **Family**
   Gentianaceae

### 5.2.3 CULTIVATION, COLLECTION, AND GEOGRAPHICAL DISTRIBUTION

*Swertia* approximately comprises 170 species, diverse genus of plants distributed across the world. *S. chirata*, an annual/biennial herb, cultivated in variety of soils with sandy loam, is rich in carbon and humus. It can be cultivated in sub-temperate regions between +1500- and 2100-m altitude. *S. chirata* is a bitter medicinal plant indigenous to temperate regions all over the world. *Swertia* genus is found at the high altitudes of western and eastern Himalayas in India, Nepal, and Bhutan regions. The complete plant of *S. chirata* has been extensively utilized and traded from India. In NMPB, it is highlighted as Medicinal Plant Species in High Trade category. Owing to its unnecessary overutilization, narrow geographical presence [31] and uncertain intrinsic challenges of seed viability as well as seed germination [32, 33], alternative methods for cultivation as well as conservation are now necessary to prevent the likely disappearance of *S. chirata*.

### 5.2.4 CHEMICAL COMPOSITION

Because of the presence of unique phytochemicals, genus *Swertia* has great therapeutic potential, primarily *S. chirata*, which has been utilized in folk medicine for different ailments. This ethnobotanical medicinal plant has been recognized typically for its extensive bitterness produced by the various phytochemicals, including amarogentin (utmost bitter component derived till yet), swertiamarin, swerchirin along with additional phytochemicals. Those have also been evidenced for its therapeutic activity [33]. Out of all the constituents extracted from *S. chirata*, key components are xanthonoids, secoiridoids, and triterpenoids [34, 35]. Among all important xanthonoid-based phytochemicals, mangiferin present important therapeutic pharmacological effects such as their capability in reducing oxidative stress, diabetes, inflammation, cancer and also possess diuretics, anti-Parkinson, and cardioprotective effects [36]. Main triterpenoidal compounds, such as ursolic acid as well as oleanolic acid, are also presented important in terms of their therapeutic potential [37]. Mangiferin, one of xanthoids, is considered one of the most pharmacologically active compounds mainly due to its anticancer and antidiabetic properties. Recently, mangiferin quantification has been done among 11 *Swertia* species from India. It was found that the level of mangiferin was found to be more in *Sarracenia minor* compared to other *Swertia* sp. obtained from the Western Ghats and the Himalayan region [38].

Recently, four xanthone-based phytochemicals, bellidifolin, methylswertianin, 1-hydroxy-3,5-dimethoxyxanthone, and decussatine, with two triterpenoidal components, 12-hydroxyoleanolic lactone as well as oleanolic acid have been isolated. Their activity against viral protein R showed that bellidifolin and oleanolic acid presented considerable activity against these viral proteins [39].

In 1973, nine tetraoxygenated xanthones have been isolated from the roots and aerial parts of *S. chirata*. Except these tetraoxygenated xanthones, several heterosides, triterpenes, and monoterpene alkaloids have been isolated from *S. chirata*. Initial pharmacological assessment in this study showed that medicinal properties attributed to the extracts could be due to the total xanthones [40]. In 1991, based on spectral and chemical evidence, new triterpenoidal swertane–based components (rarae) have been derived from *S. chirata*, with the presence of familiar triterpenoidal compounds [41]. In 1992, based on spectral evidence and chemical correlations, kairatenol, a migrated gammacerane triterpenoid, has been identified from *S. chirata* [42].

A recent report correlated contents of chemical components such as swertiamarin, amarogentin, and mangiferin in *S. chirata* with transcript levels of multiple genes of their related biosynthesis pathways. Three genes encoding PMK, ISPD, and IS of the secoiridoids biosynthesis pathway revealed increased levels, and EPSPS of mangiferin biosynthesis indicated increase in transcripts in leaf tissues [43]. Except *S. chirata*, chemical profiles of other *Swertia* species have also been explored. In 2001, in addition to the previously reported swerilactones (A to D), four new lactone-based components (swerilactones [H to K]) with a unique C-29 backbone were derived from *S. mileensis* (Qing-Ye-Dan), used in China to cure inflammation caused by virus. It has been revealed that swerilactone-based components (H to K) showed significant activity against hepatitis caused by HBV [44]. Another study demonstrated the isolation of novel lactone-based components (enantiomers), (±)-Sweriledugenin A, from *Swertia leduci* [45]. The previous study showed an isolation of five unusual lactonic enamino ketones, gentiocrucines A–E (1–5) from *Swertia macrosperma*, and *Swertia angustifolia* out of which gentiocrucines A and E showed anti-HBV activities [46].

## 5.2.5  COMMON USES

The whole herb has been utilized in curing piles, skin-associated complications, ulcers, and diabetes.

## 5.2.6  TRADITIONAL USES

*S. chirata* is commonly known as a bitter tonic in traditional system of medicine for the treatment of diabetes, digestive disorders, fever, loss of appetite, obesity, skin, and various other diseases [47].

## 5.2.7  TRADITIONAL FORMULATIONS

Earlier, the main component of an antipyretic Ayurvedic preparation known as Sudarshana powder was *S. chirata*, later replaced by *Andrographis paniculata*. Presently, the Sudarshana powder contains *A. paniculata* with 52 other bitter components (50%). The other components also have different therapeutic importance [48]. Sudarshana powder, which is a unique formula of 53 bitter components, can cure fever-related indications, including nausea, indigestion, eating disorder, and fatigue. Its utilization showed laxative (mild) response, improved the bile flow, as well as considered digestive to treat GIT complications [49–52]. Ojamin is over-the-counter herbal formulation that contains aqueous extracts of 14 herbs, including *S. chirata*. A recent study showed significant antidiabetic and antihyperlipidemic effects of Ojamin in rats. Nevertheless, the use of Ojamin with Metformin is warned [49–52]. Mahasudarshan Churna, a polyherbal Ayurvedic formulation, comprising Chirata as a main ingredient with additional 49 components, has been widely utilized in treating complications associated with hepatic tissue as well as spleen, fever, cold, and malaria [53–57]. Hyponidd, an Ayurvedic preparation comprising *Momordica charantia*, *Gymnema sylvestre*, and *S. chirata*, is used for the treatment of type 2 diabetes mellitus and polycystic ovarian disease as an insulin sensitizer. Mahasudarshan Churna has been used to treat diabetes, malarial fever, viral infection, and oxidative stress [53–57]. HPTLC analysis revealed main chemical compounds in Mahasudarshan Churna preparations were oleanolic acid, gallic acid, as well as mangiferin with other components such as quercetin, ursolic acid, and curcumin [53–57]. Di-da (*S. chiraita*, *Hebe elliptica*, *Swertia franchetiana*, and *Swertia mussotii*) Tibetan medicines in China were used for the treatment of various liver complications. *S. chirata*, *H. elliptica*, *S. franchetiana*, and *S. mussotii* are the main ingredients of Di-da. In folk medicine practices, Di-da was bitter in flavour as well as cold in nature [58]. It is beneficial in hepatic as well as gallbladder-associated complications thus is widely used in jaundice, icterohepatitis, viral hepatitis, and cholecystitis [58].

## 5.2.8  Pharmacological Uses

### 5.2.8.1  Antiviral Properties
The previous report evidenced that *S. chirata* extract showed antiviral effects against Herpes simplex virus type 1 [59]. It has been demonstrated that chloroform-soluble fraction derived from *S. chirata* showed antiviral protein R properties. Six components have been isolated as well as purified from active chloroform-soluble fraction of *S. chirata*, out of which Bellidifolin and oleanolic acid inhibited the expression of viral protein R in HeLa cells harbouring the TREx plasmid encoding the full length of viral protein R [60].

### 5.2.8.2  In Dengue
The complete herb decoction of chirata is evidenced for its effectiveness in the treatment of dengue [61].

### 5.2.8.3  In the Management of Osteoporosis
Chronic use with high-dose glucocorticoids administration has been linked to osteoporosis. Recent investigation showed that plumbagin presents considerable promising effects against dexamethasone-induced cellular damage via reducing oxidative stress, apoptosis, and osteogenic markers [62].

### 5.2.8.4  Antioxidant, Antibacterial, and Antidiabetic Effects
In earlier study, the effectiveness of this *Swertia cordata* as well as *S. chirata* in reducing oxidative stress, curing diabetes, and bacterial infections were evaluated. It was found that both plants showed promising results in terms of reducing oxidative stress, curing diabetes, and bacterial infections. Nevertheless, *S. chirata* revealed better effects than *S. cordata* [63]. A computational study showed that plumbagin inhibits FtsZ polymerization, and the binding site of plumbagin is close to the C-terminal of B. subtilis FtsZ; however, it does not exhibit any effect on the *Escherichia coli* FtsZ [64]. An analogue of plumbagin, namely SA-011, also reported for inhibiting the GTPase activity of *Bacillus anthracis* FtsZ and found to be more effective then berberine [65].

### 5.2.8.5  Anti-Metastatic Effects
*S. chirata* is a rich source of various xanthones. Recently, the cytotoxic potential of four xanthone derivatives isolated from chirata has been tested (under in vivo as well as in vitro conditions) in rodent transplanted breast cancer cell line (EAC) as well as two human breast cancer cell lines (MDA-MB-231 and MCF-7). It was found that 1,5,8-trihydroxy-3-methoxyxanthone may possibly act as an anticancer agent via augmenting ROS as well as lipid peroxidation in breast cancer cells resulting in apoptosis [66].

Another study demonstrated that anti-metastatic potential of xanthone derivates (1,5,8-trihydroxy-3-methoxy) present chirata herb, against adenocarcinoma under both in vitro and in vivo conditions. It was found that 1,5,8-trihydroxy-3-methoxy xanthone treatment effectively decreased the metastatic potential of Ehrlich ascites carcinoma–induced solid tumour via downregulating metastatic and epithelial–mesenchymal transition markers in vivo. In addition, the in vitro study demonstrated that 1,5,8-trihydroxy-3-methoxy xanthone treatment limited migratory and colony-forming capability of breast carcinoma cell line MCF-7 via downregulating metastatic and epithelial–mesenchymal transition markers [67].

### 5.2.8.5.1 Anti-SARS-CoV-2 Effects

Recently, extensive docking studies over the phytochemicals have been investigated to assess the inhibitory potential of the phytochemicals derived from *S. chirata*, Malabar nut, Tulsi, and against SARS-CoV-2 proteins. It was found that these phytochemicals (anisotine as well as amarogentin) showed promising inhibitory potential at the binding pockets of the viral proteins (anisotine: spike/Mpro proteins; amarogentin: RdRp protein) [68].

Recent findings based on docking studies demonstrated that amarogentin, alpha-amyrin, belladonnine, beta-sitosterol, kutkin, eufoliatorin, and caesalpinins showed highest affinity towards both viral spike protein and ACE2. These chemicals showed excellent pharmacokinetic report with least toxicity. Genetic network inference bases analysis showed that the following parameters have been altered during COVID-19:

- G-protein-coupled protein
- Membrane protein proteolysis
- Nuclear transport, apoptotic pathway
- Peptidyl-serine phosphorylation
- Positive regulation of ion transport
- Signalling based on kinase (B)
- Protein secretion
- Angiotensin level regulation
- Tumour necrosis factor

The target prediction study showed the majority of the components of plant bind at similar sites that have been found to be changed in SARS-Cov-2 infection [69, 70].

### 5.2.8.6   Anti-Inflammatory Effects

Amarogentin, (bitter secoiridoid glycoside) derived from *S. chirata*, has been reported for its effectiveness in leishmaniasis and carcinogenesis. A recent study suggested that amarogentin downregulates the cyclooxygenase-2 activity that can further facilitate to reduce skin carcinogenesis in an animal model. A recent molecular docking study revealed the role of amarogentin as a selective COX-2 inhibitor [71].

### 5.2.9   Mechanism of Action

The presence of antidiabetic compounds such as amarogentin, sweroside, swertiamarin, mangiferin, as well as amaroswerin in the fraction makes it effective against diabetes-induced complications [72, 73]. Chirata-derived extracts as well as its phytochemicals showed sugar-reducing effects by the following ways:

- Via decreasing level of cholesterol, low-density lipoproteins, triglycerides, as well as glucose [74].
- Via decreasing glucose and improving insulin as well as lipid levels [75].

- Via considerable decrease in cholesterol, triglycerides, glucose (fasting) levels, and increase in liver-based glycogen [76].
- Via improving intake of glucose as well as glycogen with muscular growth and the reducing level of blood glucose via increasing release of insulin [77].

## 5.2.10 Industrial Applications

### 5.2.10.1 Anti-Corrosive Agent
Recently, it was found that the *S. chirata* extract can be utilized for the corrosion resistance of carbon steel in an acidic medium [78].

### 5.2.10.2 Clastogenic Effects
Recent findings showed that in low dose of biogenic AgNPs (synthesized by using aqueous leaf extracts of *S. chirata*), bio-AgNPs can induce substantial clastogenic effects on both meristematic and reproductive plant cells. It was found that biogenic AgNPs caused chromosomal aberrations in both mitotic and meiotic cells of *Allium cepa* even at low dose [79].

### 5.2.10.3 Hepatoprotective Effects
In 2011, a study demonstrated that the administration of the plant extracts (*A. paniculata* or *S. chirata*) at 100–200 mg/kg after paracetamol insult normalized all the biochemical parameters up to the level of control. It was found that *A. paniculata* or *S. chirata* extracts provide protection against hepatotoxicity caused by paracetamol [80].

### 5.2.10.4 To Control Soft Rot Bacteria of Potato
Previously, soft rot Ecc P-138 bacterium has been considered aggressive in potatoes (from Bangladesh). Previously, an in vitro study suggested the role of *Swertia chirata* in significantly inhibiting Ecc P-138 growth. It was found that in context with the loss of weight as well as rate of infection during storage condition (22-week), the potatoes treated with *S. chirata* showed more protection against the infection caused by soft rot than control [81].

## 5.2.11 Marketed Products

Herbal preparations, including Melicon-V (ointment), Mensturyl (syrup) as well as Ayush-64, Diabecon (Edwin and Chungath; Mitra et al.) comprise chirata extract in different amounts to possess various therapeutic effects such as antipyretic, hypoglycaemic, antifungal, and antibacterial [82, 83].

## 5.2.12 Toxicity

A recent study showed that *S. chirata* extract contains ursolic acid and heavy metal along with trace elements in the recommended range as per WHO guidelines. Additionally, it was found that *S. chirata* extract and ursolic acid demonstrated considerably less inhibitory activity on rat liver microsome. Further, it was revealed that *S. chirata* extract and ursolic acid showed less inhibitory activity on CYP3A4 and CYP2D6 in comparison to known inhibitors (ketoconazole and quinidine). Thus, it was found that the traditional use of *S. chirata* could be safe in respect of both tested isozymes [84].

## 5.2.13 Clinical Data

A randomized, open label, parallel efficacy, active control, and exploratory clinical trial (CTRI/2020/05/025214) was performed to evaluate efficacy and safety of an Ayurvedic formulation (AYUSH 64) as adjunct treatment to standard of care for the management of mild-to-moderate COVID-19 patients. The outcome showed that patients showed improved clinical recovery with improvement in the symptoms, supported by in lab investigations and other radiological examination [85, 86].

Unani formulation (UNIM-401), comprising a mixture of *Fumaria parviflora*, *Terminalia chebula*, *S. chirata*, and *Psoralea corylifolia* in the proportion of 5:2:5:5 as aq. extract in dried form, showed the effectiveness against psoriasis. The previous non-inferiority randomized controlled clinical study showed that oral UNIM-401 was found to be potent among individuals with mild–acute chronic plaque psoriasis [86].

### 5.2.14 ADULTERANTS AND SUBSTITUENTS

Since *S. chirata* is critically endangered now, the recent study demonstrated the utility of other species of *Swertia*, namely, *S. paniculata* and *S. angustifolia* as likely substitutes for *S. chirata* in herbal formulations. It was also found that microwave-assisted extraction using aqueous ethanol proved to be the most efficient method for the extraction of all the pentacyclic triterpenoids compounds in different populations of *Swertia* species such as oleanolic acid, ursolic acid, betulinic acid, and lupeol, when compared with the other conventional approaches [87].

Another study showed that the newly developed optimized microwave-assisted extraction and high-performance thin-layer chromatography densitometry procedure for simultaneous estimation of three phytochemicals among five *Swertia* sp. (*S. nervosa*, *S. chirata*, *S. angustifolia*, *S. paniculata*, and *S. cordata*) from the Himalayan region, India could be efficiently executed for quality assessment herbal preparations. HPTLC findings demonstrated that *S. angustifolia* and *S. paniculata* comprised a high level of phytochemicals (swertiamarin, mangiferin, and amarogentin) and can be considered an alternative of *S. chirata*, while other species (*S. cordata* and *S. nervosa*) due to the lack of optimum level of phytochemicals could not be suggested as an alternative to *S. chirata* [88]. Recently, it has been revealed that *S. chirata* shows a low degree of genetic diversity and gene flow, and hence there is a high chance of extinction of this significant species. It was also found that among five *Swertia* species collected from the Western Himalayas, India, a high content of amarogentin and swertiamarin has been found in the *S. chirata*, followed by *S. paniculata*. This study could be valuable to discover the promising genotype and to develop conservation approaches for the top *Swertia* species [89]. The commercial name "Chirata" refers to *S. chirata* and *A. paniculate*. This vague trade name provides space for illegal practices. Considering this aspect, in 2011, exomorphic and endomorphic characteristics of *S. chirata* have been investigated, and the appropriate use of the technical terms has also been highlighted, which could be helpful in the identification and authentication of this plant [90]. At present, due to its extensive use, traditional medicine plants have been overexploited from its natural habitat, therefore now, on the verge of extinction; that's why these are highlighted as critically endangered medicinal herbs. A recent report suggested that the acute kidney injury (biopsy-proven case) is associated with the use of hyponidd tablet (containing *M. charantia*, *G. sylvestre*, and *S. chirata*) in a 60-year-old male with type-2 diabetes mellitus and hypertension [91].

### 5.2.15 MOST CITED RESEARCH

The most cited research articles reported so far are based on antioxidant, hepatoprotective, and antimalarial effects of *S. chirata*. It was found that the ethanolic extract of *S.* exhibits in vitro and in vivo antioxidant effects that could be responsible for hepatoprotective effects. Antioxidant property of *S. chirata* supports the traditional use of *S. chirata* in Tibetan medicine to cure liver complications [92]. In another highly cited research article, butanol-soluble bitter rich fraction showed that marginal activity derived from *S. chirata* showed marginal antihepatotoxic activity against carbon tetrachloride–induced liver toxicity in experimental rats [93]. The research article published in 2000 showed that 1,5-dihydroxy-3,8-dimethoxy xanthone of *S. chirata* demonstrated considerable anti-inflammatory action in acute, sub-acute, and chronic experimental models in rats [94].

### 5.2.16 PATENTS

US20130011501: In this invention, a botanical anti-fever composition comprising a therapeutically effective amount of admixture of parts or extracts of at least one plant species from each of genus *Baptista tincoria* and genus *S. chirata* was provided.

## 5.3   CHITRAK

*Plumbago zeylanica* L.

### 5.3.1   BIOLOGICAL SOURCE

*P. zeylanica* L.
   **Family**
   Plumbaginaceae

### 5.3.2   CHEMICAL COMPOSITION

In 1967, a naphthaquinone derivate chitranone has been isolated from *P. zeylanica* [95]. In 1971, biplumbagin and 3-chloroplumbagin have been isolated from *P. zeylanica* [96].

   In 1979, two quinones-based dimers of plumbagin, zeylanone, and isozeylanone were isolated from *P. zeylanica* [97]. In 1983, isoshinanolone, droserone, plumbagin, as well as naphthalenone-biplumbagin derivative have been derived from phenolic extract of *P. zeylanica* [98]. Subsequently in 1984, plumbazeylanone, a quinone, may be 5b,11a,12,12a-tetrahydro-1,7-dihydroxy-5b-(8-hydroxy-3-methyl-1,4-naphthoquinon-2-yl)-5a,12a-dimethyl-5a-dibenzo [99] fluorene-5,13:6,11-diquinone, a novel trimmer of plumbagin with an additional methyl group has been isolated from *P. zeylanica*. In 2003, two plumbagic acid glucosides, 3′-*O*-β-glucopyranosyl plumbagic acid and 3′-*O*-β-glucopyranosyl plumbagic acid methylester along with five naphthoquinones (plumbagin, chitranone, maritinone, elliptinone, and isoshinanolone), and five coumarins (seselin, 5-methoxyseselin, suberosin, xanthyletin, and xanthoxyletin) have been isolated from the roots of *P. zeylanica* [100]. In 2004, β-sitosterol, β-sitosteryl-3β-glucopyranoside, β-sitosteryl-3β-glucopyranoside-6′-*O*-palmitate, lupenone, lupeol acetate, plumbagin, and trilinolein have been isolated from the dichloromethane extract of aerial parts of *P. zeylanica* [101]. In 2008, an immunomodulatory chemical compound known as seselin ($C_{14}H_{12}O_3$; MW 228) was identified from *P. zeylanica* [102]. In 2010, six naphthoquinones were derived from *P. zeylanica* roots [103]. During the same period, 3β-hydroxylup-20(29)-ene-27, 28-dioic acid has been isolated from *P. zeylanica* methanolic extract [104]. In 2014, naphthoquinone 2 and new monoterpenoid 3 have been isolated from the stems of *P. zeylanica*. It was demonstrated that new naphthoquinone 2 demonstrated potent inhibitory activity against nuclear factor κB (NF-κB), equivalent to that of parthenolide [105]. Plumbagin present in the *P. zeylanica* L. roots retains various pharmacological properties such as anticancer, antibacterial, and anti-inflammatory. A naphthoquinone analogue and therapeutically active compound, namely plumbagin, is present in the roots of *P. zeylanica* L. where this chemical compound is accumulated and biosynthesized. Transcriptome data revealed that the co-expression of Aldo–keto reductase (PzAKR), polyketide cyclase (Pzcyclase), and cytochrome P450 (PzCYPs) transcripts, along with the polyketide synthase (PzPKS) transcripts, is responsible for its more synthesis in roots. It was also found that naphthalene derivative isoshinanolone could be a possible precursor of plumbagin synthesis via naphthoquinone biosynthesis in plants [106]. Recently, callus root culture demonstrated that sodium acetate, L-tyrosine, chitosan, yeast extract, and salicylic acid enhanced plumbagin content when added to the callus cultures [107]. Another study demonstrated that the supplementation of abiotic (yeast and malt extract) and biotic elicitors (methyl jasmonate and salicylic acid) enhanced plumbagin level substantially [108]. Recently, a bioactive compound, namely heneicosane, was firstly identified in *P. zeylanica* leaves, and antimicrobial assessment showed its excellent antimicrobial activity against *Streptococcus pneumoniae* and *Aspergillus fumigatus* [109].

### 5.3.3   COMMON USES

*P. zeylanica* roots have been used in Ethiopia for skin-related complications, while in India it has been extensively used to treat diabetes mellitus.

### 5.3.4 Traditional Uses

The common name Plumbago is originated from Latin phrase Plumbum (=lead), describing its capability to treat localized paralysis caused by lead poisoning or the capability of plant sap to make lead-dyed stains on skin. Thus, it is also known as "leadwort". The last name or species called as zeylanica means "of Ceylon". In Sanskrit it is known as "Chitrak" that suggests its implications over skin, that is, if the paste of root with water is used topically, blisters become visible immediately and further become red in colour after bursting leading to skin discolouration, thus it's called Chitrak.

### 5.3.5 Traditional Formulations

There are several Ayurvedic formulations of *P. zeylanica* Linn that have been used from decades. Aaragvadhadi, Aarogyapanchakam, Abhayadi quanthah, Abhayamodaka, Abhayarishtah, Agnikumarorasah, Agnimukham churnam, Agnitundi vati, Ajamodadi vataka, Amalakyadi, Amritashatapatalaghritam, Ashwagandharishta, Chavikadi ghrita, Chitra kathi, Chitraka ghrita, Chitraka himam, Chitraka kalkam, Chitraka phantam, Chitraka quantham, Chitraka rasayana, Chitraka swarasam, Chitrakadi avaleha, Chitrakadi Churna, Chitrakadi leha, Chitrakadi taila, Chitrakadi vati, Chitrakaharitaki, Dashamoolarishta, Drakshasava, Eladi ghrita, Hapushadyam ghrita, Khandasam churna, Kshirashatapalakam ghrita, Lauhasava, Mushkakadi, Mustadi, Panchakola, Panchatikta guggulu ghrita, Pippalyadya churna, Punarnava guggulu vati, Saptavimshatika guggulu vati, Satyadi churna and gutika, Shaddharana yoga, Shadushana, Shwetakaraviradya taila, Tejovatyadi ghrita, Trimada, Triphalasava, Varemadi, Vyoshadi guggulu vati, Yakritaplihari lauha, and Yogaraja guggulu vati [110].

*P. zeylanica* L. is a famous traditional medicinal plant that has been used extensively in the formulations of several key herbal products such as Dashmularisht, Chitrakadi vati having hepatoprotective, anticancer, neuroprotective, anti-atherogenic, and cardiotonic effects.

A polyherbal Ayurvedic formulation, namely Trimada, containing three herbs, *P. zeylanica*, *Cyperus rotundus*, and *Embelia ribes* in equivalent amount, was used to improve the performance of the digestive system and metabolism. Trimada formulation demonstrated lower cytotoxicity to human liver carcinoma cell line and demonstrated less interaction with liver microsomes. It was also discovered that Trimada formulation contains sesquiterpenes, phenolic acids, benzoquinones, triterpenes, flavonoids, and heavy metal level in the permissible limit. Recently, it was found that Trimada formulation possesses low cytotoxicity and provides insignificant interaction with CYP450 isozymes, thus, considered safe [111].

As per Ayurveda, single herb is not capable enough of attaining a required therapeutic effect. Thus, considering this aspect, recently a polyherbal formulation containing methanolic extracts of *P. zeylanica* Linn, *Datura stramonium* Linn, and *Argemone mexicana* Linn was developed to study the effect of polyherbalism on antimicrobial, antioxidant effect, and wound-healing effects of the formulation. It was found that antimicrobial and anti-inflammatory effects of polyherbal formulation incited and elevated the wound-healing process via enhanced restoration of injured site [112]. Hutabhugādi cūrṇa, which contains various herbs, including *P. zeylanica*, has been prescribed for various complications such as digestive related complications, anaemia, oedema, and piles [113]. Rasagenthi Mezhugu, a herbomineral preparation in the Siddha, is used as a remedy for all types of cancers. Earlier investigation demonstrated the efficacy of Rasagenthi Mezhugu against the cervical cancer caused by human papilloma virus and thus can be further used as potent formulation to treat human papilloma virus–induced cervical cancers [114].

### 5.3.6 Pharmacological Uses

Chitrak has been known for its potential in protecting liver, reducing inflammation, normalizing lipid profile, and treating diabetes and cancer. The previous study showed that active chemical component of *P. zeylanica* showed that improved therapeutic efficacy may be due to the micellar solubilization and resultant-enhanced partitioning of plumbagin via intestinal by Gelucire, which was manifested in the in vivo anti-inflammatory activity performed in rats [115].

### 5.3.6.1  Anticancer and Antitumour Activity

Plumbagin has been reported for causing apoptotic death in cancerous cells obtained from prostate, lungs, pancreas as well as breast, and inhibiting oxidative stress caused due to the caused overproduction of reactive oxygen species [116]. Findings obtained from previous cytotoxicity study showed that plumbagin significantly suppressed growth of tumour cell lines such as Raji, Calu-1, HeLa, and Wish [117]. Another study demonstrated physiological modifications caused by plumbagin among tongue squamous cancer cells (SCC-25). In this study, plumbagin decreased nuclear factor erythroid 2–linked factor-2 translocation at nuclear level as well as supressed upregulation of respective genes. Therefore, ROS overproduction resulted in cell cycle arrest and stemness attenuation, causing apoptosis [118]. It was also found that plumbagin increases the anticancer potential of cisplatin via elevating ROS intracellularly in tongue squamous cancer cells. In this study, it was revealed that plumbagin increases cytotoxicity, apoptosis, and autophagy caused by cisplatin among CAL-27, CD-DP (cisplatin-resistant) cells as well as CAL-27. Further it was discovered that plumbagin hinders the viability and growth of tongue squamous cell carcinoma cells via increasing the synthesis of intracellular reactive oxygen species. Plumbagin with cisplatin increased intracellular ROS, possibly by stimulating JNK and hampering AKT/mTOR signalling pathways [119]. Another study showed that compound β-sitosteryl-3β-glucopyranoside-6′-*O*-palmitate demonstrated cytotoxic activity against MCF-7 and Bowes cancer cell lines, β-sitosterol inhibited Bowes cell growth, and plumbagin was found to be cytotoxic against MCF-7 and Bowes cells [120]. Plumbagin isolated from in vitro grown plants of *Plumbago indica* L. and *P. zeylanica* L. showed anti-proliferative activity in both stomach cancer cell lines (AGS) and breast cancer cell lines (MDA-MB-231) via causing apoptotic death [121]. The previous study also suggested that plumbagin can cause apoptosis in human pancreatic cancer cells mainly via mitochondria-related pathway pursued by both caspase-dependent and caspase-independent cascades [122]. Plumbagin has also been reported for its capability in inhibiting development and metastasis of human prostate PC3-M-luciferase cells in xenograft rodent-based model [123]. An earlier report suggested the synthesis of magnesium oxide and chitosan-modified magnesium oxide nanoparticles via a green approach using leaves extract of *P. zeylanica* L. as a nucleating agent. It was found that chitosan-modified magnesium oxide nanoparticles presented a greater amount of cytotoxicity for both the bacterial and cancer cells in comparison to the magnesium oxide NPs. Additionally, it was also reported that chitosan-modified magnesium oxide nanoparticles showed less toxicity than magnesium oxide NPs [124].

### 5.3.6.2  Anti-*H. Pylori* Effects

Plumbagin, isolated from the roots of *P. zeylanica*, showed potent effects against *Helicobacter pylori* [125]. Another study showed highest inhibitory effects of *P. zeylanica* ethyl acetate extract against *H. pylori* (out of 50 screened Taiwanese folk medicinal plants), within a pH range of 1–7 [126].

### 5.3.6.3  Reversible Antifertility Activity

An earlier study revealed a reversible antifertility activity of plumbagin-free alcohol extract from *P. zeylanica* Linn root on female reproductive system and fertility of adult female Wistar rats via reducing the levels of serum progesterone, follicle stimulating hormone, and luteinizing hormone without causing any adverse toxicity [127]. Recent study showed that plumbagin increased mATPase expression, improved production of mLPO with a rise in the level of caspases-9 and -3. Plumbagin treatment also reduced sperm count and motility considerably and enhanced sperm malformations. Dose-dependent expression of the Bak, p53, and cytochrome *c* release and decrease in the expression of anti-apoptotic Bcl-2 protein considerably decreased the expression of FSH and progesterone receptors, TESK-1, and aromatase. Moreover, plumbagin strongly supressed p53-MDM2. It was concluded that plumbagin causes injury to testicular cells via an activation of mitochondrial pathway involving the p53 protein network [128].

### 5.3.6.4 Antidiabetic Effects

The previous investigation demonstrated the antidiabetic activity of plumbagin derived from the roots of *P. zeylanica* as well its impact on glucose transporter 4 translocation in streptozotocin-induced diabetic rats. It was found that Plumbagin considerably decreased the blood glucose and normalized all biochemical parameters. Moreover, plumbagin improved the hexokinase activity and reduced the level of fructose-1,6-bisphosphatase as well as glucose 6-phosphatase considerably. Experimental animals induced with diabetes plumbagin treatment showed the upregulation of protein as well as GLUT4 mRNA expression suggesting its role as a potential therapy to treat diabetic condition [129].

### 5.3.6.5 Anticancer Effects

Combined crude acetone extract obtained from *P. zeylanica* L., *Limonia acidissima* L., and *Artocarpus heterophyllus* Lam has been investigated against human breast cancer cell line (MCF-7), and in vitro and in silico molecular docking analyses were also carried out. Findings obtained from this study revealed considerable antioxidant effect, brine shrimp toxicity, and anticancer activity. Molecular docking results suggested that binding energy was high in several ligands for certain cancer receptors [130]. Recently, network pharmacology determined the mechanism of plumbagin on pancreatic cancer, anticipated molecular targets, pathways, as well as mechanisms of action. The most crucial targets for plumbagin in pancreatic cancer were found to be interleukin (IL)-6, mitogen-activated protein kinase-1, TP53, and B-cell lymphoma 2 genes [131].

### 5.3.6.6 Treatment of Myeloid Leukaemia

The previous report showed that the intraperitoneal administration of plumbagin in vivo using NB4 tumour xenograft in NOD/SCID mice led to a reduction of tumour volume without any toxicity, suggesting its potential role in the treatment of myeloid leukaemia [132].

### 5.3.6.7 Effect on Blood Coagulation

An earlier study revealed that chronic Chitrak extract treatment at a less dose extends the bleeding duration by changing adhesiveness of the platelets and the coagulation in albino rats. It could be due to the presence of active chemical compound having structure similar to vitamin K [133].

### 5.3.6.8 Nephroprotective Effects

In a previous study, kidney-protective effect of Chitrak extract (hydroalcoholic extract) in nephrotoxicity caused by cisplatin was demonstrated in rodents. It was found that high-dose treatment with this hydroalcoholic extract considerably supresses an undesirable effect of cisplatin overweight of kidney, creatinine, serum, as well as urea parameters suggesting nephroprotective activity. Additionally, the extract also showed an antioxidant effect via showing its considerable effects on glutathione peroxidase, catalase, as well as lipid peroxidation [134].

### 5.3.6.9 Hepatoprotective Effect

A recent study demonstrated that plumbagin possessed hepatoprotective activity against ischaemia-reperfusion (I/R) injury–induced insults. Findings showed that the hepatoprotective activity of plumbagin was attributed to its antioxidant, anti-inflammatory, and anti-apoptotic effects that are somewhat facilitated via reducing HMGB1 expression [135]. Plumbagin has been found to protect the liver from acute and chronic damage by inhibiting inflammation, collagen production, and decreasing macrophages and neutrophils [136].

### 5.3.7 Mechanism of Action

Effects of seselin from Chitrak extract over cell proliferation stimulated by phytohemagglutinin have been investigated in peripheral blood mononuclear cells. It was found that seselin inhibited human peripheral blood mononuclear cells proliferation-activated with phytohemagglutinin via [137]:

- Regulating development of cell cycle, IL-2 and production of IFN-γ in peripheral blood mononuclear cells (human).

- Arresting the cell cycle progression of activated human peripheral blood mononuclear cells from the G1 transition to the S phase. Suppressing IL-2 and IFN-γ production.
- Reducing the IL-2 and IFN-γ gene expression in PHA-activated human peripheral blood mononuclear cells.

Hydroxylup-20(29)-ene-27,28-dioic acid dimethyl ester from *P. zeylanica* showed anti-metastatic effect by preventing the proliferation and migration of MDA-MB-231 cells via inducing apoptosis through following mechanisms [138]:

- membrane potential of mitochondria lost with the decreased expression of Bcl-2, bad upregulation, cytochrome-*c* secretion, caspase-3 stimulation
- PARP-cleavage resulting fragmentation of DNA
- MDA-MB-231 cells adhesion inhibition to the fibronectin-treated substratum
- hindered the wound healing migration as well as MDA-MB-231 cells infiltration via the reconstitution of extracellular matrix
- reduced the release of matrix metalloproteinases (-2, -9)

Protective effect in neurodegenerative complications.

It has been known that activated microglial cells synthesize pro-inflammatory mediators, including nitric oxide and cytokines. The unnecessary production of these pro-inflammatory mediators can result in neurodegenerative. Potent naphthoquinone-based phytopharmaceutical, such as plumbagin, demonstrated anti-inflammatory effects on macrophages. It was recently studied that plumbagin substantially downregulated the expression of several cytokines such as IL-1α, G-CSF, IL-12 p40/p70, MCP-5, MCP-1, and IL-6 and thus can be potentially used for neurodegenerative complications [139].

### 5.3.7.1   Antiangiogenic and Antitumour Effects

Plumbagin modulates cellular proliferation, carcinogenesis, and radioresistance, all known to be regulated by the activation of the transcription factor NF-κB, suggesting plumbagin might affect the NF-κB activation pathway. Plumbagin (5-hydroxy-2-methyl-1,4-naphthoquinone) suppresses NF-κB activation and NF-κB-regulated gene products through the modulation of p65 and IκB-α kinase activation, leading to the potentiation of apoptosis induced by cytokine and chemotherapeutic agents [140]. Plumbagin showed anticancer effects in different cancerous cells, nevertheless the cellular targets of plumbagin have not been well identified and remained only incompletely understood. As mentioned later, various mechanism of actions have been reported so far supporting its anticancer effects. Plumbagin supressed vascular endothelial growth factor-induced migration and tube formation of human endothelial progenitor cells without affecting its viability. In addition, it was found that plumbagin supressed angiogenesis via the phospholipase C, Akt, extracellular-signal-regulated kinase, NF-κB, and hypoxia-inducible factor signalling pathways. Plumbagin is also found in considerably reduced micro vessel formation and endothelial progenitor cells-specific marker expression [140]. The previous study also demonstrated that CRM1, a nuclear export receptor whose role is changed in malignancy owing to enhanced expression and overactive transport, is considered direct cellular target of plumbagin [141]. Plumbagin has been known for its potential to stimulate ERK1/2 as well as to supress the activity of Akt in cancerous cells. A previous study demonstrated that plumbagin stimulates Src, PI3K, and NADPH oxidase. Further, it was revealed that activated PI3K or PDK1 successively stimulates Akt and Ras-Raf-MEK1/2-ERK1/2 in 3T3-L1 cells [142]. Plumbagin has been reported for its capability in modulating cellular proliferation, carcinogenesis, and radioresistance via the activation of the transcription factor NF-κB, indicating that this phytochemical could alter the NF-κB activation pathway. It was also suggested that this phytochemical supress NF-κB-regulated gene products as well as NF-κB stimulation via altering IκB-α kinase as well as p65 stimulation, resulting in an induction of apoptotic death via chemotherapeutic as well as cytokines [143].

### 5.3.7.2   In Hepatic Cancer

Plumbagin has also been reported for its capability to cause apoptosis of human hepatocellular carcinoma cells experiencing epithelial-mesenchymal change via enhancing the caspase-3 protein level and cleaving vimentin [144].

### 5.3.7.3   In Lung Cancer Management

Plumbagin reduces the osteopontin-induced invasion of human non-small cell lung cancer A549 cells via impeding the Rho-associated kinase pathway mediated by the focal adhesion kinase/AKT pathway and suppresses lung metastasis in vivo [145].

### 5.3.7.4   In Brain Tumour

Plumbagin considerably reduced glioma cell proliferation and induced cell apoptosis without any symptoms of toxicity. Also, results from this study demonstrated that these findings revealed that plumbagin could decrease FOXM-1 and can be used for the treatment of brain tumour (gliomas) [146]. It was also suggested that plumbagin stimulated the suppression of glioma cell migration, and invasion is strongly related to the decrease in the expression of MMP (-2/-9) as well as activity, and inhibition of PI3K/Akt signalling pathway activation, suggesting its role in the treatment of glioma [147].

### 5.3.7.5   In Treatment of Leukaemia

Recent study showed that plumbagin selectively kills human promyelocytic leukaemia HL-60 cells via suppression of thioredoxin reductase (contributing to its cytotoxicity) via increasing ROS, interfering with cellular redox balance, and triggering oxidative stress–based apoptotic death of HL-60 cells [148].

### 5.3.7.6   In Treatment of Polycystic Ovary Syndrome

Plumbagin has been recently reported for mitigating the pathological as well as polycystic property changes in ovary and inducing rodent ovarian granulosa cell apoptosis via PI3K/Akt/mTOR-mediated signalling pathway [149]. Previous investigation suggested that plumbagin inhibits the cell proliferation via changing the permeability of the membrane.

### 5.3.8   Industrial Applications

An earlier report revealed that compound plumbagin can be used as a natural repellent in the replacement of synthetic insecticides due to its strong enzyme inhibition activities against the malarial vector, *A. stephensi*, as a mosquito repellent. It was found that plumbagin caused more damage to the gut epithelial cells and muscles [150].

### 5.3.9   Marketed Products

Chitrakadi vati, Chitrak haritaki avaleh, Chitrakmool powder, Chitrak tablet, Chitrak-unprocessed roots, and Chitrak extract by different manufacturers.

### 5.3.10   Toxicity

Plumbagin is a valuable ingredient of TCM that is reported for its obvious toxic effects such as diarrhoea, skin rashes, and hepatic toxicity [151]. Earlier investigation based on the dermatotoxicity of *P. zeylanica* methanolic extract (80%) revealed that toxic effects of the medicinal plant could be restricted to effects such as mild irritation. Skin irritation test on rabbits using 80% methanol extract showed mild irritant effect, with an irritation index of 2.00 [152]. A recent study showed that hydroalcoholic extracts of *Glinus lotoides*, *P. zeylanica*, *Rumex steudelii*, and *Thymus schimperi* showed DNA-damaging effects when assessed by comet assay without affecting the cell viability that suggests its genotoxic potential in vitro [153]. In another study, it was demonstrated that plumbagin, without influencing the cell viability, caused considerable injury to DNA in mouse lymphoma L5178Y cells. However, a non-DNA detrimental level of plumbagin lessened the DNA injury caused via catechol, and this gives more support to notion that plumbagin reduces oxidative stress at a less dose [154].

### 5.3.11  CLINICAL DATA

A clinical study was conducted to investigate the impact of Trimad (containing *P. zeylanica* roots, *C. rotundus* tubers, *E. ribes* fruits) and Triphala in overweight and obese individuals. It was found that both formulations have shown an important role in decreasing body fat, weight as well as circumferential sizes [155]. A randomized comparative clinical trial was performed containing *P. zeylanica* with other herbs has been tested for its efficacy in treating Katigraha (low back pain). It was found that this treatment showed better effect in treating lower back pain [156]. Recently, a clinical study was performed to evaluate efficacy of herbal formulation for obesity (HFO-02 that comprises triphala powder, vrikshamla fruit powder, *Trimad* powder, and latex of Guggul) among overweight individuals. It was found that Tablet HFO-02 reduced body circumferences and skinfold thickness suggesting its possible effectiveness in obesity management [157].

### 5.3.12  ADULTERANTS AND SUBSTITUENTS

Indian *Plumbago capensis* and *Plumbago rosea* have been used as substituent of *P. zeylanica* roots. Roots belonging to these species are popular because of naphthaquinone, primarily plumbagin level along with its derivatives [158, 159]. *Baliospermum montanum* is considered a substituent of Chitrak crude drug [160].

### 5.3.13  MOST CITED RESEARCH

Screening of some Indian medicinal plants (aqueous, hexane and alcoholic extracts) for their antimicrobial potential by Ahmad et al. in 1998 is considered as the most cited research as in this research antimicrobial potential of 82 medicinal plants have been evaluated. Out of all fractions, only alcoholic extracts of *Holarrhena antidysenterica, P. zeylanica, Terminalia belerica, Emblica officinalis,* as well as *T. chebula* are found to be effective against test bacteria and all of these extracts showed no cellular toxicity [161].

### 5.3.14  PATENTS

US8067043B2: This study by Nandkishor Bapurao Managoli in 2008 was based on herbal composition containing *P. zeylanica* and other medicinal plants as ingredients for the treatment of hyperlipidaemia and the inhibition of myocardial infarction.

US10245240B2 (by Per Borgström): This invention is based on a pharmaceutical composition containing plumbagin has been developed and claimed to treat prostate carcinoma.

## 5.4  CHOTO GOKHRU

### 5.4.1  VERNACULAR NAMES

The name *Tribulus* is originated from Greek word *"tribolos"* that implies spike fruit, Gokhru, Gokshura, Chotagokhru, Land-Caltrops, Puncture Vine.

### 5.4.2  BIOLOGICAL SOURCE

It is obtained from the dried ripe seeds of *Tribulus terrestris* Linn.
    **Family**
    Zygophyllaceae

### 5.4.3 CHEMICAL COMPOSITION

The fruits of *T. terrestris* comprise key chemical components, including saponins, mainly steroidal saponins (furostanol and spirostanol type), polyphenolic compounds, and alkaloids. The saponins such as furostanol supposed to be precursors of the spiro analogues. So far, 70 compounds have been isolated and identified in *T. terrestris*. Chemical profile of *T. terrestris* is highlighted in Table 5.2.

A recent report demonstrated the presence of 26 essential and trace elements in the plant species collected in Russia and China. It was established that the elemental profile of *T. terrestris* grass alters depending on the habitat [163].

### 5.4.4 TRADITIONAL USES

*T. terrestris* fruits and roots of *T. terrestris* have been used as a folk medicine for thousands of years in China, India, Sudan, and Pakistan. The fruits of *T. terrestris* are used in various traditional systems such as traditional Chinese medicine, Ayurvedic medicine, and traditional preparation in Bulgaria for the treatment of several ailments [164]. As a folk Chinese Medication, *T. terrestris* is considered top-grade treatment in the existing Chinese pharmaceutical monograph "Shen Nong Ben Cao Jing" [165, 166]. Fruits of *T. terrestris* have been used for treating various kidney-related ailments, as diuretic, expectorant, for the treatment of itchy skin, headache, vertigo, and blockage in a milk duct as mentioned in Chinese Pharmacopoeia. Decoction of *T. terrestris* is beneficial in dry cough, inflammatory disorders, pain (lower back), pulmonary illnesses, and sciatica. *Vata Prakopa* has been used as a tonic to boost the immune system. In Indian Ayurvedic practice, *T. terrestris* fruits have been used for treating infertility, impotency, and low sexual activity. In addition, *T. terrestris* (fruits as well as roots) has also been used for cardiotonic effects [167]. In Sudan, *T. terrestris* has been used to relieve irritation as well as inflammation of the mucous membranes and used to treat inflammation of the kidneys [167]. Moreover, in Pakistan it has been used as diuretic as well as uricosuric [168]. Recent reports demonstrated that the inflammation as well as aging reducing effects of *T. terrestris* could be accountable for its anti-inflammatory and antiaging effects.

## TABLE 5.2
## Chemical Compounds Identified in *T. terrestris* [162]

| Class of Chemical Compounds | Chemical Components |
| --- | --- |
| Furostanol saponins | Neoprotodioscin, neoprototribestin, polianthoside D, protodioscin, prototribestin, terestrinin A, terestrinin B, terestrinin J–T, terestrinin D, terestrosin I, terestrosin K, tribufuroside D, tribufuroside E |
| Spirostanol saponins | 25R-5a-spirost-3,6,12-trione, 25R-spirost-4-ene-3,12-dione, 25R-spirost-4-ene-3,6,12-trione, agovoside A, dioscin, diosgenin, gitogenin, hecogenin, prosapogenin B, terestrinin U, tigogenin, tribestin, tribulosin |
| Quinic acid derivatives | 4,5-Di-*p*-*cis*-coumaroylquinic acid, 4,5-di-*p*-*trans*-coumaroylquinic acid, 5-*p*-*cis*-coumaroylquinic acid, 5-*p*-*trans*-coumaroylquinic acid |
| Flavonoids | Apiotribosides A–D, astragalin (kaempferol 3-glucoside), isoquercitrin, isorhamnetin-3-glucoside, kaempferol, kaempferol 3-rutinoside, kaempferol-3-gentiobioside, luteolin-7-*O*-β-D-glucoside, quercetin, quercetin 3,7-diglucoside, quercetin-3-gentiobioside, quercetin-3-*O*-arabinosyl galactoside, isorhamnetin-3-glucoside, quercetin-3-*O*-sophoroside-7-*O*-glucoside, rutin, tribuloside |
| Alkaloids | Harmaline, harmalol, harmane, harmine, *n*-caffeoyltyramine, norharmane, perlolyrine, tribulusterine |
| Cinnamic acid amides | Coumaroyltyramine, ferulic acid, feruloyloctopamine |

### 5.4.5 TRADITIONAL FORMULATIONS

Gokshura is one of the contentious medicinal plants in Ayurvedic medicine as well as act as an ingredient of Dashamoola, a popular traditional preparation utilized to treat different ailments. Recently, Mudabbar/*Tadbeer-e-advia* has been used for *T. terrestris* Linn fruits to detoxify, purify, and improve therapeutic action and to lessen its doses before producing the preparations as prescribed in Unani medicine. It was found that Mudabbar process perhaps can be used to enhance/modify the action of *T. terrestris* and also resulted in considerable increase in Diosgenin content [169]. *Kaamdev ghrita* (called "*Vajikarana Rasayana*") is a cow *ghee*-based classical aphrodisiac traditional preparation containing *T. terrestris* along with other medicinal plants, used to improve sexual performance, as well as to treat various sexual disorders such as premature ejaculation and infertility. Recent report showed that aphrodisiac property of Kaamdev *ghrita* could be because of antioxidant principles that appear in it [170].

A polyherbal patented as well as marketed formulation known as Speman® by The Himalaya Drug Company, India constituting a mixture of *T. terrestris* extracts with other medicinal plants that have been used to treat male sexual problems such as premature ejaculation, low sperm count, and other dysfunctions [171, 172]. Previous reports also demonstrated promising results of Speman® over andro- and gametogenic activities [173]. It was also found that the treatment with Speman® prevented the imbalance in the reproductive system caused by cyclophosphamide. Other clinical investigations also reported the doe-dependent effects of Speman® [173]. The trademarked product DA-5512 constituting *Zingiber officinale, Tapiscia sinensis, Pinus densiflora, E. officinalis, Pinus thunbergiana*, and *T. terrestris* was tested for its impact over human dermal papilla cells proliferation in vitro as well as over hair growth in C57BL/6 mice. Also, clinical investigation was conducted to evaluate the safety as well as the efficiency of DA-5512. It was found that DA-5512 can improve the growth of hairs as well as keep them healthy and thus could be used as potent treatment for hair loss [174]. A polyherbal formulation containing *T. terrestris* with other medicinal plants has been investigated for its aphrodisiac potential in *Albino* rats. It was found that polyherbal preparation showed good aphrodisiac activity on rats and could be alternatively used for various sexual dysfunctions [175]. Another study showed the effect of polyherbal preparation, MAT containing *Mucuna pruriens*, Ashwagandha, and *T. terrestris*, on sexual function in rats. It was found that MAT increased sexual function and reduced oxidative damage in male rats via Nrf2/HO-1 pathway [176]. Nano Leo is a pro-sexual nutrient preparation comprising *T. terrestris*-powdered extract with other ingredients showed improvement in sexual dysfunction. Recent study showed that *T. terrestris* extract improved sexual desire, erection, and orgasm among males suffering from erectile dysfunction [177]. In another study, a Rasayana Ghana tablet comprising *Tinospora cordifolia, E. officinalis*, and *T. terrestris* has been tested for anti-depressant and anxiolytic effects. It was found that this Ayurvedic preparation can be used in the prevention and treatment of depression and anxiety [178].

### 5.4.6 PHARMACOLOGICAL USES

#### 5.4.6.1 Antiglycation, Antioxidant, and Antiproliferative Effects

Recent investigation demonstrated that saponins containing standardized extracts of puncture vine containing herbal preparation demonstrated oxidative stress as well as cell proliferation reducing capabilities with antiglycation property in human tumour cells. In addition, it was found that saponins-enriched extract showed more antiglycation and antioxidant activity than standardized extract [179]. Another study demonstrated the comparative potential of Gum arabic and *T. terrestris* on male fertility, which showed that Gum arabic treatment showed more improved fertility in Gum arabic–treated mice Balb/c mice than *T. terrestris* [180].

#### 5.4.6.2 On Female Reproductive System

An earlier report revealed the effect of Ashwagandha and puncture vine extract over polycystic ovarian syndrome caused by letrozole in rodents. It was found that this combination normalized the oestrus cycle after being altered by letrozole. Additionally, considerable decrease in serum total cholesterol, along with a decrease in ovarian and uterine weight was normalized. Thus, the combination of *Withania somnifera* and *T. terrestris* can be effectively utilized against prolonged menstrual periods condition [181]. Recent investigation showed a stimulating effect of *T. terrestris* on the growth of differentiation protein (factor-9) expression as well as bone morphogenetic protein-15 at protein as well as mRNA stages in the oocytes/granulosa cells. Findings demonstrated that growth proteins were found to be susceptible to the chemical components of *T. terrestris* at various follicle structures [182].

### 5.4.6.3  Antihypertensive Effects

*Eucommia ulmoides* in combination with *T. terrestris* has been extensively used for therapeutic applications; however, the exact pathway underlying its efficiency has not been revealed yet. Findings from this study suggested that *E. ulmoides* in combination with *T. terrestris* showed good antihypertensive effect, which could be mediated via intestinal microbiota and their valuable metabolites [183].

### 5.4.6.4  Effect on Male Reproductive System

The previous report suggested that the ethanolic extract of *T. terrestris* possessed protective effect against cadmium-induced testicular injury possibly via an inhibiting peroxidation of testicular tissues by reducing oxidative stress as well as metal chelator effects and might be incidentally via triggering the production of testosterone from Leydig cells [184]. The other previous report suggested that the fruit extract of *T. terrestris*- normalized metronidazole caused marked alterations in spermatogenic activities, especially a suppression of oxidative stress in the testis [185]. Also, a recent study demonstrated the effect of *T. terrestris* dry extract to restore the changes caused by cyclophosphamide treatment in mice testes, perhaps because of the presence of protodioscin. It was found that *T. terrestris* prevents men-reproductive injury caused via cyclophosphamide by decreasing an oxidative degradation of lipids, ROS, as well as a carbonylation of proteins, and increasing antioxidant enzymes and 17β-HSD activity [186]. Another previous study showed that the incubation of human semen with the extract of *T. terrestris* considerably increased a motility of overall sperms, a level of advanced spermatozoa (motile), and velocity (curvilinear) with improvement in viability [187]. A recent study also showed that the treatment of Wistar rats with *T. terrestris* extracts + Malathion reduced the toxic effects of Malathion on some of the male reproductive parameters [188]. In addition, a recent study also revealed that *T. terrestris* in combination with *Anacyclus pyrethrum* treatment in male rats improved fertility parameters such as sperm count, sperm viability, motility testosterone, and luteinizing hormone levels [189]. Another study demonstrated the effect of gross saponins of *T. terrestris* on erectile function in rats resulting from type 2 diabetes mellitus. It was also revealed that gross saponins of *T. terrestris* can protect rats (resulting from type 2 diabetes mellitus) erectile function by improving penile endothelial function and preventing cavernosum fibrosis, apoptosis, which is synergistic with sildenafil [190]. Due to the evidenced anti-inflammatory property of *T. terrestris*, recently, the potential of aqueous extract and saponin fraction of *T. terrestris* shows its effects on the expression of intracellular adhesion molecule-1 (ICAM-1), vascular cell adhesion molecule-1 (VCAM-1), and E-selectin (SELE) genes in human umbilical vein endothelial cells and bone marrow endothelial cells during normal and lipopolysaccharide–induced conditions. It was found that *T. terrestris* decreases the expression of ICAM-1, VCAM-1, and E-selectin in human endothelial cells that are constantly expressed during atherosclerosis [191]. Another study suggested that Terrestrosin D from *T. terrestris* L. can inhibit bleomycin-induced inflammation and fibrosis in the lungs of mice, which could be associated with decreased inflammatory and fibrotic markers [192].

### 5.4.6.5  Anti-Inflammatory Effects

*T. terrestris* has been reported for diuretic as well as lithontriptic effects and used to treat oedema as well as urinary-based infections. Earlier reports have suggested that its capability in reducing inflammatory reaction is caused by tumour necrosis factor (TNF)-α, IL (-6,-10), NO, as well as COX-2. Recent study revealed the molecular sites as well effects of *T. terrestris* on osteoarthritis. It was found that *T. terrestris* extract treatment reduced osteoarthritis monosodium iodoacetate via downregulating COX-2, IL-6, TNF-α, as well as NO synthase-2. It was also revealed that this effect was dependent on the expression of matrix metalloproteinases-2 and -9 [193]. In a recent study, anti-inflammatory effects of tribulusamide D on lipopolysaccharide-stimulated RAW 264.7 macrophages were examined. It was found that tribulusamide D (isolated from the ethanol extract of *T. terrestris*) possessed anti-inflammatory effects by changing the expression of inflammatory mediators (nitric oxide synthase and cyclooxygenase-2 expression) and cytokines (IL-6, IL-10, and TNF-α) [194].

### 5.4.6.6 Antiurolithic Effects

The treatment of Wistar rats with an aqueous extract of *T. terrestris*, despite showing antioxidant effects, also improved renal function, repaired renal structure, increased body weight, and modulated the morphology of renal stones via considerably p38MAPK upregulation at the gene/protein level. This study demonstrated a promising antiurolithic effect of *T. terrestris* extract [195].

### 5.4.6.7 In Stroke Management

Recently, the protective effect of gross saponins of *T. terrestris* fruit over ischemic stroke in a middle cerebral artery occlusion rodent has been studied via metabolomics using GC-MS method. It was found that gross saponins of *T. terrestris* can reverse the middle cerebral artery occlusion-induced serum–metabolic abnormalities via regulating numerous metabolic pathways [196].

### 5.4.6.8 Nephroprotective Effects

Recent report demonstrated that the treatment of rats with *T. terrestris* hydroalcoholic extract showed protection against renal impairment caused by mercuric chloride perhaps because of its potential in reducing inflammation as well as oxidative stress via reducing the renal accumulation of mercury [197].

### 5.4.6.9 Antiapoptotic Effect

Recently, the antiapoptotic effect of resistance training and *T. terrestris* in the heart tissue of rats exposed to stanozolol was studied. It was found that resistance training and treatment with *T. terrestris* alone showed antiapoptotic activities. Nevertheless, resistances training with *T. terrestris* treatment, mainly at higher doses, have more desirable effects than resistance training or *T. terrestris* treatment alone on the apoptosis markers [198].

### 5.4.6.10 Antidiabetic Effects

Recent study demonstrated the effect of *T. terrestris* extract with low protodioscin content and *T. terrestris* extract with high protodioscin. It was revealed that both *T. terrestris* extracts decreased high blood glucose level in diabetic rats. Also, follicle-stimulating hormone and testosterone content were found to be higher in the *T. terrestris* extract with a high protodioscin group. It was also found that the level of testosterone could be associated with the protodioscin level in extracts and could be perhaps facilitated via FSH-mediated pathway [199]. To prevent the toxic effects induced by medications, *T. terrestris* has been reported for its protective effect against haloperidol-induced catalepsy and thus could be utilized to inhibit extrapyramidal associated undesirable effects caused by haloperidol [200]. Another report showed that the *T. terrestris* (fruit extract) treatment demonstrated protective effects against toxicity caused by cisplatin in the rodent [201]. One more report showed that hydroalcoholic extract of *T. terrestris* showed protective effects via reducing myeloperoxidase action, lipase as well as serum amylase amounts in pancreas and normalized histological parameters on cerulein-induced acute pancreatitis in mice [202].

Recent study demonstrated that hydroalcoholic extracts of *T. terrestris* root demonstrated positive quorum-quenching activity by successfully downregulating quorum sensing–controlled mechanisms such as pigment production and biofilm formation. Chemical component responsible for this action could be ß-1, 5-*O*-dibenzoyl-ribofuranose. It was also demonstrated that this chemical component neither showed bactericidal nor bacteriostatic action; however, it distresses its interaction. It was also observed that this compound antagonized quorum sensing and signalling via an inhibiting action of acyl homoserine lactone [203].

### 5.4.6.11 Anticancer Effects

*T. terrestris* (leaf and seed extracts) has shown cellular toxicity against breast cancerous cell line (MCF-7) via causing an fragmentation of DNA, induction of apoptotic death via upregulation in the expression of Bax/p53 genes as well as reduction in the Bcl2 expression. It was also found that *T. terrestris* increased caspase-3 activity and demonstrated no toxic effects against non-malignant cells. Ultimately, it was found that the *T. terrestris* extracts may possess anticancer effects via both extrinsic and intrinsic apoptotic pathways [204]. Recent study also demonstrated that Tiliroside, a chemical component of *T. terrestris* L., inhibited liver cancer development perhaps via acting as a carbonic anhydrases XII inhibitor. It was revealed that Tiliroside reduced the proliferation of liver cancer cells Hep-3B as well as SNU449 [205].

### 5.4.6.12 Effect on Oxidative Stress

It has been reported that *T. terrestris* showed antioxidant effects via reducing ROS production caused by a cellular damage of human retinal pigment epithelial cells caused by $H_2O_2$. This is done via a modulation of PI3K/Akt-Nrf-2 signalling pathway. *T. terrestris* extract considerably enhanced the cellular viability with the prevention of the apoptotic death of epithelial cells via Bcl-2, Bax, and cleaved caspase (3, 9) modulation. In addition, ethanolic extract treatment with *T. terrestris* alone considerably enhanced NQO-1, Nrf-2, HO-1, as well as GCLM m-RNA expressions in epithelial cells [206].

Another report showed that pre-treatment with *T. terrestris* (fruit extract) demonstrated a protective effect by altering lipid-based markers as well as ROS production in myocardial necrosis caused by isoproterenol in rodents [207].

### 5.4.6.13 Anti-Inflammatory Effect

Recent study demonstrated that our data suggested that *T. terrestris*, also known as Bai Jili in Chinese, displayed potent anti-inflammatory effect, and the basic mechanism was strongly related to the inhibition of Akt/MAPKs and NF-κB/iNOS-NO signalling pathways in Zebrafish. It was found that *T. terrestris* extract provides protection against lipopolysaccharide-induced inflammation in raw 264.7 macrophages [208]. Another investigation reveals that the topical application of fruit extract of *T. terrestris* improves skin inflammation in oxazolone-induced atopic dermatitis through the regulation of calcium channels, ORAI1 and TRPV3, and mast cell activation. It was also suggested that the combination of fruit extract with steroids might be a more promising and safe tactic for atopic dermatitis treatment [209].

### 5.4.6.14 Protective Effect against Ischaemia/Reperfusion Injury

Gross saponins, derived from *T. terrestris*, mainly spiral vagina steroid and snail steroid, have been used for the treatment of a variety of illnesses such as hypertension, hyperlipidaemia, platelet aggregation, and ageing. Furthermore, some findings have demonstrated that gross saponins have a substantial protective effect against ischaemia/reperfusion injury in rat hearts and brains. Tribulosin, a chemical constituent from *T. terrestris*, has been reported for its protective effects against cardiac ischaemia/reperfusion injury in rats. In this study, it was demonstrated that tribulosin provides the protection to myocardium against ischaemia/reperfusion injury via PKCε activation [210]. Recent report showed anti-leukaemic potential of *N*-feruloyltyramine derivatives from an alkaloid extract of *T. terrestris* fruits via apoptosis induction in Jurkat E6-1 cells by considerably changed expression of critical genes such as TNFR1, FADD, AIFM, CASP8, TP53, DFFA, and NFKB1. These genes are accountable for apoptotic cell death via both intrinsic and extrinsic apoptosis pathway. Additionally, Jurkat E6-1 with alkaloids extract caused cellular toxicity with increase in phosphatidylserine translocation, DNA fragmentation, as well as caspase-3 activity [211].

### 5.4.6.15   Anti-Parkinson Effects

*T. terrestris* contains a high level of flavonoids and saponins which are accountable for neuroprotective and antioxidant effects. Recent study showed anti-Parkinson's activity of *T. terrestris* methanol extract (TTME). It was found that extract modulated the mRNA expression of α-synuclein, AChE, TNF-α, and ILs in the brain homogenate with significant improvement in behavioural parameters in Parkinson's disease rats, by suggesting neuroprotection [212].

### 5.4.7   INDUSTRIAL APPLICATIONS

#### 5.4.7.1   Anticoccidial Effects

Coccidiosis that is a parasitic zoonotic infection, caused by protozoa *Eimeria tenella*, caused significant mortality, morbidity in the poultry industry. Recent investigation demonstrated that ethanol extract obtained from fruits of *T. terrestris* showed anticoccidial effects against *E. tenella* and thus can be considered in the treatment for the coccidiosis [213].

#### 5.4.7.2   For Improving the Performance of Athletes

*T. terrestris* extract has been reported as a testosterone booster in athletes for improving muscle mass, strength and performance. Moreover, the supplementation of *T. terrestris* extracts improved serum testosterone levels on male rats, primates, rabbits, and castrated rats. Recent study showed that the administration of 1250-mg capsules comprising the extract of *T. terrestris* has not shown any changes in the mass of the muscles and testosterone concentration in plasma, dihydrotestosterone, and insulin growth factor-1; however, it considerably lessened muscular injury and improved performance (anaerobic) of proficient boxers (male), which could be linked to the reduction of plasma IGF-1 binding protein-3 [214].

#### 5.4.7.3   For Oral Hygiene

Antimicrobial effects of *T. terrestris* (ethanolic fraction) alone and in a mixture with Glycyrrhiza and Shepherd's Purse have been studied against six pathogenic microorganisms. It was found that antibacterial effects of blended extract were more effective against all microorganisms than alone, which suggests the synergistic action among phytochemicals of both fractions [215].

### 5.4.8   MARKETED PRODUCTS

Nutriley spermogra capsules, natureal gokshura extract capsules, herb essential *T. terrestris* powder, Herbinox gokshura veg capsules, Himalayan organics *T. terrestris* vitamins capsules, Himalaya gokshura men's wellness.

### 5.4.9   TOXICITY

Recent acute oral toxicity study on Wistar rats showed that LD50 > 2000-mg/kg birth weight. In addition, no undesirable effect has been observed for 28 days at 750 mg/kg [216]. Earlier, acute toxicity study on animals treated with methanolic extract of *T. terrestris*, consisting phenolic/tannin, flavonoidal, steroidal/triterpenoidal, as well as anthraquinone-based compounds, showed no toxic symptoms or mortality [217]. Further, genotoxic effects of *T. terrestris* extracts [218] revealed that methanolic extract comparatively showed higher geno- and cytotoxic effects than water extract and the chloroform fraction. However, methanolic extracts have not reported for causing any injury to DNA, while water extract could cause mutations (frame shift). Recent study revealed the capability of the fruits extract of *T. terrestris* to cause DNA injury at a greater dose in cultured human peripheral blood lymphocyte. Therefore, aqueous extracts showed genotoxic effect at high doses, and therapeutic utilization must be done accordingly [219].

## 5.4.10  CLINICAL DATA

In 2000, complete herbs, including fruits of *T. terrestris*, in ghana satwa (solid water extract) form have been tested clinically for antihypertensive effect over 75 patients of either sex. It was found that *T. terrestris* Linn showed significant antihypertensive effect on both systolic/diastolic without any undesirable responses [220]. A clinical investigation showed that the treatment of patients (female) showing sexual complications, with *T. terrestris* extract (250 mg), showed an increased level of dehydroepiandrosterone. It was also found that *T. terrestris* extract can be safely and effectively used in the treatment of female sexual complications [221]. Another clinical study has been performed to evaluate the impact of Aphrodit capsules containing *T. terrestris* with four more herbs on menopause-associated symptoms. Findings showed that Aphrodit capsules were active in lessening menopausal symptoms [222]. Recently, Libicare® constituting *Trigonella foenum-graecum*, *T. terrestris*, and *Turnera diffusa* as a food supplement has been clinically (NCT04124640) tested for its effects on the improvement of sexual function in women (185 participants) with low desire and arousal levels. It was found that the effectiveness of Libicare® will increase the score of the EVAS-M scale [223]. Another clinical study has been performed to study the efficacy as well as safety of Chandrakanthi Choornam containing *T. terrestris* fruits with other herbs in patients with low sperm count (Oligospermia). Primary outcome measure is to see the changes in sperm count, motility, and morphology. Secondary outcome measure is to see the impact on the hormone level. Another clinical investigation was performed to evaluate the safety as well as effectiveness of *T. terrestris* among females with sexual desire (hypoactive) complication. It was found that the treatment with the extract of *T. terrestris* for four weeks resulted in considerable improvement in their total female sexual function index, desire, arousal, lubrication, satisfaction, and pain domains of female sexual function index and thus can be safely and effectively used in the female's treatment suffering from this complication [224]. A clinical investigation was performed to study the effect of *T. terrestris* extract over sexual satisfaction among females (post-menopausal). It was found that the syrup of *T. terrestris* increased sexual satisfaction among females (post-menopausal) [225]. In 2014, a clinical study was performed to assess the impact of alga *Ecklonia bicyclis*, *T. terrestris*, and glucosamine oligosaccharide mixture to improve sexual complications among patients with mild-to-moderate erectile complications. It was found that this type of treatment in combination with tradamix (glucosamine oligosaccharide) improves erectile and ejaculation function and sexual quality of life in patients with mild–moderate erectile dysfunction and in regard for those with moderate arterial dysfunction [226]. In 2012, clinical study has been conducted to study the effect of *Gokshura* (*T. terrestris*) in the management of *Kshina Shukra* (Oligozoospermia). It was found that *Gokshura* granules have demonstrated excellent results in the management of *Kshina Shukra*, as compared to the placebo granules [227]. Recently, Libicare®, a multi-ingredient dietary supplement comprising Fenugreek, damiana, Puncture Vine, and maidenhair tree extracts, has been tested for its efficacy in improving sexual activity among females (postmenopausal). It was found that considerable increase in sexual activity as well as its associated hormone has been observed after treatment with Libicare® [228]. Herbal-based syrup containing palm tree pollen, chotagokhru, Korean ginseng, ginger, carob, and common poppy has been recently tested (double-blind randomized clinical trial) on 100 married and healthy men for their potential in sexual experience. It was found that herbal preparation showed a considerable improvement in sexual experience of men [229]. *T. terrestris*, containing formulation called as Tradamixina, has been recently tested to study its efficacy in improving sexual activity in aged males, mainly sexual desire as well as potential erectile dysfunction, versus treatment with tadalafil. It was found that the administration of Tradamixina (twice a day) for two months improved sexual desire among aged men with no undesirable effects caused by tadalafil [230].

### 5.4.11 Adulterants and Substituents

*T. terrestris* is one of the very well-imported raw materials that are commonly utilized as a food additive in European as well as American countries. But Indian Ayurvedic Pharmacopoeia acknowledges puncture vine as *T. subramanyamii,* Goksura as well as *T. lanuginosus* also exchanged by the similar name increasing the issues of quality as well as safety. In 2010, nuclear ribosome–based RNA genes as well as internal transcribed spacer sequence have been utilized to develop species-specific DNA markers to distinguish *T. terrestris* from its adulterants [231]. *Pedalium murex* is also reported as the substituent of *T. terrestris.* In 2011, pharmacognostical investigation has been performed to discover the distinctive characteristics, both organoleptic and microscopic fruits of *T. terrestris* as well as *P. murex* [232].

### 5.4.12 Most Cited Research

Most cited research so far is entitled as an assessment of crude drugs from Yemen for antibacterial as well as cytotoxic effects in 2001 by Ali et al. Findings obtained from this study after the screening of ethanolic extracts of 20 selected plants showed that *T. terrestris* exhibited a remarkable antibacterial activity [233]. Another most cited research is Aphrodisiac effects of *T. terrestris* fraction (Protodioscin) in normal rodents as well as rodents with the removal of the testicles. It was revealed that *T. terrestris* extract demonstrated aphrodisiac effects due to an androgen boosting property of *T. terrestris* [234]. Another study demonstrated that *T. terrestris* increases some of the sex hormones, perhaps owing to the presence of protodioscin in the extract that could make it effective in the management of male erectile dysfunction among primates, rabbits, and rats [235].

### 5.4.13 Patents

US patent 20140205687 (Application Number: 14/240,738) by Mustafa Tekin, Mustafa Eraslan (2012) claimed that herbal product, comprising *T. terrestris, Avena sativa,* and *Panax ginseng,* can be used as a supplement to enhance muscle strength, body stamina, and physical performance and to cure cardiovascular complications. US6818231B2 (US10/465,068) by Brian Alexis (2003) claimed that a therapeutic composition containing *T. terrestris* with a high content of spirostanol saponin, formulated as cream, could be useful for treating bacterial, fungal, and viral infections, particularly gynaecologic infections.

## REFERENCES

1. Bhatia S. Systems for Drug Delivery: Safety, Animal, and Microbial Polysaccharides. Springer Nature, Basingstoke, UK; 2016:122–127.
2. Bhatia S. Stem cell culture. In: Introduction to Pharmaceutical Biotechnology, Volume 3: Animal Tissue Culture and Biopharmaceuticals. IOP Publishing Ltd, Bristol, UK; 2019;3:1–24.
3. Bhatia S. Organ culture. In: Introduction to Pharmaceutical Biotechnology, Volume 3: Introduction to Animal Tissue Culture Science. IOP Publishing Ltd, Bristol, UK; 2019;3:1–28.
4. Bhatia S. Animal tissue culture facilities. In: Introduction to Pharmaceutical Biotechnology, Volume 3: Animal Tissue Culture and Biopharmaceuticals. IOP Publishing Ltd, Bristol, UK; 2019;3:1–32.
5. Bhatia S. Characterization of cultured cells. In: Introduction to Pharmaceutical Biotechnology, Volume 3: Animal Tissue Culture and Biopharmaceuticals. IOP Publishing Ltd, Bristol, UK; 2019;3;1–47.
6. Bhatia S. Introduction to genomics. In: Introduction to Pharmaceutical Biotechnology, Enzymes, Proteins and Bioinformatics. IOP Publishing Ltd, Bristol, UK; 2018;2:1–39.
7. Bhatia S. Bioinformatics. In: Introduction to Pharmaceutical Biotechnology, Enzymes, Proteins and Bioinformatics. IOP Publishing Ltd, Bristol, UK; 2018;3:1–16.
8. Bhatia S. Protein and enzyme engineering. In: Introduction to Pharmaceutical Biotechnology, Enzymes, Proteins and Bioinformatics. IOP Publishing Ltd, Bristol, UK; 2018;2:1–15.
9. Bhatia S. Industrial enzymes and their applications. In: Introduction to Pharmaceutical Biotechnology, Enzymes, Proteins and Bioinformatics. IOP Publishing Ltd, Bristol, UK; 2018;2:21.

10. Bhatia S. Introduction to enzymes and their applications. In: Introduction to Pharmaceutical Biotechnology, Enzymes, Proteins and Bioinformatics. IOP Publishing Ltd, Bristol, UK; 2018;2:1–29.

11. Bhatia S. Biotransformation and enzymes. In: Introduction to Pharmaceutical Biotechnology, Enzymes, Proteins and Bioinformatics. IOP Publishing Ltd, Bristol, UK; 2018;3:1–13.

12. Bhatia S. Modern DNA science and its applications. In: Introduction to Pharmaceutical Biotechnology, Volume 1: Basic Techniques and Concepts. IOP Publishing Ltd, Bristol, UK; 2018;1(3):1–70.

13. Bhatia S. Introduction to genetic engineering. In: Introduction to Pharmaceutical Biotechnology, Volume 1: Basic Techniques and Concepts. IOP Publishing Ltd, Bristol, UK; 2018;1(3):1–63.

14. Bhatia S. Applications of stem cells in disease and gene therapy. In: Introduction to Pharmaceutical Biotechnology, Volume 1: Basic Techniques and Concepts. IOP Publishing Ltd, Bristol, UK; 2018;1:1–40.

15. Bhatia S. Transgenic animals in biotechnology. In: Introduction to Pharmaceutical Biotechnology, Volume 1: Basic Techniques and Concepts. IOP Publishing Ltd, Bristol, UK; 2018;1:1–67.

16. Bhatia S. Chapter 1: History and Scope of Plant Biotechnology. Modern Applications of Plant Biotechnology in Pharmaceutical Sciences. Academic Press, 2015:1–30.

17. Bhatia S. Chapter 2: Plant Tissue Culture, Modern Applications of Plant Biotechnology in Pharmaceutical Sciences. Academic Press, 2015:31–107.

18. Bhatia S. Chapter 3: Laboratory Organization, Modern Applications of Plant Biotechnology in Pharmaceutical Sciences. Academic Press, 2015:109–120.

19. Bhatia S. Chapter 4: Concepts and Techniques of Plant Tissue Culture Science, Modern Applications of Plant Biotechnology in Pharmaceutical Sciences. Academic Press, 2015:121–156.

20. Bhatia S. Chapter 5: Application of Plant Biotechnology, Modern Applications of Plant Biotechnology in Pharmaceutical Sciences. Academic Press, 2015:157–207.

21. Bhatia S. Chapter 6: Somatic Embryogenesis and Organogenesis, Modern Applications of Plant Biotechnology in Pharmaceutical Sciences. Academic Press, 2015:209–230.

22. Bhatia S. Chapter 7: Classical and Nonclassical Techniques for Secondary Metabolite Production in Plant Cell Culture, Modern Applications of Plant Biotechnology in Pharmaceutical Sciences. Academic Press, 2015:231–291.

23. Bhatia S. Chapter 8: Plant-Based Biotechnological Products with Their Production Host, Modes of Delivery Systems, and Stability Testing, Modern Applications of Plant Biotechnology in Pharmaceutical Sciences. Academic Press, 2015:293–331.

24. Bhatia S. Chapter 9: Edible Vaccines, Modern Applications of Plant Biotechnology in Pharmaceutical Sciences. Academic Press, 2015:333–343.

25. Bhatia S. Chapter 10: Microenvironmentation in Micropropagation, Modern Applications of Plant Biotechnology in Pharmaceutical Sciences. Academic Press, 2015:345–360.

26. Bhatia S. Chapter 11: Micropropagation, Modern Applications of Plant Biotechnology in Pharmaceutical Sciences. Academic Press, 2015:361–368.

27. Bhatia S. Chapter 12: Laws in Plant Biotechnology, Modern Applications of Plant Biotechnology in Pharmaceutical Sciences. Academic Press, 2015:369–391.

28. Bhatia S. Chapter 13: Technical Glitches in Micropropagation, Modern Applications of Plant Biotechnology in Pharmaceutical Sciences. Academic Press, 2015:393–404.

29. Bhatia S. Chapter 14: Plant Tissue Culture-Based Industries, Modern Applications of Plant Biotechnology in Pharmaceutical Sciences. Academic Press, 2015:405–417.

30. Bhatia S, Al-Harrasi A, Behl T, Anwer MK, Ahmed MM, Mittal V, Kaushik D, Chigurupati S, Kabir MT, Sharma PB, Chaugule B, Vargas-de-la-Cruz C. Anti-migraine activity of freeze dried-latex obtained from *Calotropis gigantea* Linn. *Environ Sci Pollut Res Int.* 2022;29(18):27460–27478. doi:10.1007/s11356-021-17810-x.

31. Bhat AJ, Kumar M, Negi AK, Todaria NP. Informants' consensus on ethnomedicinal plants in Kedarnath Wildlife Sanctuary of Indian Himalayas. *J Med Plants Res.* 2013;7:148–154.

32. Badola HK, Pal M. Endangered medicinal plant species in Himachal Pradesh. *Curr Sci.* 2002;83:797–798.

33. Joshi P, Dhawan V. *Swertia chirayita*—an overview. *Curr Sci.* 2005;89:635–640.

34. Kaur P, Gupta RC, Dey A, Pandey DK. Simultaneous quantification of oleanolic acid, ursolic acid, betulinic acid and lupeol in different populations of five Swertia species by using HPTLC-densitometry: comparison of different extraction methods and solvent selection. *Ind Crop Prod.* 2019;130:537–546.

35. Kaur P, Pandey DK, Gupta RC, Dey A. Simultaneous microwave assisted extraction and HPTLC quantification of mangiferin, amarogentin, and swertiamarin in Swertia species from Western Himalayas. *Ind Crop Prod.* 2019;132:449–459.

36. Pandey DK, Basu S, Jha TB. Screening of different east Himalayan species and populations of *Swertia* L. based on exomorphology and mangiferin content. *Asian Pac J Trop Biomed.* 2012;2(3):S1450–S1456.

37. Pandey DK, Kaur P. Optimization of extraction parameters of pentacyclic triterpenoids from *Swertia chirata* stem using response surface methodology. *3 Biotech.* 2018;8(3):152.

38. Kshirsagar PR, Gaikwad NB, Panda S, Hegde HV, Pai SR. Reverse phase-ultra flow liquid chromatography-diode array detector quantification of anticancerous and antidiabetic drug mangiferin from 11 Species of *Swertia* from India. *Pharmacogn Mag.* 2016 Jan;12(Suppl 1):S32–S36.

39. Woo SY, Win NN, Noe Oo WM, Ngwe H, Ito T, Abe I, Morita H. Viral protein R inhibitors from *Swertia chirata* of Myanmar. *J Biosci Bioeng.* 2019 Oct;128(4):445–449.

40. Ghosal S, Sharma PV, Chaudhuri RK, Bhattacharya SK. Chemical constituents of the gentianaceae V: tetraoxygenated xanthones of *Swertia chirata* Buch.-Ham. *J Pharm Sci.* 1973 Jun;62(6):926–930.

41. Chakravarty AK, Mukhopadhyay S, Das B, Swertane triterpenoids from *Swertia chirata*. *Phytochemistry.* 1991;30:4087–4092.

42. Chakravarty AK, Mukhopadhyay S, Masuda K, Ageta H. Kairatenol, yet another novel migrated gammacerane triterpenoid from *Swertia chirata*. *Tetrahedron Lett.* 1992;33:125–126.

43. Padhan JK, Kumar V, Sood H, Singh TR, Chauhan RS. Contents of therapeutic metabolites in *Swertia chirayita* correlate with the expression profiles of multiple genes in corresponding biosynthesis pathways. *Phytochemistry.* 2015 Aug;116:38–47.

44. Geng CA, Wang LJ, Zhang XM, et al. Anti-hepatitis B virus active lactones from the traditional Chinese herb: Swertia mileensis. *Chemistry.* 2011;17(14):3893–3903.

45. Geng CA, Chen XL, Zhou NJ, et al. LC-MS guided isolation of (±)-sweriledugenin A, a pair of enantiomeric lactones, from *Swertia leducii*. *Org Lett.* 2014;16(2):370–373.

46. Wang HL, He K, Geng CA, et al. Gentiocrucines A-E, five unusual lactonic enamino Ketones from *Swertia macrosperma* and *Swertia angustifolia*. *Planta Med.* 2012;78(17):1867–1872.

47. Aleem A, Kabir H. Review on *Swertia chirata* as traditional uses to its pyhtochemistry and phrmacological activity. *J Drug Deliv Ther [Internet]*. 2018 15 Oct;8(5-s):73–78. [cited 5 Aug. 2021].

48. Ayurveda Pharmacopia. Department of Ayurveda, Sri Lanka; 1976.

49. Nagodavithana P. Shri Sharangadara Samhita, 1st ed. Samayawardhana Book Shop (Pvt) Ltd, 2001.

50. Bhatia S, Al-Harrasi A, Behl T, Anwer MK, Ahmed MM, Mittal V, Kaushik D, Chigurupati S, Kabir MT, Sharma PB, Chaugule B, Vargas-de-la-Cruz C. Unravelling the photoprotective effects of freshwater alga *Nostoc commune* Vaucher ex Bornet et Flahault against ultraviolet radiations. *Environ Sci Pollut Res Int.* 2022 Feb;29(10):14380–14392. doi: 10.1007/s11356-021-16704-2. Epub 2021 Oct 5. PMID: 34609682.

51. Choudhari VP, Gore KP, Pawar AT. Antidiabetic, antihyperlipidemic activities and herb-drug interaction of a polyherbal formulation in streptozotocin induced diabetic rats. *J Ayurveda Integr Med.* 2017 Oct-Dec;8(4):218–225. doi: 10.1016/j.jaim.2016.11.002. Epub 2017 Nov 11. PMID: 29137853; PMCID: PMC5747493.

52. Choudhari VP, Gore KP, Pawar AT. Antidiabetic, antihyperlipidemic activities and herb-drug interaction of a polyherbal formulation in streptozotocin induced diabetic rats. *J Ayurveda Integr Med.* 2017 Oct-Dec;8(4):218–225. doi: 10.1016/j.jaim.2016.11.002. Epub 2017 Nov 11. PMID: 29137853; PMCID: PMC5747493.

53. Tambekar DH, Dahikar SB. Antibacterial activity of some Indian Ayurvedic preparations against enteric bacterial pathogens. *J Adv Pharm Technol Res.* 2011 Jan; 2(1):24–29.

54. Rajopadhye AA, Namjoshi TP, Upadhye AS. Rapid validated HPTLC method for estimation of piperine and piperlongumine in root of *Piper longum* extract and its commercial formulation. *Rev Bras.* 2012;22(6):1355–1361.

55. Chauhan S, Pundir V, Sharma A. Pharmacopeial standardization of mahasudarshan churna: a polyherbal formulation. *J Med Plants Stud.* 2013;1(2):13–18.

56. Babu PS, Stanely Mainzen Prince P. Antihyperglycaemic and antioxidant effect of hyponidd, an ayurvedic herbomineral formulation in streptozotocin-induced diabetic rats. *J Pharm Pharmacol.* 2004 Nov;56(11):1435–1442. doi: 10.1211/0022357044607. PMID: 15525451.

57. Beniwal P, Gaur N, Singh SK, Raveendran N, Malhotra V. How harmful can herbal remedies be? A case of severe acute tubulointerstitial nephritis. *Indian J Nephrol.* 2017 Nov-Dec;27(6):459–461.

58. Health Bureau of Tibet, Qinghai, Sichuan, Gansu, Yunnan, and Xinjiang. Tibetan Medicine Standards. Qinghai People's Publishing Press, Xining; 1979.

59. Verma H, Patil PR, Kolha pure RM, Gopalkrishna V. Antiviral activity of the Indian medicinal plant extract *Swertia chirata* against herpes simplex viruses: a study by in-vitro and molecular approach. *Indian J Med Microbiol.* 2008 Oct-Dec;26(4):322–326. PMID: 18974483.

60. Woo SY, Win NN, Noe Oo WM, Ngwe H, Ito T, Abe I, Morita H. Viral protein R inhibitors from *Swertia chirata* of Myanmar. *J Biosci Bioeng.* 2019 Oct;128(4):445–449. doi: 10.1016/j.jbiosc.2019.04.006. Epub 2019 May 7. PMID: 31076338.

61. Lalla JK, Ogale S, Seth S. A review on dengue and treatments. Research and reviews. *J Pharmacol Toxicol Stud.* 2014;2(4):13–23.

62. Zhang S, Li D, Yang JY, Yan TB. Plumbagin protects against glucocorticoid-induced osteoporosis through Nrf-2 pathway. *Cell Stress Chaperones.* 2015;20(4):621–629.

63. Roy P, Abdulsalam FI, Pandey DK, Bhattacharjee A, Eruvaram NR, Malik T. Evaluation of antioxidant, antibacterial, and antidiabetic potential of two traditional medicinal plants of India: *Swertia cordata* and *Swertia chirayita. Pharmacogn Res.* 2015 Jun;7(Suppl 1):S57–S62.

64. Bhattacharya A, Jindal B, Singh P, Datta A, Panda D. Plumbagin inhibits cytokinesis in Bacillus subtilis by inhibiting FtsZ assembly – a mechanistic study of its antibacterial activity. *FEBS J.* 2013 Sep;280(18):4585–4599.

65. Park HC, Gedi V, Cho JH, Hyun JW, Lee KJ, Kang J, So B, Yoon MY. Characterization and in vitro inhibition studies of Bacillus anthracis FtsZ: a potential antibacterial target. *Appl Biochem Biotechnol.* 2014 Mar;172(6):3263–3270.

66. Barua A, Choudhury P, Mandal S, Panda CK, Saha P. Therapeutic potential of xanthones from *Swertia chirata* in breast cancer cells. *Indian J Med Res.* 2020 Sep;152(3):285–295.

67. Barua A, Choudhury P, Mandal S, Panda CK, Saha P. Anti-metastatic potential of a novel xanthone sourced by *Swertia chirata* against in vivo and in vitro breast adenocarcinoma frameworks. *Asian Pac J Cancer Prev.* 2020 Oct 1;21(10):2865–2875.

68. Kar P, Kumar V, Vellingiri B, Sen A, Jaishee N, Anandraj A, Malhotra H, Bhattacharyya S, Mukhopadhyay S, Kinoshita M, Govindasamy V, Roy A, Naidoo D, Subramaniam MD. Anisotine and amarogentin as promising inhibitory candidates against SARS-CoV-2 proteins: a computational investigation. *J Biomol Struct Dyn.* 2020 Dec 11:1–11.

69. Maurya VK, Kumar S, Bhatt MLB, Saxena SK. Antiviral activity of traditional medicinal plants from Ayurveda against SARS-CoV-2 infection. *J Biomol Struct Dyn.* 2020 Oct 19:1–17.

70. Maurya VK, Kumar S, Bhatt MLB, Saxena SK. Antiviral activity of traditional medicinal plants from Ayurveda against SARS-CoV-2 infection. *J Biomol Struct Dyn.* 2020 Oct 19:1–17.

71. Shukla S, Bafna K, Sundar D, Thorat, SS. The bitter barricading of prostaglandin biosynthesis pathway: understanding the molecular mechanism of selective cyclooxygenase-2 inhibition by amarogentin, a secoiridoid glycoside from *Swertia chirayita. PLoS One.* 2014;9(6):e90637.

72. Suryawanshi S, Mehrotra N, Asthana RK, Gupta RC. Liquid chromatography/tandem mass spectrometric study and analysis of xanthone and secoiridoid glycoside composition of *Swertia chirata*, a potent antidiabetic. *Rapid Commun Mass Spectrom.* 2006;20(24):3761–3768.

73. Phoboo S, Pinto Mda S, Barbosa AC, et al. Phenolic-linked biochemical rationale for the anti-diabetic properties of *Swertia chirayita* (Roxb. ex Flem.) Karst. *Phytother Res.* 2013;27(2):227–235.

74. Ali S, Farooq M, Panhwar WA. Evaluation of hypoglycemic and hypolipidemic properties of *Swertia chirata. J Entomol Zool Stud.* 2017;5:1448.

75. Rajesh C, Holla R, Patil V, Anand AS, Prasad K. Anti-hyperglycemic effect of *Swertia chirata* root extract on indinavir treated ratsNatl. *J Physiol Pharm Pharmacol.* 2017;7:569.

76. Bhowmik A, Mosihuzzaman M, Kabir Y, Rokeya BJ. Substantial reduction in fasting glucose, cholesterol, triglycerides level and improvement in hepatic glycogen content. *Pharm Res Int.* 2018;22:1.

77. Saxena AM, Bajpai MB, Murthy PS, Mukherjee SK. Mechanism of blood sugar lowering by a swerchirin-containing hexane fraction (SWI) of *Swertia chirayita. Indian J Exp Biol.* 1993;31(2):178–181.

78. Haldhar R, Prasad D, Nguyen LTD, Kaya S, Bahadur I, Dagdag O, Kim SC. Corrosion inhibition, surface adsorption and computational studies of *Swertia chirata* extract: a sustainable and green approach. *Mater Chem Phys.* 2021;267:124613.

79. Saha N, Dutta Gupta S. Low-dose toxicity of biogenic silver nanoparticles fabricated by *Swertia chirata* on root tips and flower buds of *Allium cepa. J Hazard Mater.* 2017 May 15;330:18–28.

80. Nagalekshmi R, Menon A, Chandrasekharan DK, Nair CK. Hepatoprotective activity of *Andrographis paniculata* and *Swertia chirayita. Food Chem Toxicol.* 2011 Dec;49(12):3367–3373.

81. Rahman MM, Khan AA, Ali ME, Mian IH, Akanda AM, Abd Hamid SB. Botanicals to control soft rot bacteria of potato. *Sci World J.* 2012;2012:796472.

82. Edwin R, Chungath JI. Studies in *Swertia chirata. Indian Drugs.* 1988;25:143–146.

83. Mitra SK, Gopumadhavan S, Muralidhar TS. Effect of D-400, an ayurvedic herbal formulation on experimentally-induced diabetes mellitus. *Phytother Res.* 1996;10:433–435.

84. Ahmmed SM, Mukherjee PK, Bahadur S, Harwansh RK, Kar A, Bandyopadhyay A, Al-Dhabi NA, Duraipandiyan V. CYP450 mediated inhibition potential of *Swertia chirata*: an herb from Indian traditional medicine. *J Ethnopharmacol.* 2016 Feb 3;178:34–39.

85. http://ctri.nic.in/Clinicaltrials/showallp.php?mid1=43727&EncHid=&userName=025214

86. Khanna N, Nazli T, Siddiqui KM, Kalaivani M, Rais-ur-Rahman. A non-inferiority randomized controlled clinical trial comparing Unani formulation & psoralen plus ultraviolet A sol in chronic plaque psoriasis. *Indian J Med Res.* 2018 Jan;147(1):66–72.

87. Kaur P, Gupta RC, Dey A, Pandey DK. Simultaneous quantification of oleanolic acid, ursolic acid, betulinic acid and lupeol in different populations of five Swertia species by using HPTLC-densitometry: comparison of different extraction methods and solvent selection. *Ind Crops Prod.* 2019;130:537–546.

88. Kaur P, Pandey DK, Gupta RC, Dey A. Simultaneous microwave assisted extraction and HPTLC quantification of mangiferin, amarogentin, and swertiamarin in Swertia species from Western Himalayas. *Ind Crops Prod.* 2019;132:449–459.

89. Kaur P, Pandey DK, Gupta RC, Dey A. Assessment of genetic diversity among different population of five Swertia species by using molecular and phytochemical markers. *Ind Crops Prod.* 2019;138:111569.

90. Selvam ABD. Exomorphic and endomorphic features of *Swertia chirayita. Pharmacogn J.* 2011;3:1–6.

91. Beniwal P, Gaur N, Singh SK, Raveendran N, Malhotra V. How harmful can herbal remedies be? A case of severe acute tubulointerstitial nephritis. *Indian J Nephrol.* 2017 Nov-Dec;27(6):459–461.

92. Bhat GP, Surolia N. In vitro antimalarial activity of extracts of three plants used in the traditional medicine of India. *Am J Trop Med Hyg.* 2001;65(4):304–308.

93. Karan M, Vasisht K, Handa SS. Antihepatotoxic activity of *Swertia chirata* on carbon tetrachloride induced hepatotoxicity in rats. *Phytother Res.* 1999;13(1):24–30.

94. Banerjee S, Sur TK, et al. Assessment of the anti-inflammatory effects of *Swertia chirata* in acute and chronic experimental models in male albino rats. *Indian J Pharmacol.* 2000;32:21–24.

95. Sidhu GS, Sankaram AVB. A new biplumbagin and 3-chloroplumbagin from *Plumbago zeylanica. Tetrahedron Lett.* 1971;26: 2385–2388.

96. Sankaram AVB, Srinivasarao A, Sidhu GS. Chitranone—a new binaphthaquinone from *Plumbago zeylanica. Phytochemistry.* 1976;15(1):237–238.

97. Sankaram AVB, Rao AS, Shoolery JN. Zeylanone and isozeylanone, two novel quinones from *Plumbago zeylanica. Tetrahedron.* 1979;35:1777–1782.

98. Gunaherath GMKB, Gunatilaka AAL, Sultanbawa MUS, Balasubramaniam S. 1,2(3)-Tetrahydro-3,3′-biplumbagin: a naphthalenone and other constituents from *Plumbago zeylanica. Phytochemistry.* 1983;22:1245–1247.

99. Kamal GM, Gunaherath B, Gunatilaka AAL, Thomson RH. Structure of plumbazeylanone: a novel trimer of plumbagin from *Plumbago zeylanica. Tetrahedron Lett.* 1984;25:4801–4804.

100. Lin LC, Yang LL, Chou CJ. Cytotoxic naphthoquinones and plumbagic acid glucosides from *Plumbago zeylanica. Phytochemistry.* 2003 Feb;62(4):619–622. doi: 10.1016/s0031-9422(02)00519-8. PMID: 12560036.

101. Nguyen AT, Malonne H, Duez P, Vanhaelen-Fastre R, Vanhaelen M, Fontaine J. Cytotoxic constituents from *Plumbago zeylanica. Fitoterapia.* 2004 Jul;75(5):500–504. doi: 10.1016/j.fitote.2004.03.009. PMID: 15261389.

102. Tsai WJ, Chen YC, Wu MH, Lin LC, Chuang KA, Chang SC, Kuo YC. Seselin from *Plumbago zeylanica* inhibits phytohemagglutinin (PHA)-stimulated cell proliferation in human peripheral blood mononuclear cells. *J Ethnopharmacol.* 2008 Sep 2;119(1):67–73. doi: 10.1016/j.jep.2008.05.032. Epub 2008 Jun 5. PMID: 18577441.

103. Kishore N, Mishra BB, Tiwari VK, Tripathi V. Difuranonaphthoquinones from *Plumbago zeylanica* roots. *Phytochem Lett.* 2010;3:62–65.

104. Sathya S, Sudhagar S, Vidhya Priya M, Bharathi Raja R, Muthusamy VS, Niranjali Devaraj S, Lakshmi BS. 3β-Hydroxylup-20(29)-ene-27,28-dioic acid dimethyl ester, a novel natural product from *Plumbago zeylanica* inhibits the proliferation and migration of MDA-MB-231 cells. *Chem Biol Interact.* 2010 Dec 5;188(3):412–420.

105. Susumu O, Yoshiaki Y, Shinji T, et al. New naphthoquinone and monoterpenoid from *Plumbago zeylanica. Tetrahedron Lett.* 2014;48:6554–6556.

106. Vasav AP, Pable AA, Barvkar VT. Differential transcriptome and metabolome analysis of *Plumbago zeylanica* L. reveal putative genes involved in plumbagin biosynthesis. *Fitoterapia.* 2020 Nov;147:104761. doi: 10.1016/j.fitote.2020.104761. Epub 2020 Oct 16. PMID: 33069837.

107. Singh T, Sharma U, Agrawal V. Isolation and optimization of plumbagin production in root callus of *Plumbago zeylanica* L. augmented with chitosan and yeast extract. *Ind Crops Prod.* 2020;151:112446.

108. Roy A, Bharadvaja N. Establishment of root suspension culture of *Plumbago zeylanica* and enhanced production of plumbagin. *Ind Crops Prod.* 2019;137:419–427.

109. Vanitha V, Vijayakumar S, Nilavukkarasi M, et al. Heneicosane – a novel microbicidal bioactive alkane identified from *Plumbago zeylanica* L. *Ind Crops Prod*. 2020;154:112748.

110. Chaudhari SS, Chaudhari GS. A review on *Plumbago zeylanica* Linn. – a divine medicinal plant. *Int J Pharm Sci Rev Res*. 2015;30(2):119–127.

111. Kar A, Mukherjee PK, Saha S, Banerjee S, Goswami D, Matsabisa MG, Charoensub R, Duangyod T. Metabolite profiling and evaluation of CYP450 interaction potential of 'Trimada' – an Ayurvedic formulation. *J Ethnopharmacol*. 2021 Feb 10;266:113457. doi: 10.1016/j.jep.2020.113457. Epub 2020 Oct 9. PMID: 33039629.

112. Dev SK, Choudhury PK, Srivastava R, Sharma M. Antimicrobial, anti-inflammatory and wound healing activity of polyherbal formulation. *Biomed Pharmacother*. 2019 Mar;111:555–567.

113. Department of Indian Systems of Medicine and Homoeopathy, Ministry of Health and Family Welfare, Government of India. Part I. Department of Indian Systems of Medicine and Homoeopathy, 1st ed. Ministry of Health and Family Welfare, Government of India, New Delhi; 2003. Ayurvedic Formulary of India; p. 119.

114. Riyasdeen A, Periasamy VS, Paul P, Alshatwi AA, Akbarsha MA. Chloroform extract of Rasagenthi Mezhugu, a Siddha formulation, as an evidence-based complementary and alternative medicine for HPV-positive cervical cancers. *Evid Based Complement Alternat Med*. 2012;2012:136527.

115. Bothiraja C, Pawar AP, Dama GY, Joshi PP, Shaikh KS. Novel solvent-free gelucire extract of *Plumbago zeylanica* using non-everted rat intestinal sac method for improved therapeutic efficacy of plumbagin. *J Pharmacol Toxicol Methods*. 2012 Jul;66(1):35–42.

116. Hafeez BB, Fischer JW, Singh A, Zhong W, Mustafa A, Meske L, Sheikhani MO, Verma AK. Plumbagin inhibits prostate carcinogenesis in intact and castrated PTEN knockout mice via targeting PKCε, Stat3, and epithelial-to-mesenchymal transition markers. *Cancer Prev Res (Phila)*. 2015 May;8(5):375–386.

117. Lin LC, Yang LL, Chou CJ. Cytotoxic naphthoquinones and plumbagic acid glucosides from *Plumbago zeylanica*. *Phytochemistry*. 2003 Feb;62(4):619–622.

118. Pan ST, Qin Y, Zhou ZW, He ZX, Zhang X, Yang T, Yang YX, Wang D, Zhou SF, Qiu JX. Plumbagin suppresses epithelial to mesenchymal transition and stemness via inhibiting Nrf2-mediated signaling pathway in human tongue squamous cell carcinoma cells. *Drug Des Devel Ther*. 2015;9:5511–5551.

119. Xue D, Pan ST, Zhou X, Ye F, Zhou Q, Shi F, He F, Yu H, Qiu J. Plumbagin enhances the anticancer efficacy of cisplatin by increasing intracellular ROS in human tongue squamous cell carcinoma. *Oxid Med Cell Longev*. 2020 Mar 25;2020:5649174.

120. Nguyen AT, Malonne H, Duez P, Vanhaelen-Fastre R, Vanhaelen M, Fontaine J. Cytotoxic constituents from *Plumbago zeylanica*. *Fitoterapia*. 2004 Jul;75(5):500–504.

121. Jayanthi M, Gokulanathan A, Haribalan P, et al. Plumbagin from two Plumbago species inhibits the growth of stomach and breast cancer cell lines. *Ind Crops Prod*. 2020;146:112147.

122. Chen CA, Chang HH, Kao CY, Tsai TH, Chen YJ. Plumbagin, isolated from *Plumbago zeylanica*, induces cell death through apoptosis in human pancreatic cancer cells. *Pancreatology*. 2009;9(6):797–809.

123. Hafeez BB, Zhong W, Fischer JW, Mustafa A, Shi X, Meske L, Hong H, Cai W, Havighurst T, Kim K, Verma AK. Plumbagin, a medicinal plant (*Plumbago zeylanica*)-derived 1,4-naphthoquinone, inhibits growth and metastasis of human prostate cancer PC-3M-luciferase cells in an orthotopic xenograft mouse model. *Mol Oncol*. 2013 Jun;7(3):428–439. doi: 10.1016/j.molonc.2012.12.001. Epub 2012 Dec 14. PMID: 23273564; PMCID: PMC3625495.

124. Karthikeyan C, Sisubalan N, Sridevi M, Varaprasad K, Ghouse Basha MH, Shucai W, Sadiku R. Biocidal chitosan-magnesium oxide nanoparticles via a green precipitation process. *J Hazard Mater*. 2021 Jun 5;411:124884. doi: 10.1016/j.jhazmat.2020.124884. Epub 2021 Jan 11. PMID: 33858076.

125. Wang YC, Huang TL. High-performance liquid chromatography for quantification of plumbagin, an anti-*Helicobacter pylori* compound of *Plumbago zeylanica* L. *J Chromatogr A*. 2005 Nov 11;1094 (1–2):99–104. doi: 10.1016/j.chroma.2005.07.092. Epub 2005 Aug 15. PMID: 16257295.

126. Wang YC, Huang TL. Anti-*Helicobacter pylori* activity of *Plumbago zeylanica* L. *FEMS Immunol Med Microbiol*. 2005 Mar 1;43(3):407–412. doi: 10.1016/j.femsim.2004.10.015. PMID: 15708315.

127. Sandeep G, Dheeraj A, Sharma NK, Jhade D, Bharti A. Effect of plumbagin free alcohol extract of *Plumbago zeylanica* Linn. root on reproductive system of female Wistar rats. *Asian Pac J Trop Med*. 2011 Dec;4(12):978–984. doi: 10.1016/S1995-7645(11)60230-7. PMID: 22118035.

128. Bello IJ, Oyebode OT, Olanlokun JO, Omodara TO, Olorunsogo OO. Plumbagin induces testicular damage via mitochondrial-dependent cell death. *Chem Biol Interact*. 2021 Sep 25;347:109582. doi: 10.1016/j.cbi.2021.109582. Epub 2021 Jul 21. PMID: 34302802.

129. Sunil C, Duraipandiyan V, Agastian P, Ignacimuthu S. Antidiabetic effect of plumbagin isolated from *Plumbago zeylanica* L. root and its effect on GLUT4 translocation in streptozotocin-induced diabetic rats. *Food Chem Toxicol*. 2012 Dec;50(12):4356–4363.

130. Krishnan GS, Sebastian D, Savarimuthu I, Poovathumkal JA, Fleming AT. In vitro and in silico anti-cancer effect of combined crude acetone extracts of *Plumbago zeylanica* L., *Limonia acidissima* L. and *Artocarpus heterophyllus* Lam. *Synergy.* 2017;5:15–23.

131. Pan Q, Zhou R, Su M, Li R. The effects of plumbagin on pancreatic cancer: a mechanistic network pharmacology approach. *Med Sci Monit.* 2019 Jun 23;25:4648–4654.

132. Xu KH, Lu DP. Plumbagin induces ROS-mediated apoptosis in human promyelocytic leukemia cells in vivo. *Leuk Res.* 2010 May;34(5):658–665.

133. Vijayakumar R, Senthilvelan M, Ravindran R, Devi RS. *Plumbago zeylanica* action on blood coagulation profile with and without blood volume reduction. *Vascul Pharmacol.* 2006 Aug;45(2):86–90.

134. Rajakrishnan R, Lekshmi R, Benil PB, Thomas J, AlFarhan AH, Rakesh V, Khalaf S. Phytochemical evaluation of roots of *Plumbago zeylanica* L. and assessment of its potential as a nephroprotective agent. *Saudi J Biol Sci.* 2017 May;24(4):760–766.

135. Zaki AM, El-Tanbouly DM, Abdelsalam RM, Zaki HF. Plumbagin ameliorates hepatic ischemia-reperfusion injury in rats: role of high mobility group box 1 in inflammation, oxidative stress and apoptosis. *Biomed Pharmacother.* 2018 Oct;106:785–793.

136. Wang H, Zhang H, Zhang Y, Wang D, Cheng X, Yang F, Zhang Q, Xue Z, Li Y, Zhang L, Yang L, Miao G, Li D, Guan Z, Da Y, Yao Z, Gao F, Qiao L, Kong L, Zhang R. Plumbagin protects liver against fulminant hepatic failure and chronic liver fibrosis via inhibiting inflammation and collagen production. *Oncotarget.* 2016 Dec 13;7(50):82864–82875.

137. Tsai WJ, Chen YC, Wu MH, Lin LC, Chuang KA, Chang SC, Kuo YC. Seselin from *Plumbago zeylanica* inhibits phytohemagglutinin (PHA)-stimulated cell proliferation in human peripheral blood mononuclear cells. *J Ethnopharmacol.* 2008 Sep 2;119(1):67–73. doi: 10.1016/j.jep.2008.05.032. Epub 2008 Jun 5. PMID: 18577441.

138. Sathya S, Sudhagar S, Vidhya Priya M, Bharathi Raja R, Muthusamy VS, Niranjali Devaraj S, Lakshmi BS. 3β-Hydroxylup-20(29)-ene-27,28-dioic acid dimethyl ester, a novel natural product from *Plumbago zeylanica* inhibits the proliferation and migration of MDA-MB-231 cells. *Chem Biol Interact.* 2010 Dec 5;188(3):412–420.

139. Messeha SS, Zarmouh NO, Mendonca P, Kolta MG, Soliman KFA. The attenuating effects of plumbagin on pro-inflammatory cytokine expression in LPS-activated BV-2 microglial cells. *J Neuroimmunol.* 2017 Dec 15;313:129–137.

140. Lee HP, Chen PC, Wang SW, Fong YC, Tsai CH, et al. Plumbagin suppresses endothelial progenitor cell-related angiogenesis in vitro and in vivo. *J Func Foods.* 2019;52:537–544.

141. Liu X, Niu M, Xu X, Cai W, Zeng L, Zhou X, Yu R, Xu K. CRM1 is a direct cellular target of the natural anti-cancer agent plumbagin. *J Pharmacol Sci.* 2014;124(4):486–493. doi: 10.1254/jphs.13240fp. PMID: 24739265.

142. Yang SJ, Chang SC, Wen HC, Chen CY, Liao JF, Chang CH. Plumbagin activates ERK1/2 and Akt via superoxide, Src and PI3-kinase in 3T3-L1 cells. *Eur J Pharmacol.* 2010 Jul 25;638(1–3):21–28. doi: 10.1016/j.ejphar.2010.04.016. Epub 2010 Apr 25. PMID: 20420821.

143. Sandur SK, Ichikawa H, Sethi G, Ahn KS, Aggarwal BB. Plumbagin (5-hydroxy-2-methyl-1,4-naphthoquinone) suppresses NF-kappaB activation and NF-kappaB-regulated gene products through modulation of p65 and IkappaBalpha kinase activation, leading to potentiation of apoptosis induced by cytokine and chemotherapeutic agents. *J Biol Chem.* 2006 Jun 23;281(25):17023–17033.

144. Wei Y, Lv B, Xie J, Zhang Y, Lin Y, Wang S, Zhong J, Chen Y, Peng Y, Ma J. Plumbagin promotes human hepatoma SMMC-7721 cell apoptosis via caspase-3/vimentin signal-mediated EMT. *Drug Des Devel Ther.* 2019 Jul 15;13:2343–2355.

145. Kang CG, Im E, Lee HJ, Lee EO. Plumbagin reduces osteopontin-induced invasion through inhibiting the Rho-associated kinase signaling pathway in A549 cells and suppresses osteopontin-induced lung metastasis in BalB/c mice. *Bioorg Med Chem Lett.* 2017 May 1;27(9):1914–1918. doi: 10.1016/j.bmcl.2017.03.047. Epub 2017 Mar 20. PMID: 28359791.

146. Niu M, Cai W, Liu H, Chong Y, Hu W, Gao S, Shi Q, Zhou X, Liu X, Yu R. Plumbagin inhibits growth of gliomas in vivo via suppression of FOXM1 expression. *J Pharmacol Sci.* 2015 Jul;128(3):131–136. doi: 10.1016/j.jphs.2015.06.005. Epub 2015 Jun 25. PMID: 26154848.

147. Chen G, Yue Y, Qin J, Xiao X, Ren Q, Xiao B. Plumbagin suppresses the migration and invasion of glioma cells via downregulation of MMP-2/9 expression and inaction of PI3K/Akt signaling pathway in vitro. *J Pharmacol Sci.* 2017 May;134(1):59–67. doi: 10.1016/j.jphs.2017.04.003. Epub 2017 Apr 24. PMID: 28506595.

148. Zhang J, Peng S, Li X, Liu R, Han X, Fang J. Targeting thioredoxin reductase by plumbagin contributes to inducing apoptosis of HL-60 cells. *Arch Biochem Biophys.* 2017 Apr 1;619:16–26. doi: 10.1016/j.abb.2017.02.007. Epub 2017 Feb 27. PMID: 28249720.

149. Cai Z, He S, Li T, Zhao L, Zhang K. Plumbagin inhibits proliferation and promotes apoptosis of ovarian granulosa cells in polycystic ovary syndrome by inactivating PI3K/Akt/mTOR pathway. *Anim Cells Syst (Seoul)*. 2020 Jul 17;24(4):197–204. doi: 10.1080/19768354.2020.1790416. PMID: 33029296; PMCID: PMC7473319.

150. Pradeepa V, Senthil-Nathan S, Sathish-Narayanan S, Selin-Rani S, Vasantha-Srinivasan P, Thanigaivel A, Ponsankar A, Edwin ES, Sakthi-Bagavathy M, Kalaivani K, Murugan K, Duraipandiyan V, Al-Dhabi NA. Potential mode of action of a novel plumbagin as a mosquito repellent against the malarial vector *Anopheles stephensi*, (Culicidae: Diptera). *Pestic Biochem Physiol*. 2016 Nov;134:84–93. doi: 10.1016/j.pestbp.2016.04.001. Epub 2016 Apr 10. PMID: 27914545.

151. Yue L, Jiang N, Wu A, Qiu W, Shen X, Qin D, Li H, Lin J, Liang S, Wu J. Plumbagin can potently enhance the activity of xanthine oxidase: in vitro, in vivo and in silico studies. *BMC Pharmacol Toxicol*. 2021 Jul 17;22(1):45. doi: 10.1186/s40360-021-00511-z. PMID: 34274011; PMCID: PMC8286619.

152. Teshome K, Gebre-Mariam T, Asres K, Perry F, Engidawork E. Toxicity studies on dermal application of plant extract of *Plumbago zeylanica* used in Ethiopian traditional medicine. *J Ethnopharmacol*. 2008 May 8;117(2):236–248.

153. J. Demma, E. Engidawork, B. Hellman, Potential genotoxicity of plant extracts used in Ethiopian traditional medicine. *J Ethnopharmaco*. 2009;122:136–142.

154. Demma J, Hallberg K, Hellman B. Genotoxicity of plumbagin and its effects on catechol and NQNO-induced DNA damage in mouse lymphoma cells. *Toxicol in Vitro*. 2009;23:266–271.

155. Salunke M, Banjare J, Bhalerao S. Effect of selected herbal formulations on anthropometry and body composition in overweight and obese individuals: a randomized, double blind, placebo-controlled study. *J Herb Meds*. 2019;17–:18100298.

156. Kumar T, Sanapeti RV, Prasad BS. Evaluation of effect of poultice (Upanaha Sweda) in low back pain (Katigraha): a randomized comparative clinical trial. *Ayu*. 2019 Jul-Sep;40(3):159–163.

157. Gupte P, Harke S, Deo V, Bhushan Shrikhande B, Mahajan M, Bhalerao S. A clinical study to evaluate the efficacy of herbal formulation for obesity (HFO-02) in overweight individuals. *J Ayurveda Integr Med*. 2020 Apr-Jun;11(2):159–162.

158. Saraswathy A, Pradeep RV, Muralimanohar B, Vairamuthu S. Wound healing activity of the chloroform extract of *Plumbago rosea* Linn. and plumbagin. *Nat Prod Sci*. 2006;12:50–54.

159. Veluri R, Diwan PV. Phytochemical and pharmacological aspects of *Plumbago zeylanica*. *Indian Drugs*. 1999;36:724–730.

160. Prakash O, Jyoti AK, Kumar P, Manna NK. Adulteration and substitution in Indian medicinal plants: an overview. *J Med Plants Stud*. 2013;1:127–132.

161. Ahmad I, Mehmood Z, Mohammad F. Screening of some Indian medicinal plants for their antimicrobial properties. *J Ethnopharmacol*. 1998;62(2):183–193.

162. Ştefănescu R, Tero-Vescan A, Negroiu A, Aurică E, Vari CE. A comprehensive review of the phytochemical, pharmacological, and toxicological properties of *Tribulus terrestris* L. *Biomolecules*. 2020 May 12;10(5):752.

163. Tkachenko K, Frontasyeva M, Vasilev A, Avramov L, Shi L. Major and trace element content of *Tribulus terrestris* L. wildlife plants. *Plants (Basel)*. 2020 Dec 13;9(12):1764.

164. Pokrywka A, Obmiński Z, Malczewska-Lenczowska J, Fijałek Z, Turek-Lepa E, Grucza R. Insights into supplements with *Tribulus terrestris* used by athletes. *J Hum Kinet*. 2014 Jun 28;41:99–105.

165. Shang ZJ. *Annotation of Shen Nong Ben CaoJ ing*. Academy Press, Beijing; 2008:65.

166. Chinese Pharmacopoeia Commission. Chinese Pharmacopoeia, Vol. I. China Medical Science Press, Beijing; 2015:352.

167. Mohammed MS, Khalid HS, Osman WJA, Muddathir AK. A review on phytochemical profile and biological activities of three anti-inflammatory plants used in sudanese folkloric medicine. *Am J Pharm Tech Res*. 2014;4(4):1–14.

168. Akram M, Asif HM, Akhtar N, Shah PA, Uzair M, Shaheen G, et al. *Tribulus terrestris* Linn.: a review article. *J Med Plants Res*. 2011;5(16):3601–3605.

169. Tauheed A, Hamiduddin, Khanam S, Ali MA, Zaigham M. Comparative physicochemical evaluation of kharekhasak (*Tribulus terrestris* Linn.) before and after mudabbar process. *Pharmacogn Res*. 2017 Oct-Dec;9(4):384–389.

170. Gurav N, Gurav S, Wanjari M, Prasad S, Wayal S, Rarokar N. Development and evaluation of aphrodisiac potential of a classical ayurvedic formulation, 'Kaamdev ghrita' in rat model. *J Ayurveda Integr Med*. 2021 Apr-Jun;12(2):294–301.

171. Agarwal VK, Mittal PC. Clinical studies with Speman in cases of benign enlargement prostate. *Probe*. 1969;9:153–156. http://indianmedicine.eldoc.ub.rug.nl/root/A/375.

172. Vyas JN, Bhattachariya DD, Bhandari JR. Sexual potency disorders of the male: a clinical trial. *Probe.* 1970;9:149–153.

173. Mukram MA, Rafiq M, Anturlikar SD, Patki PS. Speman®, A proprietary ayurvedic formulation, reverses cyclophosphamide-induced oligospermia in rats. *Cent Asian J Glob Health.* 2013 Apr 2;2(1):14.

174. Yu JY, Gupta B, Park HG, Son M, Jun JH, Yong CS, Kim JA, Kim JO. Preclinical and clinical studies demonstrate that the proprietary herbal extract DA-5512 effectively stimulates hair growth and promotes hair health. *Evid Based Complement Alternat Med.* 2017;2017:4395638.

175. Sahoo HB, Nandy S, Senapati AK, Sarangi SP, Sahoo SK. Aphrodisiac activity of polyherbal formulation in experimental models on male rats. *Pharmacogn Res.* 2014 Apr;6(2):120–126.

176. Sahin K, Tuzcu M, Orhan C, Gencoglu H, Sahin N, Akdemir F, Turk G, Yilmaz I, Juturu V. MAT, a novel polyherbal aphrodisiac formulation, enhances sexual function and Nrf2/HO-1 pathway while reducing oxidative damage in male rats. *Evid Based Complement Alternat Med.* 2018 Apr 29;2018:8521782.

177. Shankhwar SN, Mahdi AA, Sharma AV, Pv K. A prospective clinical study of a prosexual nutrient: nano leo for evaluation of libido, erection, and orgasm in Indian men with erectile dysfunction. *Evid Based Complement Alternat Med.* 2020 Mar 11;2020:4598217.

178. Deole YS, Chavan SS, Ashok BK, Ravishankar B, Thakar AB, Chandola HM. Evaluation of anti-depressant and anxiolytic activity of Rasayana Ghana tablet (a compound Ayurvedic formulation) in albino mice. *Ayu.* 2011 Jul;32(3):375–379.

179. Figueiredo CCM, Gomes AC, Granero FO, Bronzel Junior JL, Silva LP, Ruiz ALTG, da Silva RMG. Antiglycation and antitumoral activity of *Tribulus terrestris* dry extract. *Avicenna J Phytomed.* 2021 May-Jun;11(3):224–237.

180. Nasir O, Alqadri N, Elsayed S, Ahmed O, Alotaibi SH, Baty R, Omer H, Abushal SA, Umbach AT. Comparative efficacy of Gum Arabic (*Acacia senegal*) and *Tribulus terrestris* on male fertility. *Saudi Pharm J.* 2020 Dec;28(12):1791–1796.

181. Saiyed A, Jahan N, Makbul SAA, Ansari M, Bano H, Habib SH. Effect of combination of *Withania somnifera* Dunal and *Tribulus terrestris* Linn on letrozole induced polycystic ovarian syndrome in rats. *Integr Med Res.* 2016 Dec;5(4):293–300.

182. Abadjieva D, Kistanova E. *Tribulus terrestris* alters the expression of growth differentiation factor 9 and bone morphogenetic protein 15 in rabbit ovaries of mothers and F1 female offspring. *PLoS One.* 2016 Feb 29;11(2):e0150400.

183. Qi YZ, Yang XS, Jiang YH, Shao LL, Jiang LY, Yang CH. Study of the mechanism underlying the anti-hypertensive effects of *Eucommia ulmoides* and *Tribulus terrestris* based on an analysis of the intestinal microbiota and metabonomics. *Biomed Res Int.* 2020 Nov 5;2020:4261485.

184. Rajendar B, Bharavi K, Rao GS, Kishore PV, Kumar PR, Kumar CS, Patel TP. Protective effect of an aphrodisiac herb *Tribulus terrestris* Linn on cadmium-induced testicular damage. *Indian J Pharmacol.* 2011 Sep;43(5):568–573.

185. Kumari M, Singh P. *Tribulus terrestris* ameliorates metronidazole-induced spermatogenic inhibition and testicular oxidative stress in the laboratory mouse. *Indian J Pharmacol.* 2015 May-Jun;47(3):304–310.

186. Pavin NF, Izaguirry AP, Soares MB, Spiazzi CC, Mendez ASL, Leivas FG, Dos Santos Brum D, Cibin FWS. *Tribulus terrestris* protects against male reproductive damage induced by cyclophosphamide in mice. *Oxid Med Cell Longev.* 2018 Aug 28;2018:5758191.

187. Khaleghi S, Bakhtiari M, Asadmobini A, Esmaeili F. *Tribulus terrestris* extract improves human sperm parameters in vitro. *J Evid Based Complementary Altern Med.* 2017 Jul;22(3):407–412.

188. Salahshoor MR, Abdolmaleki A, Faramarzi A, Jalili C, Shiva R. Does *Tribulus terrestris* improve toxic effect of Malathion on male reproductive parameters? *J Pharm Bioallied Sci.* 2020 Apr-Jun;12(2):183–191.

189. Haghmorad D, Mahmoudi MB, Haghighi P, Alidadiani P, Shahvazian E, Tavasolian P, Hosseini M, Mahmoudi M. Improvement of fertility parameters with *Tribulus terrestris* and *Anacyclus pyrethrum* treatment in male rats. *Int Braz J Urol.* 2019 Sep-Oct;45(5):1043–1054.

190. Zhang H, Tong WT, Zhang CR, Li JL, Meng H, Yang HG, Chen M. Gross saponin of *Tribulus terrestris* improves erectile dysfunction in type 2 diabetic rats by repairing the endothelial function of the penile corpus cavernosum. *Diabetes Metab Syndr Obes.* 2019 Sep 2;12:1705–1716.

191. Fereydouni Z, Amirinezhad Fard E, Mansouri K., Mohammadi Motlagh HR, Mostafaie A. Saponins from *Tribulus terrestris* L. extract down-regulate the expression of ICAM-1, VCAM-1 and E-selectin in human endothelial cell lines. *Int J Mol Cell Med.* 2020 Winter;9(1):73–83.

192. Qiu M, An M, Bian M, Yu S, Liu C, Liu Q. Terrestrosin D from *Tribulus terrestris* attenuates bleomycin-induced inflammation and suppresses fibrotic changes in the lungs of mice. *Pharm Biol.* 2019 Dec;57(1):694–700.

193. Park YJ, Cho YR, Oh JS, Ahn EK. Effects of *Tribulus terrestris* on monosodium iodoacetate-induced osteoarthritis pain in rats. *Mol Med Rep.* 2017 Oct;16(4):5303–5311.

194. Lee HH, Ahn EK, Hong SS, Oh JS. Anti-inflammatory effect of tribulusamide D isolated from *Tribulus terrestris* in lipopolysaccharide-stimulated RAW264.7 macrophages. *Mol Med Rep.* 2017 Oct;16(4):4421–4428.

195. Kaushik J, Tandon S, Bhardwaj R, Kaur T, Singla SK, Kumar J, Tandon C. Delving into the antiurolithiatic potential of *Tribulus terrestris* extract through – in vivo efficacy and preclinical safety investigations in Wistar rats. *Sci Rep.* 2019 Nov 4;9(1):15969.

196. Wang Y, Zhao H, Liu Y, Guo W, Bao Y, Zhang M, Xu T, Xie S, Liu X, Xu Y. GC-MS-based metabolomics to reveal the protective effect of gross saponins of *Tribulus terrestris* fruit against ischemic stroke in rat. *Molecules.* 2019 Feb 22;24(4):793.

197. Yadav HN, Sharma US, Singh S, Gupta YK. Effect of *Tribulus terrestris* in mercuric chloride-induced renal accumulation of mercury and nephrotoxicity in rat. *J Adv Pharm Technol Res.* 2019 Jul-Sep;10(3):132–137.

198. Arjmand A, Abedi B, Hosseini SA. Anti-apoptotic effects of resistance training and *Tribulus terrestris* consumption in the heart tissue of rats exposed to stanozolol. *Eurasian J Med.* 2021 Jun;53(2):79–84.

199. Ştefănescu R, Farczadi L, Huţanu A, Ősz BE, Măruşteri M, Negroiu A, Vari CE. *Tribulus terrestris* efficacy and safety concerns in diabetes and erectile dysfunction, assessed in an experimental model. *Plants (Basel).* 2021 Apr 10;10(4):744.

200. Nishchal BS, Rai S, Prabhu MN, Ullal SD, Rajeswari S, Gopalakrishna HN. Effect of *Tribulus terrestris* on haloperidol-induced catalepsy in mice. *Indian J Pharm Sci.* 2014 Nov-Dec;76(6):564–567.

201. Raoofi A, Khazaei M, Ghanbari A. Protective effect of hydroalcoholic extract of *Tribulus terrestris* on cisplatin induced renal tissue damage in male mice. *Int J Prev Med.* 2015 Feb 20;6:11.

202. Borran M, Minaiyan M, Zolfaghari B, Mahzouni P. Protective effect of *Tribulus terrestris* fruit extract on cerulein-induced acute pancreatitis in mice. *Avicenna J Phytomed.* 2017 May-Jun;7(3):250–260.

203. Vadakkan K, Vijayanand S, Hemapriya J, Gunasekaran R. Quorum sensing inimical activity of *Tribulus terrestris* against gram negative bacterial pathogens by signalling interference. *3 Biotech.* 2019 Apr;9(4):163.

204. Patel A, Soni A, Siddiqi NJ, Sharma P. An insight into the anticancer mechanism of *Tribulus terrestris* extracts on human breast cancer cells. *3 Biotech.* 2019 Feb;9(2):58.

205. Han R, Yang H, Lu L, Lin L. Tiliroside as a CAXII inhibitor suppresses liver cancer development and modulates E2Fs/Caspase-3 axis. *Sci Rep.* 2021 Apr 21;11(1):8626.

206. Yuan Z, Du W, He X, Zhang D, He W. *Tribulus terrestris* ameliorates oxidative stress-induced ARPE-19 cell injury through the PI3K/Akt-Nrf2 signaling pathway. *Oxid Med Cell Longev.* 2020 Jul 28;2020:7962393.

207. Sailaja KV, Shivaranjani VL, Poornima H, Rahamathulla SB, Devi KL. Protective effect of *Tribulus terrestris* L. fruit aqueous extract on lipid profile and oxidative stress in isoproterenol induced myocardial necrosis in male albino Wistar rats. *EXCLI J.* 2013 May 6;12:373–383.

208. Zhao WR, Shi WT, Zhang J, Zhang KY, Qing Y, Tang JY, Chen XL, Zhou ZY. Tribulus terrestris L. extract protects against lipopolysaccharide-induced inflammation in RAW 264.7 macrophage and zebrafish via inhibition of Akt/MAPKs and NF-κB/iNOS-NO signaling pathways. Evid Based Complement Alternat Med: eCAM. 2021;2021:6628561.

209. Kang SY, Jung HW, Nam JH, Kim WK, Kang JS, Kim YH, Cho CW, Cho CW, Park YK, Bae HS. Effects of the fruit extract of *Tribulus terrestris* on skin inflammation in mice with oxazolone-induced atopic dermatitis through regulation of calcium channels, orai-1 and TRPV3, and mast cell activation. *Evid Based Complement Alternat Med.* 2017;2017:8312946.

210. Zhang S, Li H, Yang SJ. Tribulosin protects rat hearts from ischemia/reperfusion injury. *Acta Pharmacol Sin.* 2010 Jun;31(6):671–678.

211. Basaiyye SS, Naoghare PK, Kanojiya S, Bafana A, Arrigo P, Krishnamurthi K, Sivanesan S. Molecular mechanism of apoptosis induction in Jurkat E6-1 cells by *Tribulus terrestris* alkaloids extract. *J Tradit Complement Med.* 2017 Dec 14;8(3):410–419.

212. Saleem U, Chaudhary Z, Raza Z, Shah S, Rahman MU, Zaib P, Ahmad B. Anti-Parkinson's activity of *Tribulus terrestris* via modulation of AChE, α-synuclein, TNF-α, and IL-1β. *ACS Omega.* 2020 Sep 22;5(39):25216–25227.

213. Hong S, Moon MN, Im EK, Won JS, Yoo JH, Kim O. Anti-coccidial activity of the ethanol extract of *Tribulus terrestris* fruits on *Eimeria tenella*. *Lab Anim Res.* 2018 Mar;34(1):44–47.

214. Ma Y, Guo Z, Wang X. *Tribulus terrestris* extracts alleviate muscle damage and promote anaerobic performance of trained male boxers and its mechanisms: roles of androgen, IGF-1, and IGF binding protein-3. *J Sport Health Sci.* 2017 Dec;6(4):474–481.

215. Soleimanpour S, Sedighinia FS, Safipour Afshar A, Zarif R, Ghazvini K. Antibacterial activity of *Tribulus terrestris* and its synergistic effect with *Capsella bursa-pastoris* and *Glycyrrhiza glabra* against oral pathogens: an in-vitro study. *Avicenna J Phytomed.* 2015 May-Jun;5(3):210–217.

216. Kaushik J, Tandon S, Bhardwaj R, Kaur T, Singla SK, Kumar J, Tandon C. Delving into the antiurolithiatic potential of *Tribulus terrestris* extract through – in vivo efficacy and preclinical safety investigations in Wistar rats. *Sci Rep.* 2019 Nov 4;9(1):15969.

217. EI-Shaibany A, Al-Habori M, Al-Tahami B. Anti-hyperglycaemic activity of *Tribulusterrestris* L. aerial part extract in glucose loaded normal rabbits. *Trop J Pharm Res.* 2015;14(12):2263–2268.

218. Abudayyak M, Jannuzzi AT, Özhan G, Alpertunga B. Investigation on the toxic potential of *Tribulus terrestris in vitro. Pharm Biol.* 2015;53(4):469–476.

219. Qari SH, El-Assouli SM. Evaluation of cytological and genetic effects of *Tribulus terrestris* fruit aqueous extract on cultured human lymphocytes. *Saudi J Biol Sci.* 2019 Jan;26(1):91–95.

220. Murthy AR, Dubey SD, Tripathi K. Anti-hypertensive effect of Gokshura (*Tribulus terrestris* Linn.) – a clinical study. *Anc Sci Life.* 2000 Jan;19(3–4):139–145.

221. Gama CR, Lasmar R, Gama GF, Abreu CS, Nunes CP, Geller M, Oliveira L, Santos A. Clinical assessment of *Tribulus terrestris* extract in the treatment of female sexual dysfunction. *Clin Med Insights Womens Health.* 2014 Dec 22;7:45–50.

222. Taavoni S, Ekbatani NN, Haghani H. Effect of *Tribulus terrestris*, ginger, saffron, and Cinnamomum on menopausal symptoms: a randomised, placebo-controlled clinical trial. *Prz Menopauzalny.* 2017 Mar;16(1):19–22.

223. https://clinicaltrials.gov/ct2/show/NCT04124640

224. Akhtari E, Raisi F, Keshavarz M, Hosseini H, Sohrabvand F, Bioos S, Kamalinejad M, Ghobadi A. *Tribulus terrestris* for treatment of sexual dysfunction in women: randomized double-blind placebo – controlled study. *Daru.* 2014 Apr 28;22(1):40.

225. Tadayon M, Shojaee M, Afshari P, Moghimipour E, Haghighizadeh MH. The effect of hydro-alcohol extract of *Tribulus terrestris* on sexual satisfaction in postmenopause women: a double-blind randomized placebo-controlled trial. *J Family Med Prim Care.* 2018 Sep-Oct;7(5):888–892.

226. Sansalone S, Leonardi R, Antonini G, Vitarelli A, Vespasiani G, Basic D, Morgia G, Cimino S, Russo GI. Alga *Ecklonia bicyclis*, *Tribulus terrestris*, and glucosamine oligosaccharide improve erectile function, sexual quality of life, and ejaculation function in patients with moderate mild-moderate erectile dysfunction: a prospective, randomized, placebo-controlled, single-blinded study. *Biomed Res Int.* 2014;2014:121396.

227. Sellandi TM, Thakar AB, Baghel MS. Clinical study of *Tribulus terrestris* Linn. in oligozoospermia: a double blind study. *Ayu.* 2012 Jul;33(3):356–364.

228. Palacios S, Soler E, Ramírez M, Lilue M, Khorsandi D, Losa F. Effect of a multi-ingredient based food supplement on sexual function in women with low sexual desire. *BMC Womens Health.* 2019 Apr 30;19(1):58.

229. Ebrahimpour N, Khazaneha M, Mehrbani M, Rayegan P, Raeiszadeh M. Efficacy of herbal based syrup on male sexual experiences: a double-blind randomized clinical trial. *J Tradit Complement Med.* 2020 Jan 17;11(2):103–108.

230. Iacono F, Prezioso D, Illiano E, Romeo G, Ruffo A, Amato B. Sexual asthenia: Tradamixina versus Tadalafil 5 mg daily. *BMC Surg.* 2012;12(Suppl 1):S23.

231. Balasubramani SP, Murugan R, Ravikumar K, Venkatasubramanian P. Development of ITS sequence based molecular marker to distinguish, *Tribulus terrestris* L. (Zygophyllaceae) from its adulterants. *Fitoterapia.* 2010 Sep;81(6):503–508.

232. Kevalia J, Patel B. Identification of fruits of *Tribulus terrestris* Linn. and *Pedalium murex* Linn.: a pharmacognostical approach. *Ayu.* 2011 Oct;32(4):550–553.

233. Ali NA, Jülich WD, Kusnick C, Lindequist U. Screening of Yemeni medicinal plants for antibacterial and cytotoxic activities. *J Ethnopharmacol.* 2001 Feb;74(2):173–179.

234. Gauthaman K, Adaikan PG, Prasad RN. Aphrodisiac properties of *Tribulus terrestris* extract (Protodioscin) in normal and castrated rats. *Life Sci.* 2002 Aug 9;71(12):1385–1396.

235. Gauthaman K, Ganesan AP. The hormonal effects of *Tribulus terrestris* and its role in the management of male erectile dysfunction – an evaluation using primates, rabbit and rat. *Phytomedicine.* 2008 Jan;15(1–2):44–54.

# 6 Standardization and Quality Control of Crude Drugs

*Ahmed Al-Harrasi*
Natural and Medical Sciences Research Center, University of Nizwa

*Saurabh Bhatia*
Natural and Medical Sciences Research Center, University of Nizwa
School of Health Sciences, University of Petroleum and Energy Studies

*Deepak Kaushik*
M. D. University

*Tapan Behl*
Chitkara University

*Sridevi Chigurupati*
Qassim University

## 6.1 INTRODUCTION

Owing to the scientific validation, less side effects, easy availability, cost-effectiveness, traditional and widespread utilization, multicomponent action, potent therapeutic outcomes, and convenience to use, recently, demand for crude drugs has experienced a substantial shift in recent years. In modern pharmacognosy, major emphasis has been given on assessing quality, purity, potency, safety and efficacy of crude drugs, standardized extracts, and pure phytopharmaceuticals. Standardization is defined as the method of assessing the quality and purity of crude drug materials by using certain factors such as morphological, microscopical, physical, chemical, and biological studies [1–33]. Standardization of crude drugs is a lengthy process that requires the involvement of various laboratories such as pharmacognosy, pharmacology, analytical chemistry, and microbiology; however, owing to the advancements in these fields, especially in analytical chemistry, it's now possible to standardize the crude drug material in a desirable time frame with a high degree of accuracy. Considering the safety of end consumer, crude drugs or botanicals or phytopharmaceuticals or their formulations are assessed as pharmacopeial specifications without compromising their therapeutic potential. Crude drugs or phytopharmaceuticals or botanicals or extracts are considered samples that are coded properly for their standardization and quality evaluation. Based on nature, the samples are assessed by various parameters for their identity, purity, potency, safety, and microbial limits. This is one of the most popular classes of crude drugs. Essential oils have been used for various health benefits since primordial time. Recently, the role of essential oil in treating various respiratory complications has been documented, especially for various types of respiratory infections. Natural polymers derived from plant, algal, and mammalian sources are considered another class of crude drugs that have been recently studied for their important role in different drug delivery systems. One of the most important drug delivery systems, which has been studied by using these polymers, is a nano-based carrier in order to achieve the targeted delivery, avoid wastage of

DOI: 10.1201/9781003274124-6

drug, and achieve better therapeutic effects with a satisfactory pharmacokinetic profile without any toxicity. However, owing to further benefits, crude drugs are always exploited, which considerably impact their sources and can lead to their extinction. Recent developments in plant tissue culture science have enabled us to culture plants considering their high medicinal value under in vitro condition to prevent their extinction from the nature, for genetic manipulation, and to maintain the supply of raw materials to several herbal industries. The major challenges associated with crude drugs are the lack of quality control and the violation of following standardization protocols which lead to the development of plant materials with compromised quality. This chapter demonstrates quality control measures that must be considered to maintain the quality of crude drugs [1–33].

## 6.2 IDENTIFICATION OF THE PLANT MATERIAL

Plant materials when processed offer various present challenges in terms of botanical identification, and as per the World Health Organization (WHO) guidelines, the use of incorrect species is a threat to consumer safety. According to the WHO, between 65 and 80% of the populations of developing countries currently use medicinal plants as remedies [34, 35]. Owing to the high utilization of medicinal plants, the WHO issued the Monographs on Selected Medicinal Plants volumes 1–5 from 1999 to 2010; these volumes contain a list of species with recognized medicinal benefits and the accepted means to correctly use them. In addition, many countries have their own bodies to regulate the use of medicinal plants such as Brazil that has its own agency named National Health Surveillance Agency (from the Portuguese ANVISA—Agência Nacional de Vigilância Sanitária). After processing, plant materials are generally kept in large bags or boxes, and thus, buying local herbal products presents a high chance of acquiring counterfeited, substituted, and/ or adulterated products. Owing to such reasons, a reliable procedure to verify products is required. DNA-barcoding method that relies on short, standardized regions of the genome has emerged as a more reliable technique for a species-level identification of plants [36–39].

Proper documentation and categorization of plant materials is important for the effective conservation as well as to confirm their continuous use. With the aid of conventional as well as molecular taxonomy, a plant specimen can be examined. Earlier, a taxonomist used to identify plant specimen by making a herbarium to identify the plant based on its microscopic and macroscopic features. Owing to the development of molecular taxonomy, especially the availability of molecular markers to examine genetic variation within a population (intra) and between species (inter), it's now possible to scientifically identify plant specimen more accurately. These markers can be used for inter- or intra-species variation assessment, e.g., random amplified polymorphic DNA (RAPD) and inter-simple sequence repeat (ISSR) markers, which are primarily used for intra-species variation; however, RAPD is not recommended for solving taxonomical issues, whereas single nucleotide polymorphisms (SNPs) markers, identified via fingerprinting methods such as amplified fragment length polymorphism (AFLP) or Sanger sequencing, are the most suggested [40, 41].

Thus, gene sequencing methods (such as DNA fingerprinting and DNA barcoding) save effort, time and money, which are considered more accurate. Morphological, taxonomical, microscopical, gene sequencing, and chemical investigations help in establishing data to identify plants. Each investigation has its own importance to establish chemical, genetic, morphological, taxonomical, and microscopical profiles of plants which can further help in identifying it. This database helps in distinguishing the plant from possible adulterants, substituents, foreign matter, closely related plant species, and other materials with similar properties like plants, which can ultimately prevent unnecessary adulteration and substitution of crude drugs. To identify crude drugs, usually morphological, microscopical, and phytochemical investigations are sufficient to explore their organoleptic, macroscopical, histological, and chemical characteristics. Certain reports, websites (database), and books can be followed to identify the plant specimen. However, because of the deterioration of crude drug samples, sometimes it's not possible to examine those by referring to such sources.

Primary analysis, including the evaluation of extractive as well as ash values, the determination of the percentage of purity along with safety profiles as well as microbial limit must be done.

## 6.3   DETERMINATION OF PURITY OF CRUDE DRUG MATERIALS

During crude drug quality assessment, purity is one of the most important parameters, which ensures the authenticity of the material in context with the percentage of adulterants or substituents in the given crude drug material. For the purity assessment of the given plant material or crude drug, many parameters can be assessed. Assessment of macroscopic as well as microscopic features, including histological assessment, can help in the determination of the purity. In addition, physical as well as chemical analysis of the samples by using different analytical tools such as spectroscopic as well as chromatographic techniques have been used frequently for determining the purity of the samples [42, 43].

## 6.4   ASSESSMENT OF THE POTENCY OF THE CRUDE DRUG

Potency of the given crude drug sample can be determined by assessing its therapeutic potential via different in vitro or in vivo experimental models. In Addition, some in silico approaches have been used for the identification of molecular targets and to determine the binding affinity of individual phytochemical compounds (ligand–receptor interactions). These approaches such as molecular docking, structure–activity relationship (quantitative or qualitative) based on computational assessment, pharmacophore production, functional genomic screening, and other computational tools will help not only in screening of potent phytochemicals but also determining their levels of potency by assessing their binding affinities with suitable targets [45]. Pharmacological assessment by using animal models (using different types of animals) and in vitro approaches (animal or human cell lines or organoids or tissue) have been used to assess the therapeutic effectiveness of the crude drugs [46]. Depending upon the chemical profile, therapeutic effectiveness may vary. Usually, the proportion of the most potent phytochemical determines the therapeutic effectiveness of the crude drug; however, sometime, synergistic interactions among the chemical components also determine the therapeutic efficacy. For instance, the level of tannins determines the intensity of astringent effect, the level of bitter compounds (e.g., quinine) determines the bitterness, the level of saponins determines haemolytic activity, and the level of certain alkaloidal content determines cytotoxicity, and so on.

## 6.5   SAFETY PROFILE OF CRUDE DRUGS

Safety of crude drugs is dependent on chemical profile, elemental composition, microbial content, moisture content, extent of contamination, etc. The safety profile of the crude drug or any plant materials must fulfil certain criteria of identity, purity and potency so as to be categorized as safe. Elemental compositions of the crude drugs that represent more heavy metals than permissible limits are not considered safe. Similarly, the presence of any adulterant or substituent of a crude drug or its mixing with any substandard or spurious materials also impacts the safety profile of the crude drug. Microbial spoilage or contamination also impacts the chemical nature of the crude drug that can lead to either decline in the potency or increase the toxicity of the crude drug. Changes in the storage conditions also impact the shelf life of the crude drug, which can subsequently impact the safety profile of the crude drug. Similarly, packaging materials used for the crude drug also impact its stability as well as shelf life. The type of extraction method used for extracting the suitable fraction can also impact the chemical composition that can further impact shelf life as well as safety profile of the drug. Various approaches have been used to assess the toxicity of extracts such as acute, sub-acute, and chronic toxicity. Other factors such as absolute levels of organic chlorides and organic phosphorous, levels of heavy metals, microbial or radioactive or any pollutant-related contaminants or toxic compounds, including aflatoxins, must be re-examined as per the WHO guidelines [47, 48].

## 6.6   PRESCRIBED MICROBIAL LIMITS FOR CRUDE DRUGS

Owing to the improper storage conditions, plant-based raw materials become more vulnerable to the microbial attack of both pathogenic and non-pathogenic origins. For crude drugs, limits for Enterobacteriaceae, Salmonella, and overall aerobes have been already recommended by WHO. Crude drug materials, especially those which are meant for internal use such as tragacanth, must be free from pathogenic microorganisms such as *Escherichia coli*. Maximum microbial limit for crude drugs for topical applications is always greater than that used in internal applications. Various in vitro test procedures can be utilized to detect the presence of pathogenic as well as non-pathogenic microbes in crud drug materials. Various procedures, including drying procedure, extraction method, storage conditions, inherent nature of plant materials or extracts like water retention, pH, chemical composition (e.g., some components can easily oxidize than others), can impact the level in crud drug samples [49].

## 6.7   QUALITY ASSESSMENT OF CRUDE DRUGS

Assessment of crude drug materials is based on the examination of crude drugs on the basis of their identity, quality, purity, potency, safety, efficacy, reliability, and reproducibility. Quality assessment of crud drug materials is usually performed on the basis of five different methods as mentioned below. In addition, natural products-based chemical fingerprinting has been used to determine the complex chemical profile that has now become one of the most convincing tools for the quality assessment of natural products [50]:

* Organoleptic or macroscopic characters-based assessment
* Microscopical or histological assessment
* Physical assessment
* Assessment of chemical profile
* Assessment of therapeutic efficacy

### 6.7.1   Organoleptic or Macroscopic Characters-Based Assessment

Organoleptic or macroscopic evaluation is generally performed on the basis of macroscopic features that can be assessed by using our own senses; thus, it is also called sensory evaluation. This qualitative assessment involves characteristics such as colour, odour, taste, shape, size, shape, texture, and fracture. It's also known as an approach that is based on the assessment via organs of sense or a macroscopic form of the crud drug material to assess its identity as well as quality. Two forms of the crude drugs organized (have defined shape) and unorganized (doesn't have defined shape) are usually assessed on the basis of the sensory characteristics in order to determine the quality of the crude drug. Different organs such as leaves, stems, barks, roots, stolons, rhizomes, fruits, and flowers are assessed on the basis of shape, size, taste, odour, colour, texture, fracture, etc. to determine the quality of the sample. Usually, the type of the crude drug determines the method used for the assessment, e.g., for fruits, including cardamom and seeds such as nux vomica, and characteristics such as shape, size, surface are more valuable sensory features. Similarly for the flowers, including chamomile, the arrangements of various whorls and inflorescence types are more important, whereas size and shape are more important in the case of barks, roots, rhizomes, and stolons. For the examination of leaf sample part from the characters of size and shape, venation patterns as well as margins are also considered [50–52].

### 6.7.2   Microscopical or Histological Assessment

With the aid of advancements in microscopy such as transmission and scanning electron microscopy, confocal microscopy, phase contrast microscopy, it is not only easy to study, at the cellular level, plant's inherent microscopic features such as plant development but also to study plant-based infection (plant–pathogen interactions). These advance microscopes can also be utilized to assess plant's anatomy deeper level to assess the quality of plant materials. However, still traditional approaches by using a compound microscope to study the histological characters of the plant in order to equate its characteristics with the original plant are often performed. This microscopy-based examination of plant tissues requires section cutting as well as stunning of the specimen. A fresh crude drug sample is usually preferred, and for hard tissues or an older drug, it is always advisable to soak the sample in water during the night to make it smooth enough for taking microscopical sections. Microtome-based sectioning of the plant tissue offers more thinness with reproducible results. The different sections of the plant can be investigated from different angles such as transverse (TS), longitudinal (LS), radial longitudinal (RLS), or tangential longitudinal (TLS) sections. Very thin sections are usually preferred and further exposed to staining reagents to investigate the chemical deposition of the cellular organization [53].

Frequently used staining reagents such as phloroglucinol-HCl (for lignified tissues), chlorozinc iodide (for cellulose containing tissues), tincture of alkanna (for suberized or cuticularized walls), or ruthenium red (for mucilage as well as gums containing drugs). Basic microscopic features of the plant organs such as stems, roots, and leaves vary, and this variation can be microscopically tracked via using histological images. Since decades, a compound microscope has been used (10× lower magnification and 45× higher magnification) for the microscopic analysis of the sample. For comparison, permanent glass slides or reference books can be used as standard and to compare the histological features of the sample with standard. For non-protoplasmic materials, including starch granules, aleurone grains, fixed oil, calcium oxalate crystals, mucilage, or even for certain secondary metabolites localization, a microscopic examination is usually performed. In addition, various leaf-associated parameters such as stomatal number, stomatal index, palisade ratio, vein islet, and vein termination number that are called leaf surface constants are determined by coupling camera lucida and microscope. These quality parameters act as a standard for assessing the authenticity of leaf-based crude drugs, especially for assessing any adulterants present in the sample. Moreover, the microscopic examination helps in determining the authenticity of powdered drugs which is called as powdered crude drug microscopy. Powdered characteristic of the crude drug consists of cellular matter in groups or individually which can be equated with microscopic features of the crude drugs [54].

### 6.7.3 Physical Assessment

Apart from its quality assessment on the basis of its chemical profile, crude drugs are also assessed on the basis of its physical properties of the chemical components. This type of quality assessment on the basis of physical characteristics is called physical evaluation. There are different types of physical evaluation as mentioned below [55]:

- Evaluation based on foreign material
- Evaluation based on ash values
- Evaluation based on extractive values
- Evaluation based on chromatographic methods
- Evaluation based on physical constants
- Evaluation based on spectroscopic methods.

There are various procedures that can be used for the physical assessment of crude drugs:

- Foreign materials (Foreign organic matter)
- Microbial contamination
- Moisture content
- Ash values (total ash, acid insoluble ash, water-soluble ash, sulphated ash)
- Extractive values (water-soluble extractive, ethanol-soluble extractive, petroleum ether-soluble extractive, ether-soluble extractive)
- Volatile oil determination

- Physical constants (melting, boiling, freezing points refractive index optical rotation)
- Chromatographic methods (thin-layer chromatography [TLC], high performance liquid chromatography [LC], gas chromatography, high-performance thin- layer chromatography [HPTLC])
- Spectroscopic methods (ultraviolet–visible [UV–Vis] spectroscopy, infrared [IR] spectroscopy, $H^1$ and $C^{13}$ nuclear magnetic resonance [NMR] spectroscopy, mass spectroscopy, X-ray crystallographic)

It's quite challenging to discuss all the physical parameters of crude drugs; thus, some important parameters are discussed below. Foreign matter is a substance which doesn't comply with the standard or authentic or official drugs. Foreign matter's presence could be owing to deliberate adulterating crude drugs (either to compensate the overall yield or to earn more profit from the market) or an accidental addition of a plant material that closely resembles the parent crude drug especially owing to inadequate harvesting and garbling. Nevertheless, an addition of some foreign matter (organic as well as inorganic) sometime is added intentionally to the expensive crude drugs such as saffron or agarwood oil. In addition, during the processing or storage of the crude drug, foreign matter such as insects, moulds, or animal waste material may get added to the crude drug that can deteriorate the quality of crude drugs. There are some challenges faced during the harvesting of crude drugs, as at times complete pure drug cannot be obtained. In such circumstances, official books like pharmacopoeias must be referred to in order to know the restrictions for the proportion of other plant parts, which can be allowed. However, the crude drug containing any toxic foreign matter or foreign matter beyond the permissible limit must not be considered.

### 6.7.3.1  Level of Moisture in Crude Drugs

Moisture stimulates the activity of enzymes and increases microbial growth, resulted in the deterioration of the sample. Determination of the moisture content is done via heating the crude drug sample at 105°C at constant weight and determines the overall loss in weight. There are several procedures that have been used for the determination of moisture content of crude drug sample such as

- Loss on drying procedure
- Azeotropic distillation procedure
- Karl Fischer procedure

#### 6.7.3.1.1  Loss on Drying

To determine the moisture content, several pharmacopoeias used a standard procedure. Loss on drying could be not only because of the water, but also due to the loss of volatile oil present in the crude drug sample. However, mainly it's due to the loss of the water, and moisture balance has been used to determine the variation between initial and final weight. In addition, moisture in crude drugs can also be determined using gas chromatography–LC [56].

#### 6.7.3.1.2  Azeotropic Distillation Procedure

Azeotropic distillation is performed by using Dean Stark apparatus that will help in determining moisture content. It is performed by using water with solvents, including xylene, toluene, and benzene to yield an azeotropic mixture. Distillation of such a mixture up to the azeotropic level is done. In this process, water present in the crude drug forms an azeotropic mixture with the solvent. After heating, both solvent and water distilled all together and received in a graduated tube. Owing to the density difference, water (high density then solvent) forms a lower layer, and its volume can be measured after the complete distillation [57].

#### 6.7.3.1.3  Karl Fischer Procedure

Karl Fischer procedure is commonly used for the determination of overall moisture, as it is applied to determine a small level of moisture content. A Karl Fischer reagent is made up of iodine, sulphur dioxide, and pyridine in dry methanol. In the presence of water, iodine is reduced by sulphur dioxide resulting in the loss in dark brown colouration that can be further measured by colorimetry [58, 59].

### 6.7.3.2 Ash Value Determination

A crude drug material has both organic and inorganic composition. After heating the crude drug sample unless it will convert into ash, organic matter fully burnt; however, inorganic elements still left. This inorganic elemental composition plays an important role in determining the inorganic composition of a crude drug material that is further equated with the quality of the crude drug, e.g., drugs containing more level of heavy metals are not recommended. So, every plant has its own unique elemental combination with unique proportion that can be related to its habitat as well as quality parameters. In this procedure, the weighed amount of the crude drug is subjected for incineration, where the crude drug is burnt at 450°C. This elemental composition helps in determining quality and purity of the crude drug. The following methods have been used for the determination of ash values.

#### 6.7.3.2.1 Total Ash Value Determination

To get the ash of crude drug materials, organic content and carbon present in the crude drug are subjected to heating at 450°C. Usually, the total ash comprises carbonates, phosphates, silicates, and silica. The total ash can further be utilized to investigate water-soluble and acid-insoluble ash [60, 61].

#### 6.7.3.2.2 Acid-Insoluble Ash

For deriving acid-insoluble ash, total ash can be treated with hydrochloric acid (dilute), to eliminate several inorganic salts and to produce primarily silica in the end-product of acid-insoluble ash [62].

#### 6.7.3.2.3 Water-Soluble Ash

Water-soluble ash is obtained by treating crude drug with water, and then water-soluble fraction is separated to produce water-soluble ash. During this procedure, water-insoluble salts are eliminated to determine the percentage of water-soluble ash contents.

#### 6.7.3.2.4 Sulphated Ash

Sulphated ash is derived by incineration at high temperature (around 600°C) after the treatment with dilute sulphuric acid. This will allow the conversion of all oxides and carbonates to sulphate salt [63].

### 6.7.3.3 Extractive Values

Crude drugs contain various active chemical compounds which are accountable for biological activities. These active chemical components show varied solubility in different solvents. Some are more soluble in polar solvents, whereas others show more solubility in semipolar solvents, and some also show more solubility in nonpolar solvents. Thus, extractive value is defined as overall chemical components solubilized in a given amount of single or multiple solvents. To derive the extract, different conventional as well as advanced extraction procedures have been utilized. Usually, maceration, percolation or continuous percolation, or Soxhlet extraction have been used to derive the extract by treating crud drug samples with suitable solvents. There are different types of extractive values determined to study the quality of the crude drugs as follows [64]:

- Water-soluble extractive values: It's determined for gentian, liquorice, etc., especially, for those drugs that contain water-soluble components such as tannins, sugar, mucilage, and plant acids.
- Alcohol-soluble extractive value: For the drugs that contain alcohol-soluble chemicals such as ginger and valerian.
- Petroleum-ether-soluble extractive or ether soluble extractive values: Such values are determined for such drugs that contain nonpolar solvent-preferred chemicals.

#### 6.7.3.4  Determination of Volatile Oil Content

Certain crude drugs are aromatic, and due to their aromatic characteristic, they are commercialized at large scale. Even such aromatic substances, despite inducing aromatic characteristic, also present a range of biological activities. Thus, it's important to determine the volatile oil content in an order to determine the quality and purity of the given crude drugs. Various extraction procedures ranging from simple distillation to advance distillation process have been used to extract oils from the respective plant sources. The type of plant source and extraction method determine the composition of volatile oil. To know the composition of volatile oil, especially in vapour phase, HS–GC–MS has been used. Plant materials containing essential oil cannot be assessed by soluble extractives approach as several non-volatile components can also be extracted. For the extraction of volatile oil, generally Clevenger's apparatus is used at the laboratory scale. During this process, weighed plant materials, generally fresh leaves, placed on a round bottom flask with a required amount of water along with additional 5–10% glycerine. An addition of glycerine increased the boiling point of water, which will prevent the evaporation of the water and at the same time allow the evaporation of volatile components present in the crude drug. At the end, total volatile oil collected can be measured in the graduated receiver. The design of Clevenger's apparatus can be modified for volatile oils which have greater density than water. Nature of crude drugs and the type of extraction method selected determines the ultimate chemical composition of the volatile oils and the duration of the extraction process. Nevertheless, majority of the oils are hydrodistillated in around 4–5 h. Eventually, oil derived is separated from the water followed by its drying over sodium sulphate (anhydrous) [65, 66].

#### 6.7.3.5  Determination of Physical Constants

It's important to determine the physical properties of crude drugs, especially those are present in the liquid state. Physical properties of oils (fixed as well as volatile oils) or certain phytochemicals (such as alkaloids and glycosides) of mainly active chemical components such as melting, boiling, freezing, or sublimation points are assessed to determine the quality as well as purity. After the separation of phytochemicals and oils, physical parameters are assessed. The presence of any impurity in the form of adulterant or any other foreign substances other than authenticated drug impacts physical properties of oils of phytochemicals. Variation in physical constants indicates either degradation or adulteration of the sample [67–80]. Certain physical parameters that have been used for the crude drug are mentioned below:

- Specific gravity
- Refractive index
- Optical rotation
- Rate of flow
- Melting point
- Boiling point
- Viscosity
- Density

##### 6.7.3.5.1  Detection of Refractive Index

Refractive index can be determined for liquid crude drugs such as essential and fixed oils. The refractive index values for some essential and fixed oils have already been mentioned in pharmacopoeias, and these values can be considered primary standard for detecting any impurity or adulteration or any physical changes during the storage in the sample. These unique values for the authenticated samples represent how the oil responds to and changes the path of light. Alteration in the refractive index indicates that authenticated oil has been adulterated or degraded. This is usually performed by using an Abbe refractometer (at 20°C). Refractive index is determined by calculating the ratio between the velocity of light in air phase and in the oil phase. It can be calculated by the ratio of sine of the angle of incidence by the sine of the angle of refraction.

### 6.7.3.5.2   Detection of Optical Rotation

Many oils contain optically active substances those are having ability to rotate the plane of polarized light either towards right (dextrorotatory) or left side (laevorotatory). Thus, the detection of optical rotation is important for determining the presence of optical activity as well as of active substance in the crude drug sample. When oils are adulterated or degraded, their optical activities also get impacted. Thus, it's important to determine the optical activity of the oils in an order to determine its purity as well as quality. The optical rotation of the sample can be determined by using a polarimeter (at 20°C). Optical rotation can be determined by using the following formula: angular rotation per dm of solution/grams of substance per ml of solution=$100\alpha/Ldp$, where $\alpha$ is the observed rotation, L is the length of observed layer in drug, d is the density, and p is the weight in g of substance in 100 g of solution.

### 6.7.3.5.3   Determination of Rate of Flow

Chemical components present in the crude drug sample contain numerous active components. With an aid of advancement in chromatography, chemical components present in the sample can be separated on the basis of their affinity towards solvent (mobile phase) as well as absorbent (stationary phase). For this process, initially, the extract of the respective crude drug is prepared and suitable mobile phase and stationary phase are determined. For the separation of chemical components, the sample is applied over the stationary phase (as like in TLC). Due to their different affinities towards stationary as well as mobile phase, active chemical components are separated, which form chromatogram. In the quality assessment of the crude drug, TLC has been frequently used to determine quality and purity of the drug. An Rf value (rate of flow) is the distance travelled by the solute divided by the distance travelled by the solvent. Generally, this value varies in between 0 and 1. TLC is frequently used for the quantitative assessment of crude drug as well as its respective phytochemicals. TLC chromatogram is evaluated by comparing the sample Rf value with the standard (known compounds) Rf value. These standards are called as reference standards. Chromatogram can be assessed by using spraying or visualizing reagents mainly to visualize the spots of the compounds. It can be alternatively done by UV light. Various visualizing agents are vanillin–sulphuric acid (1%) for steroids, terpenoids, and Dragendorff's reagents for alkaloidal constituents.

## 6.7.3.6   Spectroscopical Approaches for Crude Drugs

Crude drug properties are associated with their chemical profiles, especially levels of active chemical constituents in parts-per-million/parts-per-billion. These chemical components present characteristic features, due to which characterization is feasible. For quality assessment, it's important to understand chemical composition, especially structural features of chemical components that require extraction, isolation, purification, and characterization. Chromatographic and spectroscopic techniques assist in the isolation as well as characterization of the components by establishing its chromatogram or spectral fingerprint to finally compare with authenticated source. Structural features of the chemical component are based on the arrangement of atoms in the molecule and their interlinking with chemical bonds. Earlier various separation methods such as TLC, LC, gas chromatography, and capillary electrophoresis hyphenated to mass spectrometry were used for the identification, quantification, and structural elucidation of selected chemical components present in a plant sample. Such analytical methods can be used in phytochemical investigations, allowing the fingerprinting of chemical compounds. Various analytical approaches of medicinal plants validation have been reported. Chromatographic and spectroscopic ones are the most common techniques for the analysis of herbal fingerprints. However, some the procedures that have been used earlier are expensive, time-consuming, and destructive as it can affect the quality of the sample. UV and IR spectra are usually complicated and due to the superimposition and presence of several peaks, and spectra derived from these techniques are relatively more complex and thus cannot be easily interpreted. Proton ([1]H) and [13]C NMR spectral analyses present information about the number, nature, and environment of the protons and carbon skeleton in the molecule, respectively. Mass spectrum provides the information about the molecular weight and the fragmentation pattern of the compound [81].

### 6.7.3.6.1   IR Analysis

IR spectroscopy is one of the most reliable approaches that are based on the vibrations of the atoms of a molecule. This technique is accurate, non-destructive, fast, and easy to use, and a small sample is required without any reagents. The tendency of the atoms to continuously vibrate and stretch as well as bend their bonds via absorbing IR light can be utilized to understand the nature or identity of molecules of any substance. The IR region between 1430 and 910 cm$^{-1}$ is known as "fingerprint" zone where several other bending vibrations of the atoms can be observed. Comparison between the two samples (authenticated and test sample), in the fingerprint region, presents a definite identification of compounds [82].

Vibrational spectroscopic methods have been used to depict the spatial distribution of chemical components and thus utilized to detect as well as characterize chemical components of plant tissue samples up to a resolution of around 1.2 μm, applying near-IR (NIR) and mid-IR (MIR) spectroscopy. The underlying principle is the absorption of IR radiations via vibrational transitions in covalent bonds that allow a better analysis of samples, with higher resolution. The most suitable spectroscopic approach is vibrational spectroscopy, although techniques based on the spectral study cannot offer data about the chemical makeup of the sample. However, coupling between spectral and chemometric tools gives more detailed extracts information that could be beneficial for determining the authenticity of the sample. For fingerprinting analysis, it's important to study the unique information about the phytochemical, especially specific chemical interaction among atoms and behaviour of bonds that will help in a rapid identification of the sample [83].

Spectroscopic methods by means of the NIR wavelength region of the electromagnetic spectrum are used in the herbal industry to assess its quality. For crude drug analysis, NIR, MIR, Fourier transform infrared (FTIR), and attenuated total reflection (ATR) have been used. Coupling between these IR-based tools with chemometric methods could offer potent technique for quick and non-destructive technique for the analysis. These spectra-based fingerprints generated by using these tools could be more effective in determining the chemical composition of the given sample more precisely. Spectral signals generated by different chemical components present in the crude sample offer unique fingerprint that can be further utilized to determine the quality and the purity of the sample. A combination of IR with chemometric methods such as multivariate statistical approaches is beneficial for handling IR spectra. Using this approach, information of IR spectra can be extracted to study this spectral information for qualitative or quantitative purposes [84, 85].

### 6.7.3.6.2   Ultraviolet–Visible (UV–Vis) Analysis

UV spectroscopy is a simple, inexpensive, and non-destructive method that has been used in herbal industry for various types of analysis. This technique is more suitable for the analysis of liquid samples. From UV analysis, it has been noticed that peaks derived from complex composition such as liquid extract are usually more complex. This complex spectral signal due to the superimposition of peaks of different chemical components makes the data more complex to interpret. Thus, recently UV–Vis in combination with chemometric techniques has been used for assessing the quality of herbal extracts. Chemometric software and statistical approaches require mathematical and statistical techniques to obtain valuable data from complex spectra. Chemometric approaches, including principal component analysis and support vector machines, have been used to derive UV–Vis chemical fingerprint of the herbal preparation [86].

Generally, oil-based preparations (to evaluate anisidine value, peroxide value) or colourful extracts containing pigments or algal extracts containing pigments can be assessed easily. Generally, UV–Vis spectroscopy approach has been used to assess the quality as well as the purity of herbal extracts those contain colourful substances. Thus, variation in visible colours can be correlated with quality that can be assessed by using this approach. Usually in this process, an absorption spectrum of organic compounds is correlated with the quality of the sample. Crude extract or the phytochemicals containing chromophore, auxochrome, and conjugated system are solubilized in the suitable solvent. The UV–Vis spectrum representing absorption band can be assessed. Generally, non-defective as well as non-edible oils can be assessed in between 300 and 1000 nm. A 380–450 nm range belongs to carotenoid pigments, while chlorophylls and pheophytin present an absorption band at 650–700 nm. A UV spectrum of camphor displays absorption maximum at 290 nm, which represents its tendency to absorb light energy in the UV region.

### 6.7.3.6.3    Nuclear Magnetic Resonance (NMR) Spectrum

NMR-based spectral analysis presents considerable merits such as easy to use, quick sample analysis, less time taken in sample preparation, non-destructive, and reproducible technique, and due to all these advantages, quantitative analysis can be done by using a single internal standard. NMR spectroscopy offers data for rapid qualitative as well as quantitative assessment of endogenous and exogenous metabolites. For metabolite profiling of plant-based extracts, this robust, fast, and reliable technique has been used to confirm the quality of herbal products. NMR is not as precise as other analytical tools, including LC and GC–MS, but it's more non-destructive as well as reproducible and presents a far broader range. For NMR metabolomics study, only a little amount (10–50 mg) of a sample is required and this amount is sufficient to produce a 1H-NMR spectrum in 10 min, which comprises around 50–100 metabolites. And it's generally feasible to detect 10–20 established components in this region [87]. The NMR spectra are documented in deuterated NMR solvents by using an NMR spectrometer (initially at 400 MHz or greater). Spectral data analysis is usually based on dividing the spectral data into separate regions (roughly 0.01–0.04 ppm in width). These distinct areas, in turn, are converted into a list of values for each spectrum. These days, NMR spectral analysis is done with the help of advance algorithms that line up the peaks, remove unwanted variation, the use of all data points [88]. Several statistical algorithms tools such as multivariate have been used to study NMR spectra [89]. Coupling of NMR with various chromatographic techniques and chemometric methods has been done for the effective separation and structural elucidation of unknown compounds present in the sample [90–93].

### 6.7.3.6.4    Mass Spectral Analysis

Mass spectrometry (MS) can be used to assess valuable structural data of phytochemicals present in herbal extracts. A general procedure of MS includes the bombardment of molecules with electrons to ionize them during the process of fragmentation. Fragmentation patterns suggest the mass-to-charge (m/e) value. Using MS, high-throughput screening could be done to screen lead components from extract. These days, novel chinmedomics MS-based approaches have been used for screening active components. Tandem MS/MS approaches have been used in the identification of metabolites. In herbal industry, MS has been utilized to assess the structural features of active chemical components present in plant extract to offer lead components. An MS tool has been effectively utilized for determining correct mass, molecular formula, intensities, and respective peaks of the active chemical components. To improve sensitivity, selectivity, and efficacy, MS coupling with chromatographic techniques could be done, including LC, capillary electrophoresis, and gas chromatography. One of the merits for detecting lead chemical components is screening tool called high-throughput efficacy. Coupling such as MS–MS helps in a quantitative assessment of known components and coupling with an analyser time-of-flight (TOF) provides more accuracy as well as sensitivity with a structural determination of unidentified compounds [94, 95].

### 6.7.3.6.5    X-Ray Diffraction

X-ray diffraction is a robust, direct, rapid, and non-destructive method for complete structure determination. X-ray diffractometry has been used to determine the structural features of a crystalline compound present in the extract. X-ray diffraction peaks obtained from crystalline substance present in the extract helps in the identification of the structural features, as it occurs in nature in a 3D model [96].

### 6.7.4    Chemical Assessment

The chemical assessment of different active phytochemicals is important to know the chemical nature of the extract as well as to establish their chemical profile. However, depending upon the form of extract (solid or liquid), respective procedures are employed to affirm their chemical as well as physical nature. For oils, parameters like saponification value, iodine value, acidity, and peroxide value are used to determine the purity as well as quality of oils. Such parameters have been extensively utilized to assess purity, total unsaturated components, and level of free fatty acid. For certain oils, specific gravity, kinematic viscosity, heating value, density, flash point, pour point, and cetane number have also determined. In certain cases, chemical components present in the plant extract are derivatized, which could be further separated and recovered for further analysis. Several alkaloidal compounds, organic acids, and phenolic components are initially converted to water-soluble salts and then recovered for further analysis. Also, based on the chemical nature, an organic chemical compound can be converted to chelate or colourful complex to allow its detection via colorimetric methods. One of the quickest, inexpensive, and the most conventional approaches to detect the presence of active chemical components in the herbal extract is proximate chemical analysis or preliminary phytochemical screening. In phytochemical screening, different chemical tests have been performed to detect the presence of different phytochemical or secondary metabolites [97–102].

### 6.7.4.1    Phytochemical Screening (for Primary Metabolites Study)

#### 6.7.4.1.1    Carbohydrates Analysis

##### 6.7.4.1.1.1    For Sugar Determination—
Barfoed's test: A test sample in a diluted form is treated with a Barfoed's reagent (cupric acetate with acetic acid) followed by heating in a boiling water bath, which results in the formation of red cupric oxide. This suggests the presence of monosaccharide sugars.

Fehling's solution test: A test sample in diluted form is treated with an equal amount of Fehling's solution A and B followed by heating for some time. This leads to the formation of cuprous oxide brick red ppt, suggesting the presence of reducing sugars.

Molisch's test: A test solution in diluted form is subjected to α-naphthol (alcoholic solution) treatment (few drops) followed by treatment with concentrated sulphuric acid (by adding from the side of the test tube). This results in the formation of a ring (purple) at the intersection lower than the aqueous layer, suggesting the presence of sugars.

Osazone formation: A test sample in diluted form is subjected to the treatment with phenylhydrazine hydrochloride, sodium acetate, and acetic acid, which results in the formation of yellow crystals of osazone. This suggests the presence of sugars (glucose and fructose form the same osazone and glucosazone).

Seliwanoff's test: This test has been used for the estimation of pentose sugars. A test solution (diluted form) is treated with a crystal of resorcinol (1,3-isomer of benzenediol) and subjected to heating followed by the treatment with concentrated hydrochloric acid. This has resulted in the pink colour formation, suggesting the presence of ketose sugars.

Pentose sugar test: A test sample in diluted form is subjected to heating followed by the treatment with phloroglucinol. Afterwards, an equivalent amount of hydrochloric acid has been added. An addition of acid turns solution into red colour which suggests the presence of pentose sugars.

Furfural test: It's a test used to determine the difference among fructose and glucose. In this test, a test solution in diluted form is subjected to heating followed by the treatment with phosphoric acid (a drop). An addition of acid turns the solution into a colourless solution. Afterwards, a small piece of filter paper pre-treated with aniline (10%) in acetic acid is placed over the test tube. Fumes released from the solution convert the paper colour into red, which indicates the following.

##### 6.7.4.1.1.2    For Polysaccharides—
Cellulose: Cellulose, a polysaccharide that forms cell wall and is found in woody materials. A thin section of a sample when treated with a drop of phloroglucinol

and hydrochloric acid presents red colour, suggesting the presence of cellulose. Similarly, when thin section of the sample is treated with iodine solution (0.1 M) and sulphuric acid (80%), it gives blue or bluish violet colour.

Inulin: This water-soluble storage polysaccharide is stored as a food reserve in various plants. A test solution of the sample is treated with $\alpha$-naphthol and sulphuric acid solution resulting in the formation of reddish-brown colour, which indicates the presence of inulin.

Lignin test: This polysaccharide-based natural organic compound contains pentose sugar and forms cell-wall constituent of many crude drugs. The test sample (thin section) when treated with phloroglucinol and hydrochloric acid turns the stained portion pink, indicating the presence of lignin. Similarly, when the test sample in a form of the thin cutting is stained with a safranin reagent, it leads to the formation of the pink stain. Another procedure which is often used is based on the treatment of the thin section with thionine solution for almost 15 min followed by washing with ethanol. Formation of bluish violet stain indicates the presence of lignified cells.

Mucilages: Mucilages are the highly viscous and stocky polysaccharide solution produced by the plants. To test the presence of mucilage sample is usually treated with polycationic dye, a ruthenium red reagent. This treatment presents the development of pink colour. Another procedure involves the treatment of the sample with a thionine reagent for 15 min followed by washing with ethanol. This type of treatment resulted in the development of reddish-purple colour, which indicates the presence of mucilage.

Pectin: This polysaccharide is diversely present in citrus plants containing galacturonic acid. In this test, a thin section of the sample is treated with a toluidine blue O reagent. This type of treatment turns the blue colour to pink or reddish-purple colour.

Starch: Starch is a polysaccharide that is deposited intracellularly as starch grains. Structurally, it's made up of amylose as well as amylopectin. A test sample solution is allowed to treat it with iodine solution (weak form). Treatment with iodine turns the solution colour into blue which suggests the presence of starch. When the same solution is subjected to heating blue colour formed vanishes and once more appears after cooling. The ratio of amylose and amylopectin impacts the colour of the solution. Deep blue coloration often occurs because of the presence of amylose, whereas amylopectin gives purple colour with iodine solution.

Suberin: This polysaccharide forms cuticular cell wall in some crude drugs. A thin section of the test sample when treated with Sudan red solution results in the formation of reddish-orange colour due to the presence of suberized cells.

### 6.7.4.1.2   For Protein Analysis

Biuret test: In this procedure, a test solution is treated with diazonium salt that leads to the development of reddish-orange colour because of the presence of peptide linkage of proteins. Biuret test always comes negative for the sample containing amino acids but not protein.

For testing protein in aleurone grains: In plants, proteins are stored in the form of aleurone grain also called aleurone. Plant cells comprising aleurone form the aleurone layer. For testing protein in the sample, test solution is treated with an alcoholic iodine solution that results in the formation brownish or yellowish-brown colour, indicating the presence of protein. Similarly, to test protein in the sample, a test solution is treated with alcoholic iodine solution, resulting in the formation of brown or yellowish-brown colour, suggesting the presence of proteins. Another procedure includes the treatment of the sample with few drops of alcoholic picric acid solution that results in the formation of yellow colour, suggesting the presence of protein. In one more method, a test sample is treated with mercuric nitrate solution, which results in the formation of brick red colour, indicating the presence of protein.

### 6.7.4.1.3   For Testing Amino Acids

#### 6.7.4.1.3.1   Ninhydrin Test   In this test, a dilute solution of the sample is treated with a ninhydrin reagent and is further subjected to heating resulting in the formation of purple colour that suggests the presence of amino acids.

*6.7.4.1.3.2   Millon's Test*   In this test, a dilute solution of the sample is treated with a Millon's reagent that leads to the formation of precipitates, indicating the presence of amino acids.

*6.7.4.1.3.3   Test for Tyrosine*   A test solution of the sample is allowed to treat with a Millon's reagent, resulting in the formation of dark red colour, which suggests the presence of tyrosine.

*6.7.4.1.3.4   Test for Cysteine*   A test solution of the sample is allowed to treat with a sodium hydroxide solution (40%) followed by the treatment with lead acetate solution (10%). This solution is further subjected to boiling. The formation of black precipitate suggests the presence of cysteine.

*6.7.4.1.3.5   Test for Tryptophan*   A test solution of the sample is allowed to treat with glyox-ylic acid followed by the treatment with concentrated sulphuric acid, resulting in the formation of reddish-purple ring observed at the intersection of the two layers, suggesting the presence of tryptophan.

*6.7.4.1.3.6   Test for Histidine*   A test solution of the sample is treated with diazonium salt, result-ing in the formation of reddish-orange colour, suggesting the presence of histidine.

### *6.7.4.1.4   Fixed Oils and Fats*

By using a nonpolar solvent such as petroleum ether, fixed oils and fats present in the plant sample can be extracted. Since both fixed oils and fats are triglycerides of long-chain fatty acids, they can be tested via the presence of glycerol.

*6.7.4.1.4.1   Saponification Test*   A test sample is allowed to treat with alcoholic potassium hydroxide (0.5 N) followed by the treatment with a phenolphthalein reagent. Afterwards, a solution is subjected to heating which results in the formation of soap, owing to a neutralization of alkali, representing the pres-ence of fatty compounds. In another approach, a test sample is treated with a copper sulphate solution (1%) followed by the treatment with sodium hydroxide (10%) solution. This leads to the formation of a blue colour solution owing to the presence of glycerine. Also, at the same time, it leads to the forma-tion of cupric hydroxide (imparts blue colour) that remains in a solubilized state owing to its solubility in glycerine. In another procedure, a test solution is treated with a small amount of sodium hydrogen sulphate resulting in the production of a pungent smell, representing the presence of glycerine.

### 6.7.5   PHYTOCHEMICAL SCREENING (FOR SECONDARY METABOLITES STUDY)

A plant has in-built bio-machinery to produce an array of secondary metabolites from primary metabolites. Secondary metabolites are produced by using sequential intermediate steps of bioconversion of primary metabolite by using several key enzymes. The expression of these enzymes is controlled by their respective genes. The expression of these genes is controlled by certain biotic and abiotic factors. Thus, an ultimate pace of biosynthetic pathways determines the yield of secondary metabolites. Or the rate of expression of these genes, enzymes, and other rate of involvement of other intermediates in the biosynthetic reaction determines the rate of reaction as well as its yield. Some characteristics of the phytochemical test for some important secondary metabolites are listed below.

### 6.7.5.1   Alkaloids

Alkaloids are nitrogenous basic organic compounds that impart cytotoxic, bitterness, and other therapeutic properties to the plants and can be qualitatively detected via various phytochemical tests. There are three different types of alkaloids:

- True alkaloids: These are the organic compounds that comprise nitrogen in the heterocy-clic ring and derive from amino acids.

- Protoalkaloids: These are the organic compounds that comprise nitrogen but not in the heterocyclic ring and just like true alkaloids they also derive from amino acids.
- Pseudo or false alkaloids: These are the organic compounds that comprise nitrogen in the heterocyclic ring but do not derive from amino acids.

Since protoalkaloids and true alkaloids are derived from amino acids, they can be easily precipitated by alkaloidal reagents. Nevertheless, some pseudoalkaloids, e.g., caffeine, don't give positive results for alkaloids, they are derived from other pathways without an involvement of amino acid. The most common alkaloid-based tests are listed below:

Dragendorff's reagent: It's a solution of potassium iodide, bismuth nitrate, and tartaric acid that produce orange precipitates in the presence of alkaloids.

Mayer's reagent: It's a solution of mercuric chloride and potassium iodide that produce yellow precipitate with the reagent in the presence of alkaloids.

Wagner's reagent: It's an iodine solution that gives brown to a reddish-brown precipitate with the reagent in the presence of alkaloids.

Hager's reagent: It's a picric acid (saturated solution) in the cold water that gives a characteristic crystalline precipitate with reagents in the presence of alkaloids.

Miscellaneous tests for alkaloids give colourful reactions with the following test:

- Bertrand test: By using 5% silicotungstic acid.
- Iodoplatinate test: By using platinum chloride in a mixture of hydrochloric acid and potassium iodide.
- Kraut's test: By using a modified Dragendorff's reagent.
- Marme's test: By using a potassium cadmium iodide solution.
- Marquis test: By using a formaldehyde in concentrated sulphuric acid.
- Scheibler's test: By using a phosphotungstic acid or its sodium salt.
- Sonnenschein's test: By using phosphomolybdic acid.
- Tannic acid test: By using 5% tannic acid.

After the identification of alkaloids phytochemical nature of the extracted fraction is confirmed, a further sample is subjected to identify the type of alkaloid present in the sample. To determine the type of alkaloid, some of the most common tests that have been used so far are enlisted below.

### 6.7.5.1.1  Test for Tropane Alkaloids
Crude drugs containing tropane alkaloids, including hyoscine, hyoscyamine, atropine, cocaine, have been assessed by using the following phytochemical tests:

Schär's test: A test solution of a sample is treated with concentrated sulphuric acid, resulting in the formation of green colour, suggesting the presence of tropane alkaloids.

Vitali Morin test: A test solution is treated with fuming nitric acid followed by the treatment with alcoholic potassium hydroxide, results in the formation of purple colour initially and then turns into red and ultimately disappears, suggesting the presence of tropane alkaloids.

Gerrard's test: Test solution of a sample is treated with mercuric chloride (alcoholic solution) resulting into formation of red colour, suggesting the presence of tropane alkaloids.

### 6.7.5.1.2  Test for Imidazole Alkaloids
Imidazole alkaloids such as pilocarpine that contain an imidazole nucleus can qualitatively be examined by following tests:

Helch's violet colour test: It's a solution of hydrogen peroxide, potassium dichromate, and dilute sulphuric acid that gives blue colour in the presence of imidazole alkaloids.

### 6.7.5.1.3   Test for Quinoline Alkaloids

Quinoline alkaloids containing crude drugs have quinoline derivatives such as quinine, quinidine, cinchonine, cinchonidine which can qualitatively be examined by the following tests:

Thalleioquin test: It's a solution of bromine water and strong ammonia solution that gives emerald, green colour in the presence of quinoline alkaloids.

Herapathite test for quinine: It's a solution of glacial acetic acid, ethanol, concentrated sulphuric acid and iodine which gives green colour with a quinine containing solution.

### 6.7.5.1.4   Test for Opium Alkaloids

Opium alkaloids containing drugs contain organic compounds with an opium nucleus such as morphine and codeine.

Sulphomolybdic acid test: Test solution containing morphine gives violet colour with sulphomolybdic acid indicates the presence of opium alkaloids.

### 6.7.5.1.5   Test for Indole Alkaloids

Alkaloids containing an indole nucleus such as ergotamine, reserpine, strychnine, vinblastine, vincristine are called indole alkaloids. The various followings tests have been performed to detect the presence of indole alkaloids as follows:

van Urk reagent test: It's a solution of *p*-dimethyl amino benzaldehyde and sulphuric acid that presents blue colour in the presence of indole alkaloids.

Miscellaneous test: A test solution containing indole alkaloids (e.g., strychnine) presents violet colour with sulphuric acid, whereas brucine gives yellow colour with nitric acid.

### 6.7.5.1.6   Test for Steroidal Alkaloids

These are the crude drugs such as solanum and veratrum that contain organic compounds those are having nitrogen in steroidal molecules.

Formaldehyde and sulphuric acid test: A test solution containing steroidal alkaloids, e.g., solasodine gives instant red colour with sulphuric acid and formaldehyde.

### 6.7.5.1.7   Test for Purine Alkaloids (Pseudoalkaloids)

This class of alkaloids falls in pseudoalkaloid category such as caffeine, theophylline, theobromine.

Murexide test: A test solution containing a sample is treated with concentrated nitric acid followed by the evaporation and further treatment with ammonium hydroxide solution, resulting in purple colour. The appearance of the purple colour indicates the presence of purine alkaloids.

## 6.7.5.2   Glycosides

Glycosides are the organic compounds that contain one sugar part, namely glycone part, and one aglycone part (non-sugar part). The aglycone part is mainly responsible for biological activity; however, because of the presence of the sugar part, it usually gives the positive results against the Molisch test. Nevertheless, these organic compounds don't give positive results against chemical tests of reducing or nonreducing sugars. For such reasons, glycosides are initially treated with suitable hydrolysing reagents such as alkali, acids, or enzymatic solution to hydrolyse into two separate parts (glycone and aglycone) that can further show the positive findings against chemical tests of the reducing sugars depending upon their chemical nature. The aglycone part of glycosides can be identified by certain specific chemical tests as mentioned below.

### 6.7.5.2.1   Test for Anthraquinone Glycosides

Crude drugs such as senna, cascara, rhubarb, and aloe containing organic compounds those are having anthraquinone nucleus show positive results against the following tests:

Borntrager's test: In this test, a solution containing sample initially extracts with a nonpolar solvent such as chloroform and then is treated with ammonia. An ammonia layer shows pink to red colour, suggesting the existence of anthraquinone glycosides in the test solution.

Modified Borntrager's test: One of the drawbacks of Borntrager's chemical test is that it shows negative results against anthraquinone glycosides (C-type). Thus, to circumvent this, ferric chloride and hydrochloric acid are used. In this test, the sample is treated with a ferric chloride solution followed by the treatment with hydrochloric acid. This solution is subjected to heating on water bath and then is filtered followed by its extraction with carbon tetrachloride. A carbon tetrachloride layer is mixed with water and then is treated with ammonia (dilute). The appearance of cherry red colour in an ammonia layer suggests the presence of anthraquinone glycosides.

### 6.7.5.2.2 Cardiac Glycosides

Cardiac glycosides are the group of the diverse organic compounds that contain steroidal nucleus, which are broadly classified as cardenolides and bufadienolides. These categories are divided on the basis of the presence of a 5- or 6-membered unsaturated lactone ring. Such organic compounds (both classes) present a positive test against all cardiac glycoside's tests.

Balget's test: In this test, leaf lamina is treated with sodium picrate solution that leads to the formation of yellow or orange colour. This colour variation suggests the presence of cardiac glycosides.

Kedde's test: A test solution containing crude drugs is initially extracted with chloroform and then is evaporated till dryness is followed by the treatment with ethanol and 3,5-dinitrobenzoic acid. A solution derived from the above procedure is further treated with a sodium hydroxide solution that results in the development of purple colour, suggesting the presence of cardiac glycosides that contain unsaturated lactone.

Keller Kiliani test: This test is used for testing deoxysugars. A test solution containing a sample is treated with acetic acid, followed by the treatment of ferric chloride with a gradual addition of sulphuric acid (conc.) from the side. This reaction results in the appearance of a reddish-brown colour at the junction, which converts into blue colour after a while, which indicates the presence of cardiac glycosides.

Legal's test: A test solution containing a crude drug sample is treated with pyridine, followed by the subsequent treatment with sodium nitroprusside as well as sodium hydroxide, leads to the development of pink to red colour, which suggests the presence of cardiac glycosides.

Liebermann–Burchard test: A test solution containing a sample is treated with acetic anhydride, followed by the treatment with concentrated sulphuric acid from the side of the test tube, resulting in the appearance of red colour that turns blue and then shows green appearance, which indicates the presence of cardiac glycosides.

Liebermann's reaction: A test solution containing a sample is treated with chloroform, followed by the treatment with an equivalent amount of acetic anhydride and at last with an addition of sulphuric acid (concentrated). This results in the development of blue colour, which indicates the presence of cardiac glycosides.

Raymond's test: A test solution containing a crude drug sample is treated with hot methanol sodium hydroxide, which leads to the development of violet colour, suggesting the presence of cardiac glycosides.

### 6.7.5.2.3 Steroidal and Triterpenoidal Glycosides

Terpenes as well as steroid containing compounds are synthesized via the same route and thus they present all test positive, however, with slight variation due to the different structural features.

Liebermann–Burchard test: In this test, a sample solution is treated with an equivalent amount of acetic anhydride, followed by a treatment with sulphuric acid (concentrated). An addition of sulphuric acid must be done slowly from the side of the test tube to allow a reaction mixture to give results properly. Sap green colour formation indicates the presence of steroids and steroidal glycosides, whereas brown colour suggests the presence of triterpenoids.

Salkowski's test: A test sample solution is treated with sulphuric acid (concentrated). This acidic treatment if forms red in a lower layer indicates the presence of steroids, while the appearance of violet colour suggests the presence of cardiac glycosides.

### 6.7.5.2.4   Saponin Glycosides

These are the triterpene-based compounds that are generally toxic in nature and have the tendency to form froth. The following tests can be used to detect the presence of saponin glycosides.

Foam test: A test solution with a defined volume in a test tube is shaken for 30 s, leading to the formation of stable foam in the presence of saponin glycosides.

Haemolysis test: In this test, a defined volume of a blood sample is treated with a test sample in a normal saline solution. In the presence of saponins, haemolysis of the blood cells occurs.

### 6.7.5.2.5   Cyanogenic Glycosides

Cyanogenic glycosides are organic compounds that release hydrogen cyanide when hydrolysed.

Grignard reaction test: This test is often called sodium picrate test in which a filter paper strip is placed into a picric acid solution (10%), followed by the treatment with a sodium carbonate solution (10%) with subsequent draining with water. Place a little sample of the test sample in a conical flask and moisten it with water. The paper soaked with sodium picrate is placed over the mouth of the flask. After moistening, if the test sample contains cyanogenic glycosides, it produces hydrocyanic acid vapours that change the colour of the paper from yellow to brick or reddish brown due to the reaction between sodium picrate and hydrocyanic acid vapours, resulting in the formation of sodium isopurpurate.

Cuprocyanide test: A filter paper strip is initially treated with a test solution followed by the treatment with copper sulphate, resulting in the formation of distinct stain.

Ferriferrocyanide test: A test solution is treated with ferric sulphate followed by the treatment with ferric chloride as well as hydrochloric acid, resulting in the formation of Prussian blue colour, indicating the presence of cyanogenic glycosides.

Mercuric nitrate test: A test solution when treated with mercuric nitrate forms black precipitate of metallic mercury from the solution of mercuric nitrate in the presence of a cyanogenic glycosides solution.

### 6.7.5.2.6   Flavonoids and Flavonoid Glycosides

Flavonoids (C6–C3–C6) are the polyphenolic antioxidant compounds derived from phenylpropanoids and subclassified as flavones, isoflavones, flavanones, flavonols, chalcones, aurones, etc.

Alkali test: A test sample solution when treated with sodium hydroxide solution shows yellow colour, which is converted into colourless upon an addition of dilute acid, suggesting the presence of flavonoids.

Shinoda test: A test solution of a sample when treated with magnesium turning followed by the treatment with hydrochloric acid (concentrated) shows pink colour, suggesting the presence of flavonoids.

Zinc dust test: A test solution is treated with zinc dust followed by the treatment with hydrochloric acid (concentrated) resulting in red colour, which suggests the presence of flavonoids.

Leucoanthocyanidins: A test solution is treated with hydrochloric acid (2N), followed by the heating, resulting in a red colour solution that suggests the presence of leucoanthocyanidins. However, the development of yellowish-brown colour suggests the presence of catechins.

### 6.7.5.2.7   Coumarin Glycosides

Coumarins are the organic compounds that contain o-coumaric acid lactone and are usually found in glycosides.

Fluorescence test: A test sample solution containing a crude drug sample is extracted with ethanol. A spot of extract obtained is applied over the Whatman filter paper. This spot is examined under UV light. The appearance of blue, blue-green, or violet fluorescence suggests the presence of coumarins.

## 6.7.5.3   Tannins

Tannins often are called proanthocyanidins. These are the complex organic compounds containing polyphenolic nucleus. These compounds are divided as hydrolysable and condensed tannins. Based on their hydrolysis products, hydrolysable tannins are divided into gallotannins and ellagitannins.

Owing to their more complex structural arrangements, condensed tannins are not easily hydrolysed. These organic compounds can be detected by various chemical tests as mentioned below.

### 6.7.5.3.1  *Goldbeater's Skin Test*

In this test, a small piece of Goldbeaters skin is treated with HCL (2%), followed by washing with distilled water. A Latheron test sample is spotted for 5 min followed by rinsing with distilled water and a subsequent treatment with ferrous sulphate solution (1%). The appearance of brown or black colour over goldbeaters skin indicates the presence of tannin.

Ferric chloride test: A test solution is treated with ferric chloride solution, resulting in the development of blue colour, which indicates the presence of hydrolysable tannins, while the condensed tannins give green colour.

Gelatin test: A test solution is treated with gelatin (1%), followed by the treatment with sodium chloride (10%). The formation of precipitates indicates the presence of tannins.

Phenazone test: A test solution is treated with sodium acid phosphate (0.5 g), followed by the treatment with a phenazone solution (2%), which results in the formation of coloured precipitate, suggesting the presence of tannins.

Catechin test: Place a wooden matchstick in a test solution, dry it, and treat it with concentrated hydrochloric acid.

This is followed by heating matchstick near the flame. This turns matchstick into pink colour because of the presence of condensed tannins, which produces phloroglucinol.

Chlorogenic acid test: A test solution is treated with ammonia, followed by an exposure to air. The development of green colour suggests the presence of chlorogenic acid.

### 6.7.5.4  Polyphenols

Polyphenolic compounds show high reactivity against acidic or basic reagents. Polyphenolic compounds present in different aqueous or alcoholic extracts of the test solution show colourful reactions against some reagents.

Acetic acid test: A test solution treated with acetic acid when presents red colour indicates the presence of polyphenolic compounds.

Nitric acid test: A test solution treated with nitric acid when presents reddish to yellow colour indicates the presence of polyphenolic compounds.

Iodine test: A test solution treated with an iodine solution when presents transient yellow coloration indicates the presence of polyphenolic compounds.

Potassium dichromate test: A test solution treated with potassium dichromate when presents red precipitate indicates the presence of polyphenolic compounds.

Ammonium hydroxide test: A test solution treated with ammonium hydroxide when presents red colour indicates the presence of polyphenolic compounds.

Potassium ferricyanide test: A test solution treated with potassium ferricyanide when presents red colour indicates the presence of polyphenolic compounds.

Potassium permanganate test: A test solution containing polyphenolic compounds when treated with potassium permanganate (dilute) is readily decolourized, which suggests the presence of polyphenolic compounds.

Miscellaneous test: A test solution when treated with ammonium hydroxide followed by the treatment with silver nitrate solution (10%) with subsequent heating resulting in the formation of white precipitate indicates the presence of polyphenolic compounds.

### 6.7.5.5  Essential Oils

Essential oils are the class of crude drugs present in the liquid state; however, chemically, both contain volatile and non-volatile components. Not all essential oils contain volatile principles. Oils those contain volatile compounds are called volatile oils.

Tincture alkanna test: A thin section of the crude drug sample is treated with few drops of tincture alkanna, oil globules shows red colour, suggesting the presence of volatile oil.

Sudan red III test: A thin section of the crude drug sample is treated with Sudan red III solution, a deposition of essential oil presents in the crude drug sample shows red colour, suggesting the presence of volatile oil.

### 6.7.6 THERAPEUTIC EFFICACY ASSESSMENT

Biological assessment includes an in vitro or in vivo assessment of potency of the crude drugs on microorganism (bacterial or fungal strains), experimental animals and health volunteers. Since mere physical and chemical assessment is not sufficient to establish therapeutic efficacy and safety profile of the crude drug, biological assessment plays an important role, and its assessment can be done by various in vitro and in vivo approaches.

## REFERENCES

1. Bhatia S, Sardana S, Sharma A, Vargas De La Cruz CB, Chaugule B, Khodaie L. Development of broad spectrum mycosporine loaded sunscreen formulation from *Ulva fasciata* Delile. *Biomedicine (Taipei)*. 2019;9(3):17.
2. Bhatia S, Sardana S, Senwar KR, Dhillon A, Sharma A, Naved T. In vitro antioxidant and antinociceptive properties of *Porphyra vietnamensis*. *Biomedicine (Taipei)*. 2019;9(1):3.
3. Bhatia S, Sharma K, Nagpal K, Bera T. Investigation of the factors influencing the molecular weight of porphyran and its associated antifungal activity. *Bioact Carbohydr Diet Fibre*. 2015;5:153–168.
4. Bhatia S, Al-Harrasi A, Behl T, Anwer MK, Ahmed MM, Mittal V, Kaushik D, Chigurupati S, Kabir MT, Sharma PB, Chaugule B, Vargas-de-la-Cruz C. Unravelling the photoprotective effects of freshwater alga Nostoc commune Vaucher ex Bornet et Flahault against ultraviolet radiations. *Environ Sci Pollut Res Int*. 2021 Oct 5. doi: 10.1007/s11356-021-16704-2. Epub ahead of print. PMID: 34609682.
5. Bhatia S, Sharma K, Sharma A, Nagpal K, Bera T. Anti-inflammatory, analgesic and antiulcer properties of *Porphyra vietnamensis*. *Avicenna J Phytomed*. 2015;5(1):69–77.
6. Bhatia S, Kumar V, Sharma K, Nagpal K, Bera T. Significance of algal polymer in designing amphotericin B nanoparticles. *Sci World J*. 2014;2014:564573.
7. Bhatia S, Rathee P, Sharma K, Chaugule BB, Kar N, Bera T. Immuno-modulation effect of sulphated polysaccharide (porphyran) from *Porphyra vietnamensis*. *Int J Biol Macromol*. 2013;57:50–56.
8. Bhatia S, Garg A, Sharma K, Kumar S, Sharma A, Purohit AP. Mycosporine and mycosporine-like amino acids: a paramount tool against ultra violet irradiation. *Pharmacogn Rev*. 2011;5(10):138–146.
9. Bhatia S, Sharma K, Namdeo AG, Chaugule BB, Kavale M, Nanda S. Broad-spectrum sun-protective action of Porphyra-334 derived from *Porphyra vietnamensis*. *Pharmacogn Res*. 2010;2(1):45–49.
10. Bhatia S, Sharma K, Dahiya R, Bera T. Modern Applications of Plant Biotechnology in Pharmaceutical Sciences. Academic Press, Elsevier, Cambridge, MA; 2015:164–174.
11. Bhatia S. Nanotechnology in Drug Delivery: Fundamentals, Design, and Applications. CRC Press, Boca Raton, FL; 2016:121–127.
12. Bhatia S, Goli D. Leishmaniasis: Biology, Control and New Approaches for Its Treatment. CRC Press, Boca Raton, FL; 2016:164–173.
13. Bhatia S. Natural Polymer Drug Delivery Systems: Nanoparticles, Plants, and Algae. Springer Nature, Basingstoke, UK; 2016:117–127.
14. Bhatia S. Introduction to Pharmaceutical Biotechnology, Volume 2: Enzymes, Proteins and Bioinformatics. IOP Publishing Ltd, Bristol, UK; 2018:1.
15. Bhatia S. Introduction to Pharmaceutical Biotechnology, Volume 1: Basic Techniques and Concepts. IOP Publishing Ltd, Bristol, UK; 2018:2.
16. Bhatia S. Introduction to Pharmaceutical Biotechnology, Volume 3: Animal Tissue Culture Technology. IOP Publishing Ltd, Bristol, UK; 2019:3.
17. Bhatia S. Natural Polymer Drug Delivery Systems: Nanotechnology and Its Drug Delivery Applications. Springer International Publishing, Switzerland; 2016:1–32.
18. Bhatia S. Natural Polymer Drug Delivery Systems: Nanoparticles Types, Classification, Characterization, Fabrication Methods and Drug Delivery Applications. Springer International Publishing, Switzerland; 2016:33–93.

19. Bhatia S. Natural Polymer Drug Delivery Systems: Natural Polymers vs Synthetic Polymer. Springer International Publishing, Switzerland; 2016:95–118.

20. Bhatia S. Natural Polymer Drug Delivery Systems: Plant Derived Polymers, Properties, and Modification & Applications. Springer International Publishing, Switzerland; 2016:119–184.

21. Bhatia S. Natural Polymer Drug Delivery Systems: Marine Polysaccharides Based Nano-Materials and Its Applications. Springer International Publishing, Switzerland; 2016:185–225.

22. Bhatia S. Systems for Drug Delivery: Mammalian Polysaccharides and Its Nanomaterials. Springer International Publishing, Switzerland; 2016:1–27.

23. Bhatia S. Systems for Drug Delivery: Microbial Polysaccharides as Advance Nanomaterials. Springer International Publishing, Switzerland; 2016:29–54.

24. Bhatia S. Systems for Drug Delivery: Chitosan Based Nanomaterials and Its Applications. Springer International Publishing, Switzerland; 2016:55–117.

25. Bhatia S. Systems for Drug Delivery: Advance Polymers and Its Applications. Springer International Publishing, Switzerland; 2016:119–146.

26. Bhatia S. Systems for Drug Delivery: Advanced Application of Natural Polysaccharides. Springer International Publishing, Switzerland; 2016:147–170.

27. Bhatia S. Systems for Drug Delivery: Modern Polysaccharides and Its Current Advancements. Springer International Publishing, Switzerland; 2016:171–188.

28. Bhatia S. Systems for Drug Delivery: Toxicity of Nanodrug Delivery Systems. Springer International Publishing, Switzerland; 2016:189–197.

29. Bhatia S, Sharma K, Bera T. Structural characterization and pharmaceutical properties of porphyran. *Asian J Pharm* 2015;9:93–101.

30. Bhatia S, Sharma A, Sharma K, Kavale M, Chaugule BB, Dhalwal K, Namdeo AG, Mahadik KR. Novel algal polysaccharides from marine source: porphyran. *Pharmacogn Rev*. 2008;4:271–276.

31. Bhatia S. Nanotechnology in Drug Delivery Fundamentals, Design, and Applications: Part 1: Protein and Peptide-Based Drug Delivery Systems. Apple Academic Press, Palm Bay, FL; 2016:50–204.

32. Bhatia S. Nanotechnology in Drug Delivery Fundamentals, Design, and Applications: Part 2: Peptide-Mediated Nanoparticle Drug Delivery System. Apple Academic Press, Palm Bay, FL; 2016:205–280.

33. Bhatia S, Al-Harrasi A, Behl T, Anwer MK, Ahmed MM, Mittal V, Kaushik D, Chigurupati S, Kabir MT, Sharma PB, Chaugule B, Vargas-de-la-Cruz C. Anti-migraine activity of freeze dried-latex obtained from *Calotropis gigantea* Linn. *Environ Sci Pollut Res Int*. 2022 Apr;29(18):27460–27478. doi: 10.1007/s11356-021-17810-x. Epub 2022 Jan 4. PMID: 34981370.

34. Altemimi A, Lakhssassi N, Baharlouei A, Watson DG, Lightfoot DA. Phytochemicals: extraction, isolation, and identification of bioactive compounds from plant extracts. *Plants (Basel)*. 2017 Sep 22; 6(4):42. doi: 10.3390/plants6040042. PMID: 28937585; PMCID: PMC5750618.

35. Palhares RM, Gonçalves Drummond M, Dos Santos Alves Figueiredo Brasil B, Pereira Cosenza G, das Graças Lins Brandão M, Oliveira G. Medicinal plants recommended by the world health organization: DNA barcode identification associated with chemical analyses guarantees their quality. *PLoS One*. 2015 May 15;10(5):e0127866. doi: 10.1371/journal.pone.0127866. PMID: 25978064; PMCID: PMC4433216.

36. Salmerón-Manzano E, Garrido-Cardenas JA, Manzano-Agugliaro F. Worldwide research trends on medicinal plants. *Int J Environ Res Public Health*. 2020 May 12;17(10):3376. doi: 10.3390/ijerph17103376. PMID: 32408690; PMCID: PMC7277765.

37. Anand U, Jacobo-Herrera N, Altemimi A, Lakhssassi N. A comprehensive review on medicinal plants as antimicrobial therapeutics: potential avenues of biocompatible drug discovery. *Metabolites*. 2019 Nov 1;9(11):258. doi: 10.3390/metabo9110258. PMID: 31683833; PMCID: PMC6918160.

38. Fitzgerald M, Heinrich M, Booker A. Medicinal plant analysis: a historical and regional discussion of emergent complex techniques. *Front Pharmacol*. 2020 Jan 9;10:1480. doi: 10.3389/fphar.2019.01480. PMID: 31998121; PMCID: PMC6962180.

39. Osathanunkul M, Madesis P, de Boer H. Bar-HRM for authentication of plant-based medicines: evaluation of three medicinal products derived from Acanthaceae species. *PLoS One*. 2015;10(5):e0128476. Published 2015 May 26. doi:10.1371/journal.pone.0128476

40. Ganie SH, Upadhyay P, Das S, Prasad Sharma M. Authentication of medicinal plants by DNA markers. *Plant Gene*. 2015 Dec;4:83–99. doi: 10.1016/j.plgene.2015.10.002. Epub 2015 Oct 22. PMID: 32289060; PMCID: PMC7103949.

41. Nicod JC, Largiadèr CR. SNPs by AFLP (SBA): a rapid SNP isolation strategy for non-model organisms. *Nucleic Acids Res*. 2003 Mar 1;31(5):e19. doi: 10.1093/nar/gng019. PMID: 12595568; PMCID: PMC149841.

42. Katiyar C, Gupta A, Kanjilal S, Katiyar S. Drug discovery from plant sources: an integrated approach. *Ayu.* 2012 Jan;33(1):10–19. doi: 10.4103/0974-8520.100295. PMID: 23049178; PMCID: PMC3456845.

43. Ichim MC, Häser A, Nick P. Microscopic authentication of commercial herbal products in the globalized market: potential and limitations. *Front Pharmacol.* 2020 Jun 9;11:876. doi: 10.3389/fphar.2020.00876. PMID: 32581819; PMCID: PMC7295937.

44. Kumar A, Mishra DC, Angadi UB, Yadav R, Rai A, Kumar D. Inhibition potencies of phytochemicals derived from sesame against SARS-CoV-2 main protease: a molecular docking and simulation study. *Front Chem.* 2021 Oct 8;9:744376. doi: 10.3389/fchem.2021.744376. PMID: 34692642; PMCID: PMC8531729.

45. Mortensen A, Sorensen IK, Wilde C, Dragoni S, Mullerová D, Toussaint O, Zloch Z, Sgaragli G, Ovesná J. Biological models for phytochemical research: from cell to human organism. *Br J Nutr.* 2008 May;99(E Suppl 1):ES118–ES126. doi: 10.1017/S0007114508965806. PMID: 18503732.

46. Calixto JB. Efficacy, safety, quality control, marketing and regulatory guidelines for herbal medicines (phytotherapeutic agents). *Braz J Med Biol Res.* 2000 Feb;33(2):179–189. doi: 10.1590/s0100-879x2000000200004. PMID: 10657057.

47. Satake M. [The utilization and safety of medicinal plants and crude drugs]. *Kokuritsu Iyakuhin Shokuhin Eisei Kenkyusho Hokoku.* 1998;(116):13–29. Japanese. PMID: 10097510.

48. Rajeshwari P, Raveesha K. Mycological analysis and aflatoxin B1 contaminant estimation of herbal drug raw materials. *Afr J Tradit Complement Altern Med.* 2016 Aug 12;13(5):123–131. doi: 10.21010/ajtcam.v13i5.16. PMID: 28487902; PMCID: PMC5416630.

49. Li Y, Shen Y, Yao CL, Guo DA. Quality assessment of herbal medicines based on chemical fingerprints combined with chemometrics approach: a review. *J Pharm Biomed Anal.* 2020 Jun 5;185:113215. doi: 10.1016/j.jpba.2020.113215. Epub 2020 Mar 2. PMID: 32199327.

50. Simmler C, Chen SN, Anderson J, Lankin DC, Phansalkar R, Krause E, Dietz B, Bolton JL, Nikolic D, van Breemen RB, Pauli GF. Botanical integrity: Part 2: traditional and modern analytical approaches. *HerbalGram.* 2016 Spring;109:60–64. PMID: 30287984; PMCID: PMC6168214.

51. Pharmacognostical studies. *Prog Drug Res.* 2016;71:5–10. PMID: 26939259.

52. Joharchi MR, Amiri MS. Taxonomic evaluation of misidentification of crude herbal drugs marketed in Iran. *Avicenna J Phytomed.* 2012 Spring;2(2):105–112. PMID: 25050238; PMCID: PMC4075662.

53. Atkinson JA, Wells DM. An updated protocol for high throughput plant tissue sectioning. *Front Plant Sci.* 2017 Oct 4;8:1721. doi: 10.3389/fpls.2017.01721. PMID: 29046689; PMCID: PMC5632646.

54. Techen N, Crockett SL, Khan IA, Scheffler BE. Authentication of medicinal plants using molecular biology techniques to compliment conventional methods. *Curr Med Chem.* 2004 Jun;11(11):1391–1401. doi: 10.2174/0929867043365206. PMID: 15180573.

55. Siraj J, Belew S, Suleman S. Ethnobotanical assessment and physicochemical properties of commonly used medicinal plants in Jimma Zone, Southwest Ethiopia: traditional healers based cross-sectional study. *J Exp Pharmacol.* 2020 Dec 21;12:665–681. doi: 10.2147/JEP.S267903. PMID: 33376416; PMCID: PMC7762448.

56. Mary NY. Determination of moisture in crude drugs by gas-liquid chromatography. *J Pharm Sci.* 1967 Dec;56(12):1670–1672. doi: 10.1002/jps.2600561231. PMID: 5588725.

57. Zhen Z, Wang H, Yue Y, Li D, Song X, Li J. Determination of water content of crude oil by azeotropic distillation Karl Fischer coulometric titration. *Anal Bioanal Chem.* 2020 Jul;412(19):4639–4645. doi: 10.1007/s00216-020-02714-5. Epub 2020 May 30. PMID: 32474722.

58. Cowley PS, Evans FJ, Michaels HJ. Moisture determination in crude drugs by the Karl Fisher method. *Planta Med.* 1968 Dec;16(4):388–394. doi: 10.1055/s-0028-1099925. PMID: 5734376.

59. Larsson W, Jalbert J, Gilbert R, Cedergren A. Efficiency of methods for Karl Fischer determination of water in oils based on oven evaporation and azeotropic distillation. *Anal Chem.* 2003 Mar 15;75(6):1227–1232. doi: 10.1021/ac026229+. PMID: 12659179.

60. Ajazuddin, Saraf S. Evaluation of physicochemical and phytochemical properties of Safoof-E-Sana, a Unani polyherbal formulation. *Pharmacognosy Res.* 2010 Sep;2(5):318–322. doi: 10.4103/0974-8490.72332. PMID: 21589760; PMCID: PMC3093045.

61. Rao Y, Xiang B. Determination of total ash and acid-insoluble ash of Chinese herbal medicine Prunellae Spica by near infrared spectroscopy. *Yakugaku Zasshi.* 2009 Jul;129(7):881–886. doi: 10.1248/yakushi.129.881. PMID: 19571524.

62. Kim D, Kim B, Yun E, Kim J, Chae Y, Park S. Statistical quality control of total ash, acid-insoluble ash, loss on drying, and hazardous heavy metals contained in the component medicinal herbs of "Ssanghwatang", a widely used oriental formula in Korea. *J Nat Med.* 2013 Jan;67(1):27–35. doi: 10.1007/s11418-012-0640-4. Epub 2012 Mar 15. PMID: 22418854.

63. Arambewela LS, Arawwawala LD. Standardization of *Alpinia calcarata* Roscoe rhizomes. *Pharmacognosy Res.* 2010 Sep;2(5):285–288. doi: 10.4103/0974-8490.72324. PMID: 21589752; PMCID: PMC3093041.

64. Chandel HS, Pathak AK, Tailang M. Standardization of some herbal antidiabetic drugs in polyherbal formulation. *Pharmacognosy Res.* 2011 Jan;3(1):49–56. doi: 10.4103/0974-8490.79116. PMID: 21731396; PMCID: PMC3119272.

65. Zarshenas MM, Samani SM, Petramfar P, Moein M. Analysis of the essential oil components from different *Carum copticum* L. samples from Iran. *Pharmacognosy Res.* 2014 Jan;6(1):62–66. doi: 10.4103/0974-8490.122920. PMID: 24497745; PMCID: PMC3897012.

66. da Silva CE, da Costa WF, Minguzzi S, da Silva RC, Simionatto E. Assessment of volatile chemical composition of the essential oil of *Jatropha ribifolia* (Pohl) Baill by HS-SPME-GC-MS using different fibers. *J Anal Methods Chem.* 2013;2013:352606. doi: 10.1155/2013/352606. Epub 2013 Nov 24. PMID: 24371539; PMCID: PMC3859261.

67. Evans WC. Trease and Evans' Pharmacognosy. Saunders an Imprint of Elsevier, 2005:41–47.

68. Alikhan I, Khanum A. Medicinal and Aromatic Plants of India. Ukaaz Publication, 2005:133–134.

69. The Indian Pharmacopoeia. Govt. of India Publication, New Delhi; 1996. Anonymous.

70. The wealth of India: a dictionary of Indian raw materials and industrial products (industrial products—part I). *Ind Med Gaz.* 1949 Oct;84(10):476–477. PMID: PMC5189551. https://www.ncbi.nlm.nih.gov/pmc/articles/PMC5189551/

71. Nandkarni KM. Indian Materia Medica. Popular Prakashan Limited, Bombay; 1976:1291.

72. Zafar R. Practical Pharmacognosy. CBS Publishers and Distributors Pvt Ltd, January 1, 2007.

73. Shah BN, Seth AK. Textbook of Pharmacognosy and Phytochemistry (2nd Edn.). CBS Publishers and Distributors Pvt Ltd; 2020;2.

74. Kokate CK. Practical Pharmacognosy. Nirali Prakashan, Pune; 2017;17:37–84.

75. Handa SS, Kapoor VK. Textbook of Pharmacognosy. Pragati Publisher, Meerut; 2021.

76. Khandelwal KR. Practical Pharmacognosy. Pragati Books Pvt. Ltd, 2008.

77. Zafar R. Practical Pharmacognosy. CBS Publishers and Distributors Pvt Ltd, 2007.

78. Rastogi RP, Mehrotra BN. Compendium of Indian Medicinal Plant. Central Drug Research Institute Lucknow, National Institute of Science, New Delhi; 1999:147.

79. Kokate CK. Practical Pharmacognosy. Vallabh Prakashan, 1994:115–121.

80. Kokate CK, Purohit AP, Gokhale SB. Pharmacognosy. CBS Publishers and Distributors, 2005:169.

81. Fatima K, Mahmud S, Yasin H, Asif R, Qadeer K, Ahmad I. Authentication of various commercially available crude drugs using different quality control testing parameters. *Pak J Pharm Sci.* 2020 Jul;33(4):1641–1657. PMID: 33583798.

82. Huck C. Infrared spectroscopic technologies for the quality control of herbal medicines. In: Evidence-Based Validation of Herbal Medicine. 2015:477–493. doi: 10.1016/B978-0-12-800874-4.00022-2. Epub 2015 Apr 3. PMCID: PMC7149424.

83. Li Y, Shen Y, Yao CL, Guo DA. Quality assessment of herbal medicines based on chemical fingerprints combined with chemometrics approach: a review. *J Pharm Biomed Anal.* 2020 Jun 5;185:113215. doi: 10.1016/j.jpba.2020.113215. Epub 2020 Mar 2. PMID: 32199327.

84. Li YZ, Min SG, Liu X. [Applications of near-infrared spectroscopy to analysis of traditional Chinese herbal medicine]. *Guang Pu Xue Yu Guang Pu Fen Xi.* 2008 Jul;28(7):1549–1553. Chinese. PMID: 18844158.

85. Rooney JS, McDowell A, Strachan CJ, Gordon KC. Evaluation of vibrational spectroscopic methods to identify and quantify multiple adulterants in herbal medicines. *Talanta.* 2015 Jun 1;138:77–85. doi: 10.1016/j.talanta.2015.02.016. Epub 2015 Feb 16. PMID: 25863375.

86. Li Y, Shen Y, Yao CL, Guo DA. Quality assessment of herbal medicines based on chemical fingerprints combined with chemometrics approach: a review. *J Pharm Biomed Anal.* 2020 Jun 5;185:113215. doi: 10.1016/j.jpba.2020.113215. Epub 2020 Mar 2. PMID: 32199327.

87. Kim HK, Choi YH, Verpoorte R. NMR-based plant metabolomics: where do we stand, where do we go? *Trends Biotechnol.* 2011;29:267–275.

88. Ward J, Beale MH. NMR spectroscopy in plant metabolomics. In: Plant Metabolomics, Biotechnology in Agriculture and Forestry Series (eds. K. Saito, R.A. Dixon, L. Willmitzer). Springer, 2006; 57:81–91.

89. Wang YL, Tang HR, Holmes E, Lindon JC, Turini ME, Sprenger N, Bergonzelli G, Fay LB, Kochhar S, Nicholson JK. Biochemical characterization of rat intestine development using high-resolution magic-angle-spinning 1H NMR spectroscopy and multivariate data analysis. *J Proteom Res.* 2005; 4(4):1324–1329.

90. Mahrous EA, Farag MA. Two dimensional NMR spectroscopic approaches for exploring plant metabolome: a review. *J Adv Res*. 2015 Jan;6(1):3–15. doi: 10.1016/j.jare.2014.10.003. Epub 2014 Oct 18. PMID: 25685540; PMCID: PMC4293671.

91. Wolfender JL, Ndjoko K, Hostettmann K. The potential of LC-NMR in phytochemical analysis. *Phytochem Anal*. 2001 Jan–Feb;12(1):2–22. doi: 10.1002/1099-1565(200101/02)12:1<2::AID-PCA552> 3.0.CO;2-K. PMID: 11704957.

92. Valentino G, Graziani V, D'Abrosca B, Pacifico S, Fiorentino A, Scognamiglio M. NMR-based plant metabolomics in nutraceutical research: an overview. *Molecules*. 2020 Mar 23;25(6):1444. doi: 10.3390/molecules25061444. PMID: 32210071; PMCID: PMC7145309.

93. Hussin M, Abdul Hamid A, Abas F, Ramli NS, Jaafar AH, Roowi S, Majid NA, Pak Dek MS. NMR-based metabolomics profiling for radical scavenging and anti-aging properties of selected herbs. *Molecules*. 2019 Sep 3;24(17):3208. doi: 10.3390/molecules24173208. PMID: 31484470; PMCID: PMC6749213.

94. Zhang A, Sun H, Wang X. Mass spectrometry-driven drug discovery for development of herbal medicine. *Mass Spectrom Rev*. 2018 May;37(3):307–320. doi: 10.1002/mas.21529. Epub 2016 Dec 23. PMID: 28009933.

95. Demarque DP, Dusi RG, de Sousa FDM, Grossi SM, Silvério MRS, Lopes NP, Espindola LS. Mass spectrometry-based metabolomics approach in the isolation of bioactive natural products. *Sci Rep*. 2020 Jan 23;10(1):1051. doi: 10.1038/s41598-020-58046-y. PMID: 31974423; PMCID: PMC6978511.

96. Mino Y, Usami H, Ota N, Takeda Y, Ichihara T, Fujita T. Inorganic chemical approaches to pharmacognosy. VII. X-ray fluorescence spectrometric studies on the inorganic constituents of crude drugs. (5). The relationship between inorganic constituents of plants and the soils on which they are grown. *Chem Pharm Bull (Tokyo)*. 1990 Aug;38(8):2204–2207. doi: 10.1248/cpb.38.2204. PMID: 2279283.

97. Markham RK. Techniques of Flavonoid Identification. Academic Press, London; 1982:15–45.

98. Harborne JB. A Guide to Modern Techniques of Plant Analysis (3rd Edn.). Chapman and Hall, London, New York; 1998. Phytochemical methods.

99. Hostettmann K, Martson A. Saponin. University Press, Cambridge; 1995:233.

100. Clarke EGC. Isolation and Identification of Drugs, Vol. 2. Pharmaceutical Press, London, UK; 1975:905.

101. Santos SC, de Mello JC. Taninos. In: Farmacognosia: da planta ao medicamento (6th Edn.) (eds. CM Simoes, G Schenkel, G Gosmann, JC de Mello, LA Mentz, PR Petrovick). UFRGS, Porto Alegre; 2007:615–656.

102. Pearson D. Edinburgh (7th Edn.). Churchill Livingstone, New York; 1976. The chemical analysis of foods.

For Product Safety Concerns and Information please contact our EU
representative  GPSR@taylorandfrancis.com
Taylor & Francis Verlag GmbH, Kaufingerstraße 24, 80331 München, Germany